International Association of Geodesy Symposia

Ivan I. Mueller, Series Editor

International Association of Geodesy Symposia

Ivan I. Mueller, Series Editor

Global and Regional Geodynamics

Symposium No. 101

Edinburgh, Scotland, August 3–5, 1989

Convened and Edited by

P. Vyskocil
C. Reigber
P.A. Cross

Springer Science+Business Media, LLC

P. Vyskocil
International Center on Recent
    Crustal Movements
C5-25066 ZDIBY, 98
County Praha-Vychod
Prague
Czechoslovakia

C. Reigber
German Geodetic Research Institute
Marstallplatz 8
D-8000 München 22
Federal Republic of Germany

P.A. Cross
University of Newcastle upon Tyne
Department of Surveying
Newcastle upon Tyne NE1 7RU
United Kingdom

Series Editor
Ivan I. Mueller
Department of Geodetic Science & Surveying
The Ohio State University
Columbus, OH 43210-1247
USA

For information regarding previous symposia volumes contact:
Secretáire Général
Bureau Central de l'Association Internationale de Géodésie
138, rue de Grenelle
75700 Paris
France

Library of Congress Cataloging-in-Publication Data
Global and regional geodynamics / P. Vyskocil, C. Reigber, P.A.
Cross, editors.
    p.   cm. — (International Association of Geodesy symposia ;
symposium 101)
    Proceedings of a symposium held within the General Meeting of the
International Association of Geodesy in Edinburgh, Scotland, August
3–5, 1989.
    Includes bibliographical references.
    1. Geodesy — Congresses.   2. Geodynamics — Congresses.
I. Vyskocil, P. (Pavel)   II. Reigber, Ch.   III. Cross, P.A.
IV.   International Association of Geodesy. General Meeting (1989 :
Edinburgh, Scotland)   V. Series.
QB275.G56   1990
526′.1 — dc20   90-9522

Printed on acid-free paper.

Camera-ready copy provided by the editors.

9 8 7 6 5 4 3 2 1

ISBN 978-0-387-97265-7   ISBN 978-1-4615-7109-4 (eBook)
DOI 10.1007/978-1-4615-7109-4

Foreword

A General Meeting of the IAG was held in Edinburgh, Scotland, to commemorate its 125th Anniversary. The Edinburgh meeting, which attracted 360 scientific delegates and 80 accompanying persons from 44 countries, was hosted jointly by the Royal Society, the Royal Society of Edinburgh and the University of Edinburgh.

The scientific part of the program, which was held in the Appleton Tower of the University, included the following five symposia:

Symposium 101	Global and Regional Geodynamics
Symposium 102	GPS and Other Radio Tracking Systems
Symposium 103	Gravity, Gradiometry and Gravimetry
Symposium 104	Sea Surface Topography, the Geoid and Vertical Datums
Symposium 105	Earth Rotation and Coordinate Reference Frames

All together there were 90 oral and 160 poster presentations. The program was arranged to prevent any overlapping of oral presentations, and thus enabled delegates to participate in all the sessions.

The 125th Anniversary Ceremony took place on August 7, 1989, in the noble surroundings of the McEwan Hall where, 53 years earlier, Vening-Meinesz gave one of the two Union Lectures at the 6th General Assembly of the IUGG. The Ceremony commenced with welcome speeches by the British hosts. An interlude of traditional Scottish singing and dancing was followed by the Presidential Address given by Professor Ivan Mueller, on 125 years of international cooperation in geodesy. The Ceremony continued with greetings from representatives of sister societies, and was concluded by the presentation of the Levallois Medal to Professor Arne Bjerhammar. The 125th Anniversary was also commemorated by an exhibition entitled *The Shape of the Earth*, which was mounted in the Royal Museum of Scotland. An abbreviated version of the President's speech and the list of all participants are included in the proceedings of Symposium 102.

A social program enabled delegates to experience some of the hospitality and culture of both Edinburgh and Scotland, as well as provided an opportunity to explore the beautiful City of Edinburgh and the surrounding countryside.

A Scottish Ceilidh on the last night concluded a pleasant week, which was not only scientifically stimulating, but also gave delegates and accompanying persons an opportunity to renew *auld acquaintances* and make new ones.

The International Association of Geodesy and the UK Organizing Committee express their appreciation to the local organizers of the General Meeting, especially to Dr. Roger G. Hipkin and Mr. Wm. H. Rutherford, for their tireless efforts in running the meeting to its successful conclusion.

Commencing with these symposia the proceedings of IAG-organized scientific meetings will be published by Springer Verlag Inc., New York from author-produced camera-ready manuscripts. Although these manuscripts are reviewed and edited by IAG, their contents are the sole responsibility of the authors, and they do not reflect official IAG opinion, policy or approval.

V. Ashkenazi
A. H. Dodson
UK Organizing Committee

Ivan I. Mueller
President, International
Association of Geodesy

Preface

Symposium 101 was convened as a joint meeting of International Association of Geodesy Commission VII on Recent Crustal Movements (CRCM) and IAG Commission VIII on International Coordination of Space Techniques for Geodesy and Geodynamics (CSTG).

CRCM has as its primary scientific objectives (a) the further improvement of the complex analysis of the relevant crustal motion results currently available, aimed especially at the modeling of geodynamical processes in local or regional scale in order to understand the mechanisms of dynamical driving forces, and (b) the continuous improvement of methods for monitoring recent crustal movements as well as data processing and analysis. CSTG has as its goal the development of links between the various groups engaged in the field of space geodesy and geodynamics by various techniques, and the coordination of these groups. Also it is charged with the role of elaborating and proposing projects implying international cooperation, following their progress and reporting on their advancement and results.

At the XIXth IAG General Assembly in Vancouver, the CRCM and CSTG jointly agreed on the need for a stepwise improvement in their cooperation on a global and regional scale in order to extend the information on recent crustal movements within and among the "classical" areas such as Europe, North America and the Pacific area, and to establish the proper basis for monitoring crustal movements in Africa and South America.

Symposium 101 at the IAG General Meeting in Edinburgh was a direct result of this improved cooperation and was an opportunity to present (a) the most recent results on global and regional geodynamics derived from existing cooperative programs using space techniques, (b) improvements and new ideas in space reduction techniques, (c) the most recent results achieved by terrestrial techniques, and (d) theoretical studies of geodynamic modeling, geological and geophysical interpretation and analysis. It focused on the following topics:

1. Plate motions and deformations from space measurements.
2. Regional deformation by space techniques.
3. Tracking and modeling requirements for geodynamic missions.
4. Regional and local crustal movements by terrestrial techniques.
5. Statistical assessment and interpretation of results.

The success of the symposium is illustrated by the large number of high quality papers presented. The majority of these are included as written contributions to this volume.

P. Vyskocil   Ch. Reigber   P.A. Cross

Contents and Program

Regional Dynamics

Modeling of Deformation

Deformation Studies by GPS

STATION POSITIONS AND PLATE MOTION FROM LAGEOS LONG ARC LLA8903

M. M. Watkins, R. J. Eanes, B. D. Tapley and B. E. Schutz
Center for Space Research, The University of Texas at Austin
Austin, Texas 78712 U.S.A

ABSTRACT

Recent long-arc solutions for the Lageos orbit, station positions, Earth rotation parameters and dynamical force modeling parameters have yielded improvements in implied tectonic network motion. The latest such effort includes all Lageos laser tracking data from May 1976 through early 1989, a single continuous arc of over 12.5 years. Baseline rates are determined for the entire network, including North and South America, the Pacific basin, Australia, with special attention paid to the Eurasian fixed sites for this presentation.

INTRODUCTION

The determination of geophysical and geodetic parameters at The University of Texas Center for Space Research using Lageos laser ranges involves a combination of long- and short-arc techniques (Tapley et al., 1985). The long arcs provide the starting point for such research and are used primarily to study long period perturbations in the Lageos orbit. The short arcs are constructed from the residuals of the long arc orbit and form the basis for most geodetic work. Both techniques will be briefly described in the context of the most recent UT/CSR long arc solution, LLA8903, as well as the resulting geodetic parameters in the form of baselines and angular velocities for several major plates.

METHOD

The most recently computed Lageos long arc solution at the University of Texas, LLA8903, spans the period from May 7, 1976 (shortly after the Lageos launch) to February 28, 1989. Within that span were some 360,000 three-minute normal points in 37,000 passes from 103 tracking sites. Approximately the last two years of this span were composed of quick-look data. The nominal force model for this long arc was similar to that used for other recent UT/CSR long arcs and is summarized in Table 1 (Tapley et al., 1988a; Schutz et al., 1989a). An item to be noted is the use of a gravity model, PTGF3, which was developed at UT/CSR as a part of TOPEX model improvement (Tapley et al., 1989b). The adjusted parameters in LLA8903 were J_2, J_3, a small subset of the ocean tide model and an empirical along-track acceleration every 15

1

days. The root mean square (rms) residual after these adjustments was 28 cm.

The short-arc processing used the residuals from LLA8903 to compute mean orbit element adjustment at *n*-day intervals and station coordinate adjustments every 60 days, where *n* varied from 15 to 3. This variable arc length (15 to 3 days) was a new feature of the post-processing of the LLA8903 residuals that had not been used previously at UT/CSR. The need for such a technique arises from the fact that the data quality and quantity has been quite inhomogeneous over the nearly 13 years spanned by LLA8903. The sparser (and less accurate) early data will not support the adjustment of Earth rotation and orbit parameters at a finer grid than 15 days, whereas the recent data allows a greater resolution. Figure 1 demonstrates these trends clearly. For LLA8903, three discrete interval lengths were chosen: 15-day adjustments from 1976 to 1979, 6-day adjustments from 1980 to 1983, and 3-day adjustments from 1983 to 1989.

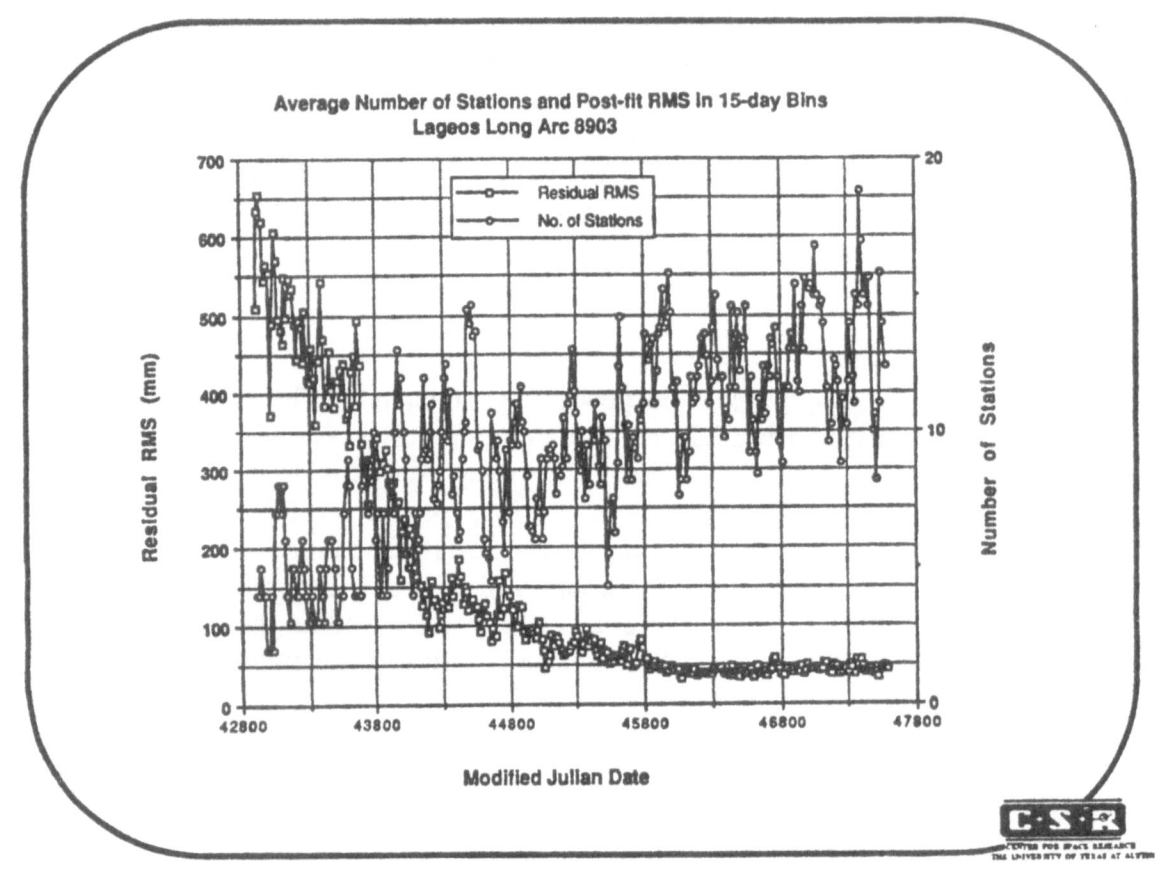

FIGURE 1

2

The variations in data quality have also been modeled by an improved weighting algorithm involving two factors. The first is the gradual increase in data quality with time, modeled as a four part linear model, and the second is the variation in quality from station to station in the network, modeled as a station dependent noise floor which is added in an rms sense to the linear weighting described above. The net result is a weighting which more accurately reflects the inhomogeneities (both temporally and spatially) in data quality over the 13 years. The rms residual after these short arc adjustments was 8 cm over the entire data span, 3–5 cm in the 3-day intervals for the later years.

An alternative method of estimating the station motions is to estimate an angular velocity for the major plates containing tracking sites (North America, South America, Pacific, Eurasia, India-Australia, and Africa). This is carried out in place of the 60-day station coordinate adjustments, and an implicit assumption of rigid, horizontal plate motion is made (Watkins, 1985).

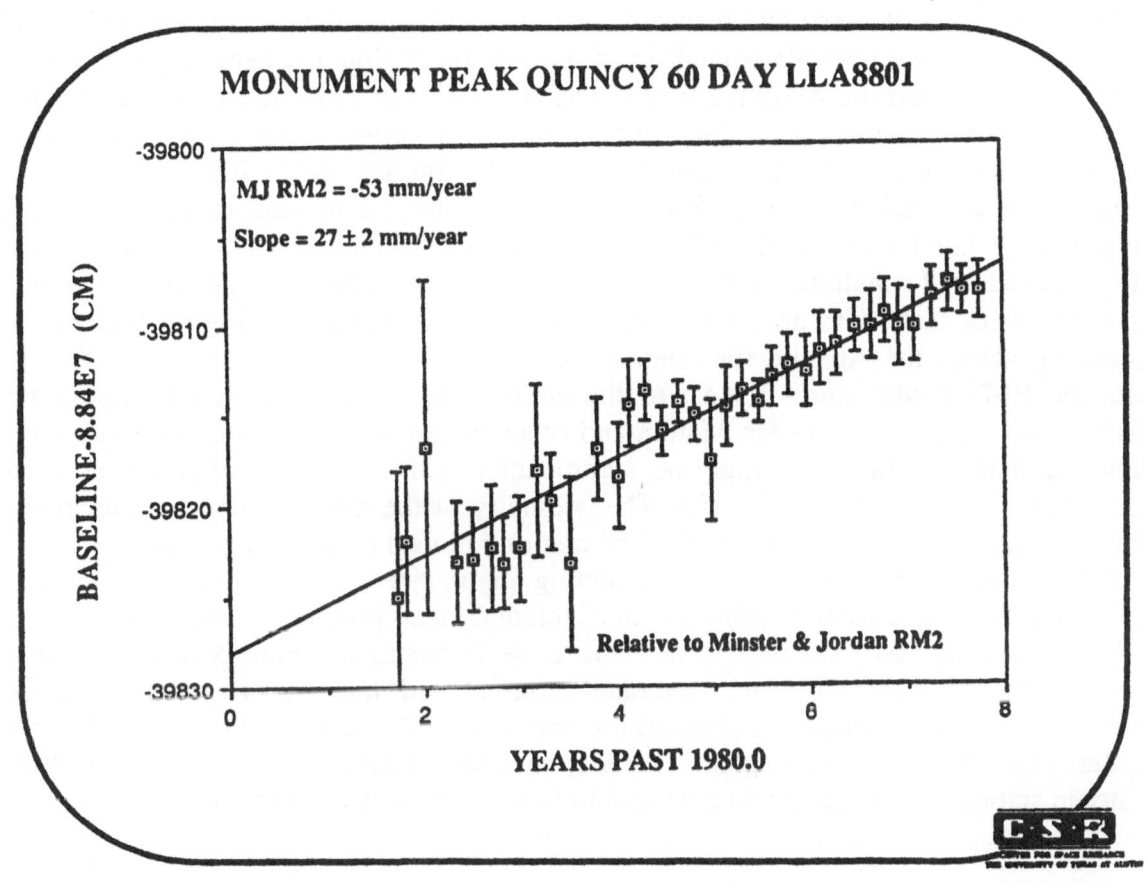

FIGURE 2

RESULTS

The time series of station coordinates obtained from the short arc processing was transformed into a corresponding set of vector baselines. A sample of such a baseline time series for the scalar length component from the previous UT/CSR long arc is presented as Figure 2. This particular plot is for the well-known Monument Peak-Quincy (or SAFE) baseline across the San Andreas Fault in southern California. The points have been plotted relative to the predicted rate of Minster and Jordan RM2 (1978) and show the widely reported 2 cm/yr difference between the observed and predicted rates. Grouping rates obtained in a similar manner for several areas of the Earth yields the tectonic motion maps presented in Figures 3–6. Figure 3 summarizes motions in the Pacific basin, and the most striking results are the large discrepancy along the Simosato (Japan)-Hollas (Hawaii) baseline on the order of 15 mm/yr but the very small discrepancy along the Simosato-Yaragadee (Australia) baseline, constraining the geometry of the deformation of the plate containing the Simosato site. Also of interest is the 10 mm/yr rate observed between Hollas and Monument Peak, which may indicate deformation of the Pacific plate between these two stations. The rates from various sites to Arequipa indicate possible deviations of the motion of South American plate from the RM2 prediction, but the Arequipa station has larger instrument noise than many current state-of-the-art stations, and the uncertainties in those rates may be slightly underestimated. Figure 4 is a similar plot for North America, indicating the San Andreas discrepancy as well as general spreading across the Basin and range province as indicated by the 10 mm/yr ± 4 Quincy-McDonald Observatory (Texas) baseline rate. The intraplate deformations in the eastern U.S. seem considerably smaller, on the order of 3 mm/yr or less, with an uncertainty at about the same level. Figure 5 shows the spreading across the Atlantic for selected stations. The observed rates seem consistent with the RM2 predictions for most of the continental (European) sites except Matera (Italy). This difference between Matera and other European sites is seen more clearly in Figure 6. The baseline rates imply no significant motion between any two sites except those rates involving the Matera site. The geometry of the rates implies a vector motion of Italy which may be consistent with deformation induced by the African plate.

Table 2 summarizes the results of estimating angular velocities of five major plates of the Earth. The adjustment of Africa is not included since it contains only a few stations (all in North Africa) with little data. The table indicates the considerable agreement between the long-term geologic average motions and the modern 13-year average motions, particularly when it is realized that the LLA8903 results include several stations in deformation zones and neglects vertical motions. Preliminary solutions limited to cratonic stations have been performed and indicate improved agreement.

4

Table 1. UTOPIA models.

Reference frames

Model	UTOPIA
Conventional Inertial System (CIS)	J2000 S.I. units
Precession	1976 IAU
Nutation	1980 IAU
Planetary ephemerides	JPL DE-200
Conventional Terrestrial System (CTS)	CSR89L(02) Schutz et al., 1989
Polar motion	CSR89L(02) Schutz et al., 1989
UT1	CSR89L(02) Schutz et al., 1989
Plate motion	Minster-Jordan AM1-2 Epoch: 1983.0
C_{21}, S_{21}	MERIT standards

Force models

Model	UTOPIA
GM	398600.4405 km^3/sec^2
Geopotential	PTGF3
n-body	Sun, Moon and all planets except Pluto
Solid Earth tides (frequency dependent)	$k_2 = 0.3$ $\delta = 0°$
Ocean tides	CSR8801 (Lageos derived)
Relativity	Central body perturbation, Lense-Thirring geodesic precession

Force models (continued)

Model	UTOPIA
Solar radiation	Solar constant = 4.5605E-6 nt/m^2 at 1 AU
	Conical shadow model for Earth and moon $A_e = 6402000$ m $A_m = 1738000$ m
Earth radiation pressure	Albedo and infrared 2nd degree zonal model
Empirical drag	15-day tangential acceleration $\sim -3 \times 10^{-12}$ m/sec^2

Measurement models

Model	UTOPIA
Laser range	Instantaneous range or full light-time computation
Troposphere	Marini & Murray tropospheric correction
Geometric tides	$h_2 = 0.6090$ $l_2 = 0.0825$ $\delta = 0°$
Ocean loading	None
Relativistic correction	Time delay due to Earth

Solution and integration models

Procedure	UTOPIA
Least square solution	Givens/Gentleman
Numerical integration	Encke method Krogh-Shampine-Gordon fixed-step and order for 2nd order diff. eq. Lundberg et al., 1989

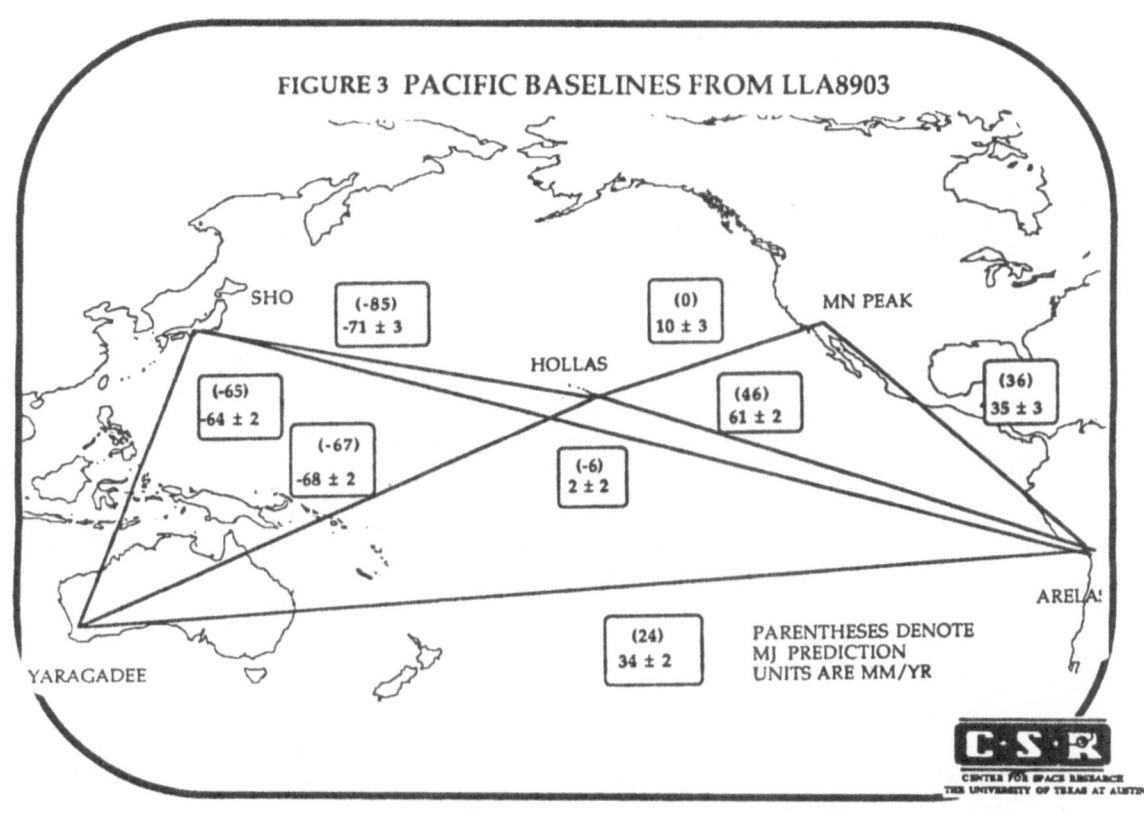

FIGURE 3 PACIFIC BASELINES FROM LLA8903

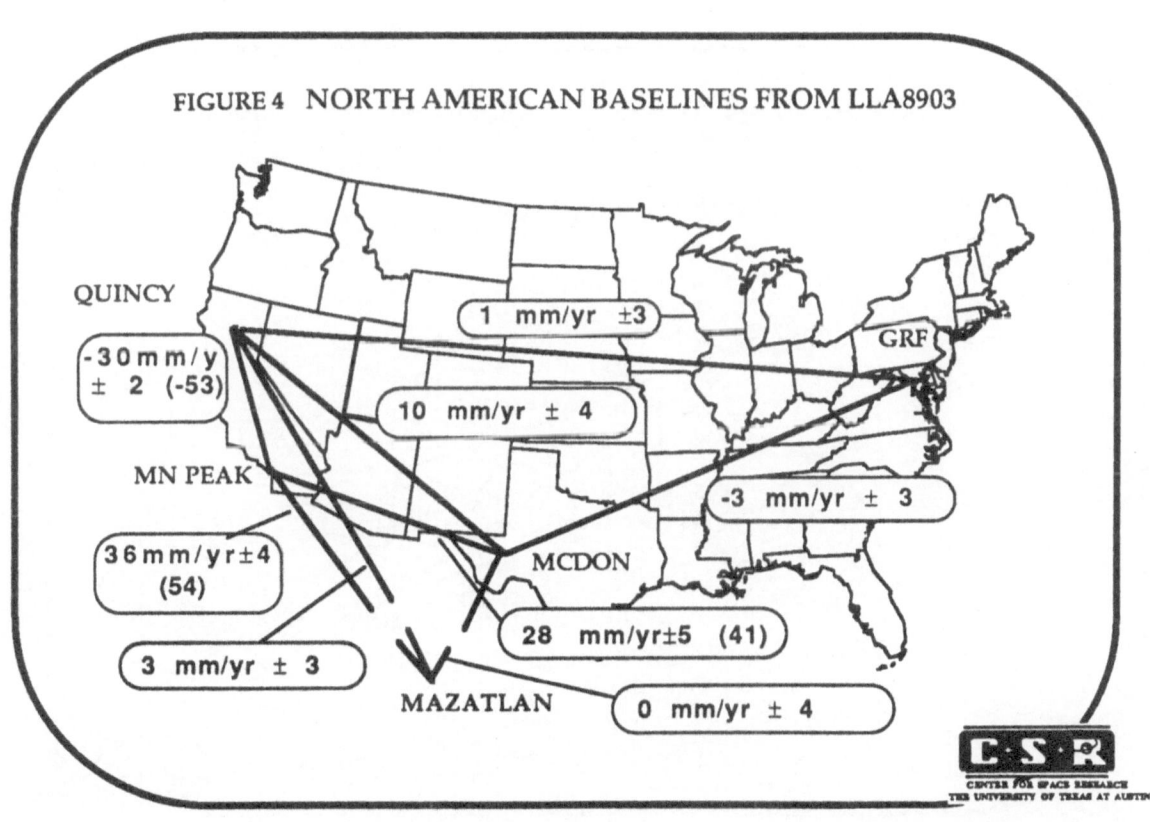

FIGURE 4 NORTH AMERICAN BASELINES FROM LLA8903

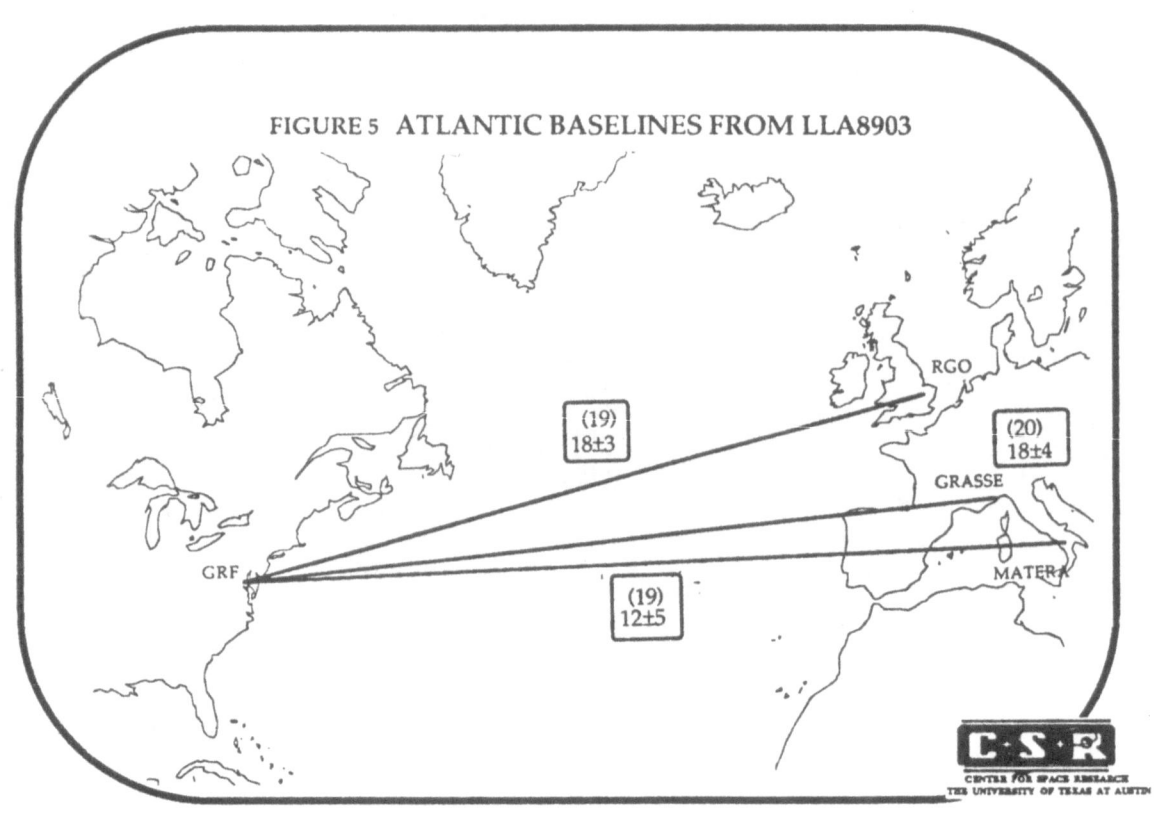

FIGURE 5 ATLANTIC BASELINES FROM LLA8903

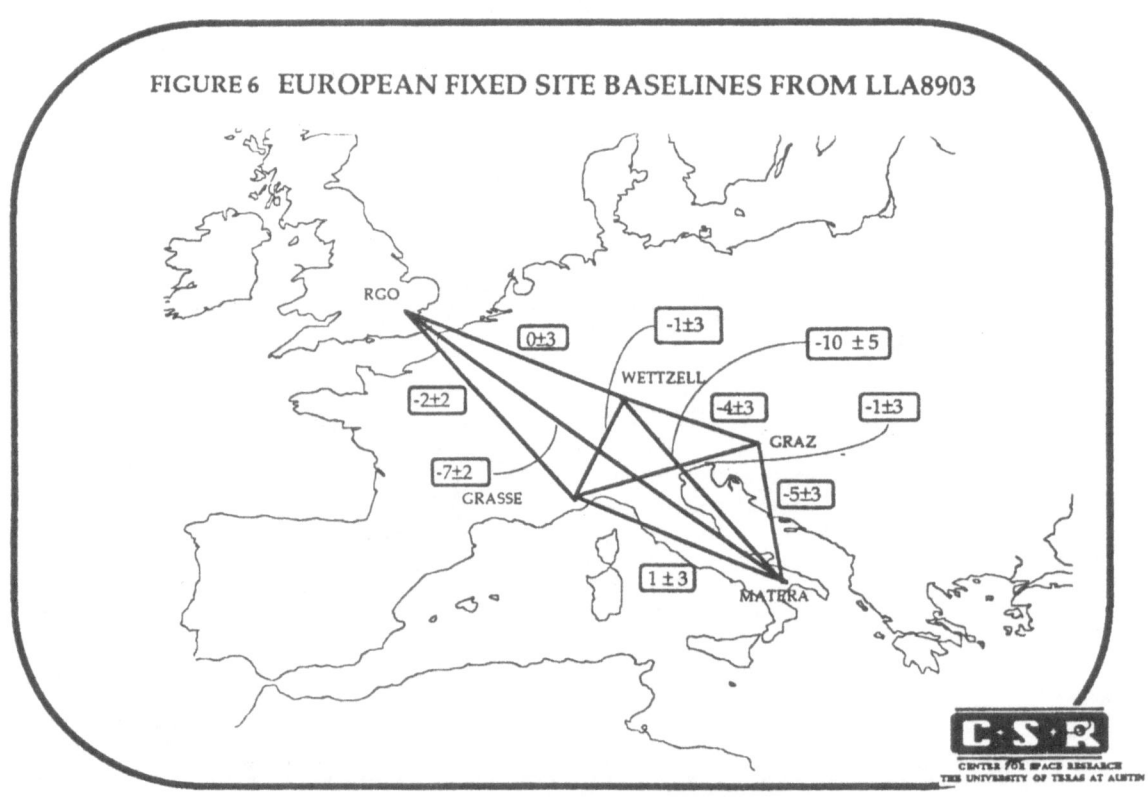

FIGURE 6 EUROPEAN FIXED SITE BASELINES FROM LLA8903

8

Table 2. Global plate velocities.

Plate pair	Latitude (°N)	Longitude (°E)	Velocity (°/MY)
NA–PA	51.1	−74.2	0.875
EU–NA	46.4	129.9	0.280
NA–IN	−23.4	−126.0	0.727
EU–PA	66.6	−86.1	0.963
IN–PA	60.8	1.1	1.110
NA–SA	−17.1	−47.0	1.190
Uncertainties	2°–8°	2°–7°	0.015–0.035°/MY
Minster and Jordan RM2			
NA–PA	48.8	−73.9	0.852
EU–NA	65.9	132.4	0.231
NA–IN	−34.2	−133.3	0.795
EU–PA	60.6	−78.9	0.977
IN–PA	60.7	−5.8	1.246
NA–SA	25.6	−53.8	0.167

CONCLUSIONS

The results obtained from LLA8903 indicate that the increased data span and processing enhancements over earlier UT/CSR long arcs have yielded improved estimates for geodetic parameters and their uncertainties. The technique of satellite laser ranging continues to demonstrate its ability to monitor plate motions at the several millimeter per year level, and the accuracy of both the data and data processing methods continues to improve.

Acknowledgments. This research was supported partially by the NASA/Goddard Space Flight Center under Contract No. NAG5-1118. Additional computing resources were provided by the University of Texas System Center for High Performance Computing.

REFERENCES

Lundberg, J. B, Schutz, B. E., Fields, R. K. and Watkins, M. M. (1989). The application of Encke's method to long arc orbit determination solutions, proceedings of AAS/AIAA Astrodynamics Specialist Conference, Stowe, Vermont, August 7–10, 1989.

Minster, J. B. and Jordan, T. H. (1978). Present-day plate motions, *J. Geophys. Res.* 83(B11), 5331–5354.

Schutz, B. E., Watkins, M. M., Eanes, R. J. and Tapley, B. D. (1989a). Global plate motions derived from Lageos laser ranging, presented at Spring Meeting of the

American Geophysical Union, Baltimore, Maryland, May 1989.

Schutz, B. E., Eanes, R. J. and Watkins, M. M. (1989b). Contribution to *IERS Annual Report for 1988*, Central Bureau of IERS – Observatoire de Paris, Paris, France, June 1989.

Tapley, B. D., Schutz, B. E. and Eanes, R. J. (1985). Station coordinates, baselines, and Earth rotation from Lageos laser ranging: 1976–1984, *J. Geophys. Res.*, **90**(B11), 9235–9248.

Tapley, B. D., Schutz, B. E., Eanes, R. J. and Watkins, M. M. (1988a). Analysis of a twelve-year Lageos long arc, *Eos Trans. AGU*, **69**(16).

Tapley, B. D., Shum, C. K., Yuan, D. N., Ries, J. C. and Schutz, B. E. (1988b). An improved model for the Earth's gravity field (TEG–1), *Proceedings of Chapman Conference on Progress in the Determination of the Earth's Gravity Field*, Ft. Lauderdale, Florida. September 1988.

Watkins, M. M. (1985). Plate Tectonics from Satellite Laser Ranging, *CSR-TM-85-01*, Center for Space Research, The University of Texas at Austin, Austin, Texas.

PLATE MOTIONS DERIVED FROM THE
DGFI 89 L03 SOLUTION

Ch. Reigber, W. Ellmer, H. Müller, E. Geiss, P. Schwintzer, F.H. Massmann
Deutsches Geodätisches Forschungsinstitut/Abt. I
Munich, Fed. Rep. of Germany

ABSTRACT

As part of DGFI's reprocessing program of all laser range measurements to LAGEOS back to 1980, the five-years interval 1983 to 1987 has been finalized. The full set of 19708 data passes in that period was carefully screened and compressed into two-minute normal points. This compressed data information was used to simultaneously estimate orbital parameters, a few dynamic model corrections, five-day Earth rotation parameters, geocentric station coordinates for a common epoch and horizontal station motions for the entire network.

This paper focusses on the description of the directly estimated horizontal station position shifts of those 17 stations which have a good data record over this five-year period and the resulting inter- and intraplate motions. Some first - of course preliminary - results are also presented for the motion of the mobile laser sites in the Aegean area, which had their first reoccupation in 1987.

INTRODUCTION

Since almost 13 years now the LAGEOS satellite has served as the principal laser ranging target, ideally suited for investigations on the Earth rotation and the relative motion of points on the Earth's surface. The tracking network has substantially been improved over the years and it is still improving, not only in terms of the number of stations and better geographical distribution, but also in terms of higher data rates and improved single shot accuracies. It has also been demonstrated by the various analysis groups that models and estimation procedures need to be improved if the high laser range accuracy is intended to be fully exploited. During the last 18 months the German Geodetic Research Institute has spent quite some effort in extending the capabilities and improving algorithms and models of its parameter estimation program DOGS. All full rate data of LAGEOS underwent a new finescreening process. Until July 1989 two-minute normal points were ready for the period 1983 through 1987. The upgraded DOGS software with improved and more consistent models was used to estimate in the dynamic mode various reference frame and dynamic model parameters in a multi-year solution. This paper describes the results obtained from the multi-year solution for the motion in latitude and longitude of those stations which had a good data record over the considered period. From these values spherical distance rates between stations on five lithospheric plates are computed. First results are also presented from the first repeated measurents on five laser pads in the Aegean area in the course of the WEGENER/MEDLAS project activities. Motion values are still very unreliable, but the general tendency of the horizontal motion is with the exception of one station in reasonable agreement with deformation models in this particular region.

METHOD OF DATA REDUCTION

Laser ranging data to LAGEOS exist since its launch in May 1976. We have used for this study all two-minute normal point data regenerated at DGFI for the period 1983 - 1987. This set of 255596 normal point data was created out of a total of 37 million full rate data points gathered from 80 fixed and mobile laser sites around the world in that period. For 23 nearby stations good terrestrially measured eccentricities were available to combine them reliably. Thus the total number of independent stations remaining in the solution were 57. The five years data set was processed in the dynamic mode with the DGFI Orbit and Parameter Estimation Program DOGS. This program comprises a data preprocessing module, an orbit determination module and a parameter estimation module. All conservative and non-conservative force field components and kinematical model parameters required for cm-geodesy are implemented in the orbit determination module. The main characteristics of the nominal model used for this analysis are given in table 1. This model is an improvement over the model used in former studies, such as (Reigber et al., 1988).

Table 1:   DOGS LAGEOS Model MMD2

Reference Frame

CIS	-	mean equator and equinox of J2000.0
Precession	-	IAU 1976 (MERIT standards)
Nutation	-	IAU 1980 + Herring 1987 correction
Planetary Ephemerides	-	DE200/LE200 ephemerides
Initial Pole Series	-	BIH87C02S homog. series (corr. to mean pole)
UT1 Parameter	-	BIH87C02S homog. series
Initial Station Coordinates	-	DGFI 87L03
Initial Station Velocities	-	Zero
Rotation of Coordinate System	-	stations 7086, 7090, 7105, 7109, 7110, 7122, 7210, 7810, 7834, 7835, 7838, 7839,7840, 7907, 7939

Dynamical Model

GM	-	3.98600440 E+14 m³ s⁻²

GM   - $3.98600440\ \mathrm{E}{+}14\ \mathrm{m}^3\ \mathrm{s}^{-2}$

Semi Major Axis of Earth   -   6378136 m

Flattening l/f   -   298.257810

Rotation of Earth   -   0.7292115 E-4 rad/s

Gravity Model of Earth   -   GEM-T1

C(2,1), S(2,1)   -   applied, computed with respect to zero mean pole

Gravity Model of Moon   -   Ferrari 77 (4 x 4)

Third Body   -   Sun, Moon, Jupiter, Venus, Saturn, Mars, Mercury

Air Drag   -   not applied

Albedo   -   Heurtel analytical model

Ocean Tides   -   GEM-T1 (32 waves)

Solid Earth Tides   -   MERIT standards (Wahr model), without Honkasalo term (permanent tide)

Relativistic Motion Equation   -   harmonic isotropic

Measurement Model

Marini-Murray Refraction Model   -   MERIT standards

Solid Earth Tides Displacement   -   MERIT standards (Wahr model)

Ocean Loading Site Displacement   -   Schwiderski model

Pole Tide   -   applied

Normal point elimin. criterion   -   eliminated when 3 or less single shots per normal point

The DOGS orbit computation module is used to form from these data for each 30 day period a normal equation system for the arc-dependent and arc-independent parameters (c.f. table 2). After reduction of the normals for the arc-dependent parameters the 60 monthly systems were combined to the final normal equation system, which is then inverted. Contrary to previous solutions (Reigber et al., 1988), in which the time dependency of the station coordinates was taken into account by computing after a time interval Δt always a new station position solution (step function: abrupt coordinate change when going from epoch t_i to epoch $t_i + \Delta t$; typical interval $\Delta t = 3$ months, 6 months, 12 months) in the DGFI89L03 solution station coordinates are estimated at a reference epoch t_r and simultaneously u individual station dependent motion parameters $p\,(t - t_r)$ are adjusted. (assumption: coordinates of a station change linearly with time).

This last approach has the advantage that over an arbitrarily long analysis interval the computation can be performed in a consistent reference system (the orientation of the terrestrial reference frame needs only to be introduced once - at the reference epoch). The adjusted coordinates and motion values are directly usable for interpretation and intercomparison with geological models.

When trying to estimate a reference frame from laser ranging data only, a singularity remains in the orientation of the system. In order to get rid of this datum defect, we have introduced in this DGFI89L03-solution the Bender/Goad (1979) condition for the position, that is the three rotation angles between the initial and adjusted coordinates of the selected set of stations (c.f. table 1) are forced to become zero in the average.

Table 2.   Adjusted Parameters in DGFI89L03 Solution

Arc-dependent parameters:

- State vector at epoch: every month
- Solar radiation pressure scaling factor: every 15 days
- Empirical along track acceleration: every 15 days

Arc-independent parameters:

- Gravity field parameters: GM
- ERP parameters: x_p, y_p, LOD every 5 days
- Station positions: φ_i, λ_i, h_i epoch values (1984.0)
- Horizontal station motions: $\dot{\varphi}_i$, $\dot{\lambda}_i \cos \varphi_i$

As far as the drift is concerned, with the second set of equations in table 3 we have forced the rotational drift rates about the x-, y-, z-axes to be zero with respect to the given set of the Minster and Jordan AM0-2 (1978) horizontal station position drift rates. This datum fixation enables the direct intercomparison of the adjusted geodetic rates of change with those as inferred from the AM0-2 geologic model because both sets then are expressed in the same reference system.

In order to constrain the relation between the CTS and CIS one length of day value was fixed to the initial one in the solution.

Table 3.   Constraints introduced

Orientation of CTS

–   Position (c.f. BENDER/GOAD): $\alpha_X = 0, \quad \alpha_Y = 0, \quad \alpha_Z = 0$

- $$\sum_{i=1}^{m} w_i \left(Rd\lambda_i \cos\phi_i \sin\phi_i \cos\lambda_i - Rd\phi_i \sin\lambda_i \right) = 0$$

- $$\sum_{i=1}^{m} w_i \left(Rd\lambda_i \cos\phi_i \sin\phi_i \sin\lambda_i - Rd\phi_i \cos\lambda_i \right) = 0$$

- $$\sum_{i=1}^{m} w_i \left(Rd\lambda_i \cos^2\phi_i \right) = 0$$

–   Drift: $\dot{\alpha}_X = c_1, \quad \dot{\alpha}_Y = c_2, \quad \dot{\alpha}_Z = c_3$

- $$\sum_{i=1}^{m} w_i \left(Rd\dot{\lambda}_i \cos\phi_i \sin\phi_i \cos\lambda_i - Rd\dot{\phi}_i \sin\lambda_i \right) = c_1$$

- $$\sum_{i=1}^{m} w_i \left(Rd\dot{\lambda}_i \cos\phi_i \sin\phi_i \sin\lambda_i - Rd\dot{\phi}_i \cos\lambda_i \right) = c_2$$

- $$\sum_{i=1}^{m} w_i \left(Rd\dot{\lambda}_i \cos^2\phi_i \right) = c_3$$

RESULTS

The DGFI89L03 solution consists of 2138 adjusted parameters, of which 100 are motion values in longitude and latitude for a total of 50 stations. The rms fit of this five-years solution is 6.4 cm. Here the results for only the horizontal station motions are given.

For 15 stations the station velocity could be estimated with a formal error of smaller than 5 mm/yr. The actual computed motion values for these 15 stations, belonging to five different plates, are given in table 4. These values are compared with the predicted values according to the M/J AM0-2 and Nuvel (De Mets et al., 1989) models in the same table. Considerable differences in the direction and/or size of the motions is found for stations Simosato (7838), Huahine (7121), Quincy (7109),

Table 4.

Station	$\dot\phi$	$\dot\lambda\cos\phi$	$\sigma_{\dot\phi}$	$\sigma_{\dot\lambda\cos\phi}$	$\dot\phi$	$\dot\lambda\cos\phi$	$\dot\phi$	$\dot\lambda\cos\phi$
		DGFI89L03			MJ AM0-2		NUVEL	
EURA								
1181 POTSDM	3.0	2.6	.6	.6	1.6	1.9	1.1	1.9
7810 ZIMMER	1.2	2.7	.4	.5	1.7	1.9	1.2	1.9
7834 WETZEL	1.3	1.3	.1	.2	1.6	2.0	1.1	1.9
7835 GRASSE	.3	1.5	.2	.2	1.7	2.0	1.2	1.9
7839 GRAZ	1.5	1.9	.1	.1	1.6	2.1	1.0	2.0
7840 HEARST	2.0	1.8	.1	.1	1.8	1.6	1.4	1.7
7939 MATERA	1.9	1.4	.1	.1	1.6	2.2	1.0	2.0
AEGEAN								
7510 ASKITS	2.6	.3	.9	1.0	-	-	-	-
7512 KATAVI	-6.3	-9.6	2.0	2.1	-	-	-	-
7515 DIONYS	-5.5	.6	1.0	1.0	-	-	-	-
7517 ROUMEL	-3.8	-.2	.8	.8	-	-	-	-
7525 CHRYSO	-1.3	-.4	1.2	1.2	-	-	-	-
NOAM								
7086 FTDAVS	-.2	-.5	.1	.2	-1.0	-1.3	-.3	-1.7
7105 GSFC	.6	-1.0	.1	.2	.2	-1.7	.7	-1.9
7109 QUINCY	-1.4	-3.3	.1	.1	-1.6	-1.4	-1.0	-1.8
7122 MAZTLN	-.8	-1.4	.1	.1	-1.1	-1.1	-.4	-1.5
PCFC								
7110 MNPEAK	1.1	-4.1	.1	.1	2.8	-4.8	2.8	-4.8
7121 HUAHIN	-.9	-11.4	.4	.4	3.7	-9.1	3.7	-9.1
7210 MAUI	3.0	-7.1	.1	.1	3.7	-6.8	3.7	-6.8
INDI								
7090 YARAGA	6.3	3.1	.1	.1	6.0	4.3	6.0	3.7
SOAM								
7907 AREQUI	1.1	1.0	.1	.1	1.0	-.7	1.2	-1.1
EURA(JAPAN)								
7838 SIMOSA	-.9	-.8	.2	.2	-1.5	2.4	-1.7	1.3

Figure 1.   Horizontal Station Motions from DGFI89L03

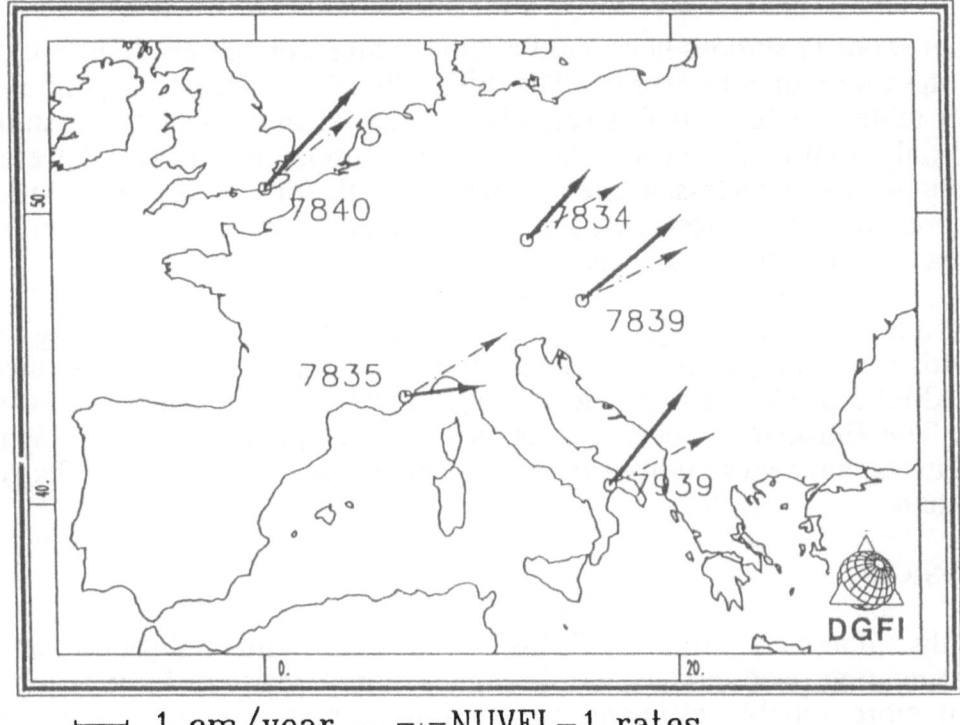

⊢——⊣ 1 cm/year   - - -NUVEL-1 rates

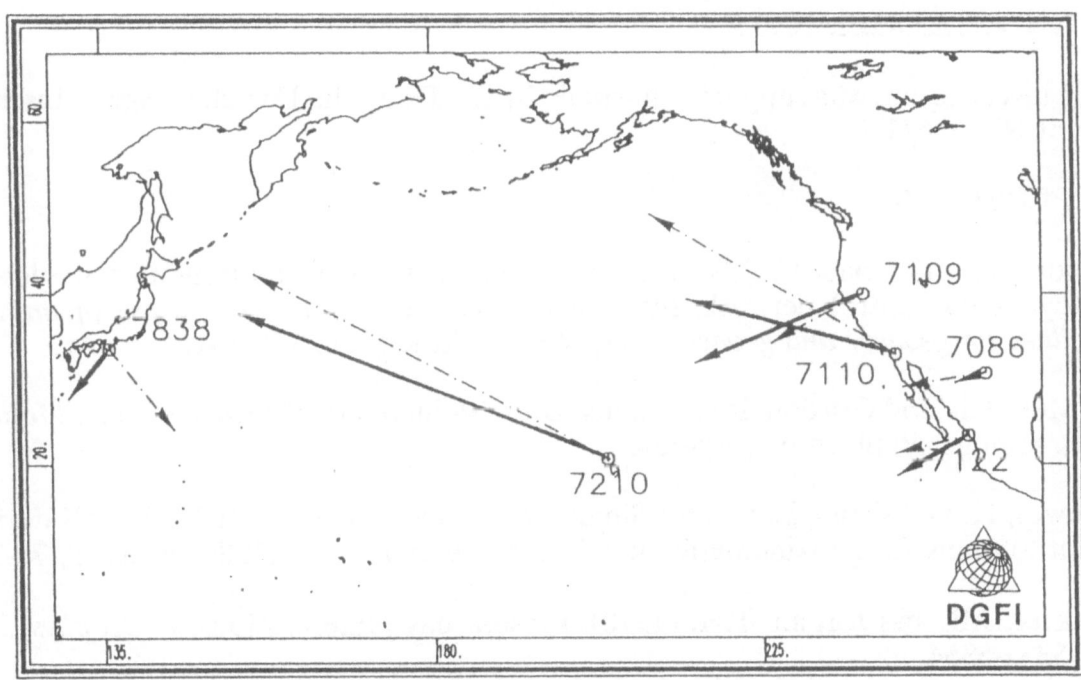

⊢——⊣ 1 cm/year   - - -NUVEL-1 rates

17

Arequipa (7907) and Grasse (7835) c.f. figure 1. Data problems are expected for stations Huahine and Grasse.

For another set of 12 stations horizontal motion values were determined with formal errors in the range of 5 to 50 mm. The WEGENER/MEDLAS Aegean network stations in table 4 belong to this set. The horizontal motion of these stations is estimated only from two relatively short tracking campaigns and can therefore not be very reliable. Nevertheless, as seen from figure 1 the observed N-S extension and E-W compression is in reasonable agreement with geodynamic deformation models in this area (Drewes and Geiss, 1986).

From the station motions the geodesic rates between those stations having the best data records were computed and are summarized in table 5. Major discrepancies with the AMO-2 predictions exist for all lines having as one endpoint the station Simosato. The European geodesic rates as given in figure 1 imply no significant motion between any two stations if the three sigma criterion is taken as the realistic error estimate.

CONCLUSION

The results obtained from DGFI89L03 indicate that through enhanced preprocessing, improved models and through a more consistent form of parameter estimation more reliable estimates for station motions are achievable. Further processing of older and newer LAGEOS data will continue to improve the results.

ACKNOWLEDGEMENTS

This investigation was supported partially by the Deutsche Forschungsgemeinschaft, Grant RE 536/1-2.

REFERENCES

Bender, P. and Goad, C. (1979): Probable LAGEOS contributions to a worldwide geodynamics control network. In: Veis and Lievieratos (eds) *The use of artificial satellites for geodesy and geodynamics*, Vol. II, Athen, pp. 145-161.

DeMets, Ch. and Gordon, R.G., Argus, D.F., Stein, S. (1989): *Current Plate Motions*. Pers. Comm., Publ. in preparation.

Drewes, H. and Geiss, E. (1986): Simulation study on the use of MEDLAS derived point motions for geokinematic models. *Adv. in space res.*, (1986) 6: no. 9, 71-74.

Minster, J.B. and Jordan, T.H. (1978): Present-day plate motions. *J. Geophys. Res.*, 83: 5331-5354.

Reigber, Ch.; Schwintzer, P., Müller, H.; Barth, W.; Massmann, F.H. (1988): The terrestrial reference frame underlying the Pre-ERS-1 Earth Model determination. *manuscripta geodaetica* (1988)13: 349-358.

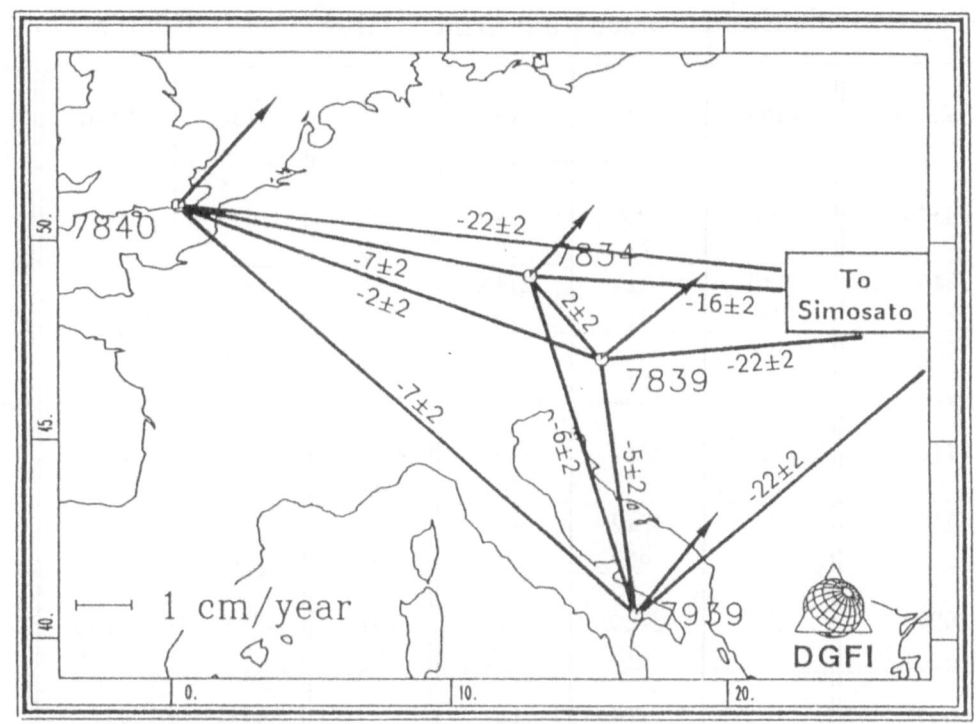

Figure 2. Spherical Distance Rates Between Fixed European SLR Sites
as Resulting from DGFI89L03 Solution (mm/yr)

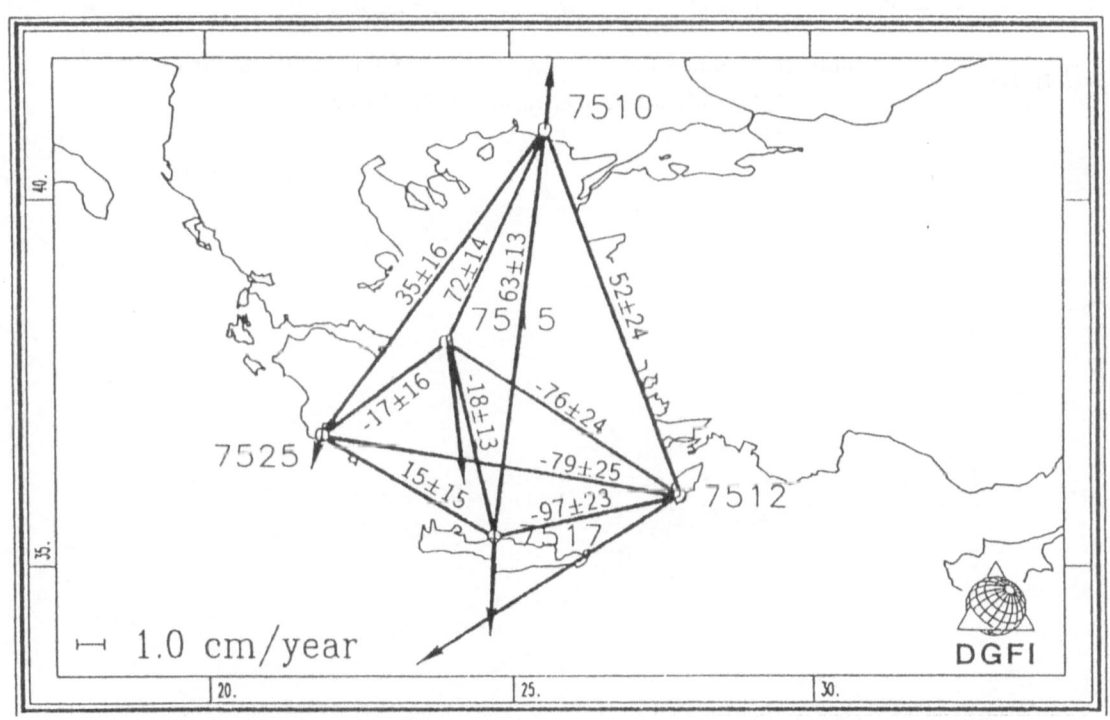

Spherical Distance Rates Between   Aegean SLR Sites
as Resulting from DGFI89L03 Solution (mm/yr)

19

Table 5. Spherical Distance Rates from DGFI89L03, bottom: M/J AM0-2 rates

ST. NR.	7105	7186	7122	7907	7110	7210	7090
7834	9±2 21	7±2 21	17±2 21	1±2 20	10±2 1	-31±2 -41	-30±2 -26
7840	14±2 22	10±2 21	21±2 22	10±2 19	15±2 6	-22±2 -28	-39±2 -27
7939	6±2 23	3±2 22	14±2 22	4±2 24	7±1 -4	-39±2 -43	-18±2 -15
7110	23±2 16	38±2 41	34±1 55	37±1 42	- -	11±1 0	-88±1 -103
7210	3±2 16	49±2 34	50±2 48	82±1 66	11±1 0	- -	-99±2 -103
7090	-79±2 -88	-50±2 -70	-48±2 -58	70±2 62	-88±1 -103	-99±2 -103	- -
7907	-3±2 -6	-2±2 -11	3±1 -12	- -	37±1 42	82±1 66	70±2 62
7838	2±2 -6	10±2 -11	7±2 -12	12±2 -25	-28±2 -65	-68±2 -99	-81±2 -77

Figure 3. Spherical Distance Rates (mm/yr) from DGFI89L03 Solution
(Values in Brackets inferred from M/J AMO-2 Model)

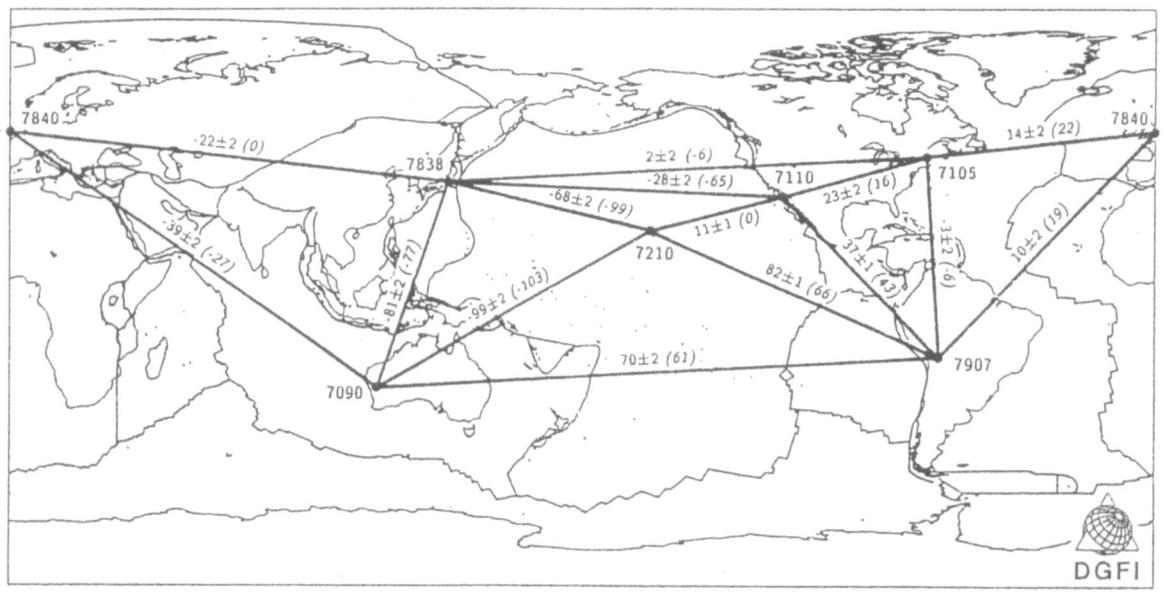

PLATE MOTIONS AND DEFORMATION FROM LAGEOS

David E. Smith, Ronald Kolenkiewicz
NASA, Goddard Space Flight Center, Laboratory for Terrestrial Physics,
Code 920, Greenbelt, MD 20771 USA

Peter J. Dunn, Mark H. Torrence, John W. Robbins,
Steven M. Klosko, Ronald G. Williamson
ST Systems Corp., Lanham, MD 20706 USA

Erricos C. Pavlis, Nancy B. Douglas
University of Maryland, College Park, MD USA

Susan K. Fricke
RMS Technologies Inc., Lanham, MD 20706 USA

INTRODUCTION

LAGEOS laser ranging observations collected by the global tracking network between May 1976 and December 1988 have been analysed to yield a comprehensive geodetic parameter solution. Some of the stations in the participating network have a continuous tracking record over the full LAGEOS mission lifetime and can be used to monitor positions in a limited network for over 12 years. However, the introduction of several new stations of improved precision has allowed determination of relative positions to centimeter accuracy for each quarter of as year solution since the beginning of 1980. A nine year history of these three-dimensional positions yields horizontal inter-station baseline rates to an accuracy of a few mm/yr as well as vertical displacement rates to similar accuracy. Comparisons of the measurements of plate motion between sites centrally located within the plates with those predicted by current geological models agree to better than 9 mm/yr RMS. The phenomena of post-glacial rebound is considered as an explanation for the vertical motion.

The determination of geodetic (and geophysical) information from laser ranging observations relies on maintainence of a unique reference frame throughout the nine year tracking history and the appropriate modeling of the dynamics involved. Our approach is based on the dynamic method which simultaneously determines an accurate description of the LAGEOS' orbit. A more detailed discussion of this analysis can be found in (Smith et al., 1989a, Smith et al., 1989b, and Christodoulidis et al., 1985). The SL7.1 results are presented in the form of SLR velocity vectors and the intersite geodesic rates which are further compared to those implied by the geologic plate motion models.

RESULTS

Figure 1a shows the decreasing behaviour of the monthly orbital RMS of fit for LAGEOS. In the early years this RMS was around 30 cm, with a fair amount of scatter. As tracking accuracy and our modeling capabilities improved and as more stations were added to the network, this RMS and its scatter have decreased considerably. For the last few years the RMS has been consistently under 5 cm. Figure 1b reflects a similar pattern for monthly GM values. Before 1980, considerable variations and large uncertainties prevailed, but with increased data, the value of GM stabilized around a mean of 398600.4408 km^3s^{-2}. This suggests that the original variation was not a real phenomena, but is rather an artifact of data quality and quantity.

Table 1 and Figure 2 show the station velocities and vector motions for 17 SLR stations. The determination of site velocities is based on the observed intersite geodesic rates and the adoption of the motion for two reference stations. The solution utilizes a least squares estimation procedure. In Figure 2, the black arrows with error ellipses are based upon the LAGEOS SL7.1 analysis. Vectors predicted by AM0-2 are shown as halftoned arrows.

Figures 3a and 3b show motion vectors for RGO, Grasse, Wettzell, Potsdam, Graz and Matera superimposed on simplified tectonic maps of Europe. Figure 3a show both the SL7.1 vectors (solid black arrows) with their associated error ellipses and the Minster and Jordan AM0-2 (1978) vectors (halftoned arrows) for each site. The AM0-2 vector shown for Matera is based upon the assumption of African motion for the Adriatic region. Geophysical evidence has been reported which supports this assumption (Mantovani et al., 1989). Our results for Matera indicate better agreement between SLR results and geologic model predictions under the African motion assumption as opposed to assuming Eurasian behavior for Matera. Figure 3b shows the same sites and their residual motion vectors between SL7.1 and AM0-2. This analysis indicates good agreement between geodetic and geologic motion vectors for stations centrally located within Europe, and, with the exception of Matera, the fixed stations in Europe appear to move as a single tectonic block.

Figure 4a and 4b are similar maps for the SLR sites located in North America. Quincy, Mazatalan, Platteville, McDonald, and Greenbelt are assumed to lie on the North American plate, with Monument Peak on the Pacific plate. It should be kept in mind that some of these sites are located in diffuse zones of tectonic deformation and may be affected by local and regional tectonics. Figure 4b show the residual motion vectors between SL7.1 and AM0-2. It can be seen from these figures that there is less agreement between vector motion in the Basin and Range and along the San Andreas. A more detailed discussion of geodesic baselines across North America (Smith et al., 1989) suggests that Basin and Range spreading is observed at 5 - 10 mm/yr, and that motion across the San Andreas Fault system is 20 mm/yr less than predicted by rigid plate models. In Figure 4b this effect can be deduced from the residual motion vectors as well.

Figure 5a and 5b show the correlation of the geodesic interplate rates for the stations centrally located on major plates (Yaragadee, Orroral, Hawaii, Huahine, RGO, Wettzell, Graz, Greenbelt, McDonald, and Mazatalan) obtained from our LAGEOS analysis and those predicted using the AM0-2 and NUVEL-1 (DeMets et al., 1989) models respectively. These figures indicate that the agreement between SLR observations over the last decade and the linear rates predicted from million year averages developed from magnetic anomaly profiles, transform fault azimuths and Earthquake slip vectors, is quite striking. Furthermore, the SLR results agree equally well with the more recent model NUVEL-1 which contains substantially more data than AM0-2. For these SLR sites which are located in the stable plate interiors, the motion predicted from the geological models is very similar.

REFERENCES

Christodoulidis, D. C., D. E. Smith, R. Kolenkiewicz, S. M. Klosko, M. H. Torrence and P. J. Dunn, Observing Tectonic Plate Motions and Deformations from Satellite Laser Ranging, *J. Geophys. Res.*, V.90, pp.9249-9263, 1985.

DeMets, C., R. G. Gordon, D. F. Argus, and S. Stein, Current Plate Motions, *Geophys. J. International*, in press, 1989.

Minster, J. B. and T. H. Jordan, Presnt-day Plate Motions, *J. Geophys. Res.*, V.83, pp.5331-5354, 1978.

Mantovani, E., D. Babbucci, D. Albarello and M. Mucciarelli, Deformation Pattern in the Central Mediterranean and Behavior of the African/Adriatic Promontory, *Tectonophysics*, in press, 1989.

Smith, D. E., R. Kolenkiewicz, P. J. Dunn, M. H. Torrence, J. W. Robbins, S. M. Klosko, R. G. Williamson, E. C. Pavlis, N. B. Douglas and S. K. Fricke, Tectonic Motion and Deformation from Satellite Laser Ranging to LAGEOS, *J. Geophys Res.*, in press, 1989a.

Smith, D. E., R. Kolenkiewicz, B. H. Putney, P. J. Dunn, S. M. Klosko, E. C. Pavlis, J. W. Robbins, M. H. Torrence, R. G. Williamson, and S. K. Fricke, A geodetic Earth motion model derived from LAGEOS observations: GSFC-SL7, *NASA Tech. Memo.*, in preparation, 1989b.

Table 1.
SLR Station Velocity Model

Station Name	SLR Model Azimuth (°)	Rate (mm/yr)	Error Ellipse Parameters S. Major (mm/yr)	S. Minor (mm/yr)	Orient. (°)	M/J AM0-2 Model Azimuth (°)	Rate (mm/yr)
Quincy	254	22	2.6	1.6	-12	221	21
McDonald Obs.	245	24	2.9	1.9	-11	234	17
Mazatlan	251	16	2.6	2.0	-23	225	15
Platteville	231	13	9.6	5.1	-7	239	19
Greenbelt	277	18	*	*	*	277	18
Monument Pk.	288	44	2.6	1.6	-17	300	55
Huahine	288	92	6.7	2.7	-52	297	80
Hawaii	299	77	*	*	*	299	77
Wettzell	35	21	4.1	2.1	38	50	26
RGO	35	22	4.4	2.0	22	43	24
Graz	46	27	4.6	2.3	44	52	26
Matera	29	25	4.2	2.1	43	54	27
Simosato‡	278	10	4.2	3.0	-46	123	28
Easter Island	99	81	4.2	2.4	-62	96	90
Arequipa	37	12	3.4	2.0	69	326	12
Orroral	20	68	4.5	2.8	1	22	56
Yaragadee	28	63	3.5	2.4	-18	35	74

* Greenbelt and Hawaii are constrained to move as M/J AM0-2.

‡ Although the plate upon which Simosato resides is under question, we have assumed Eurasian in our analysis.

QRA1BAN1 solution - 890308

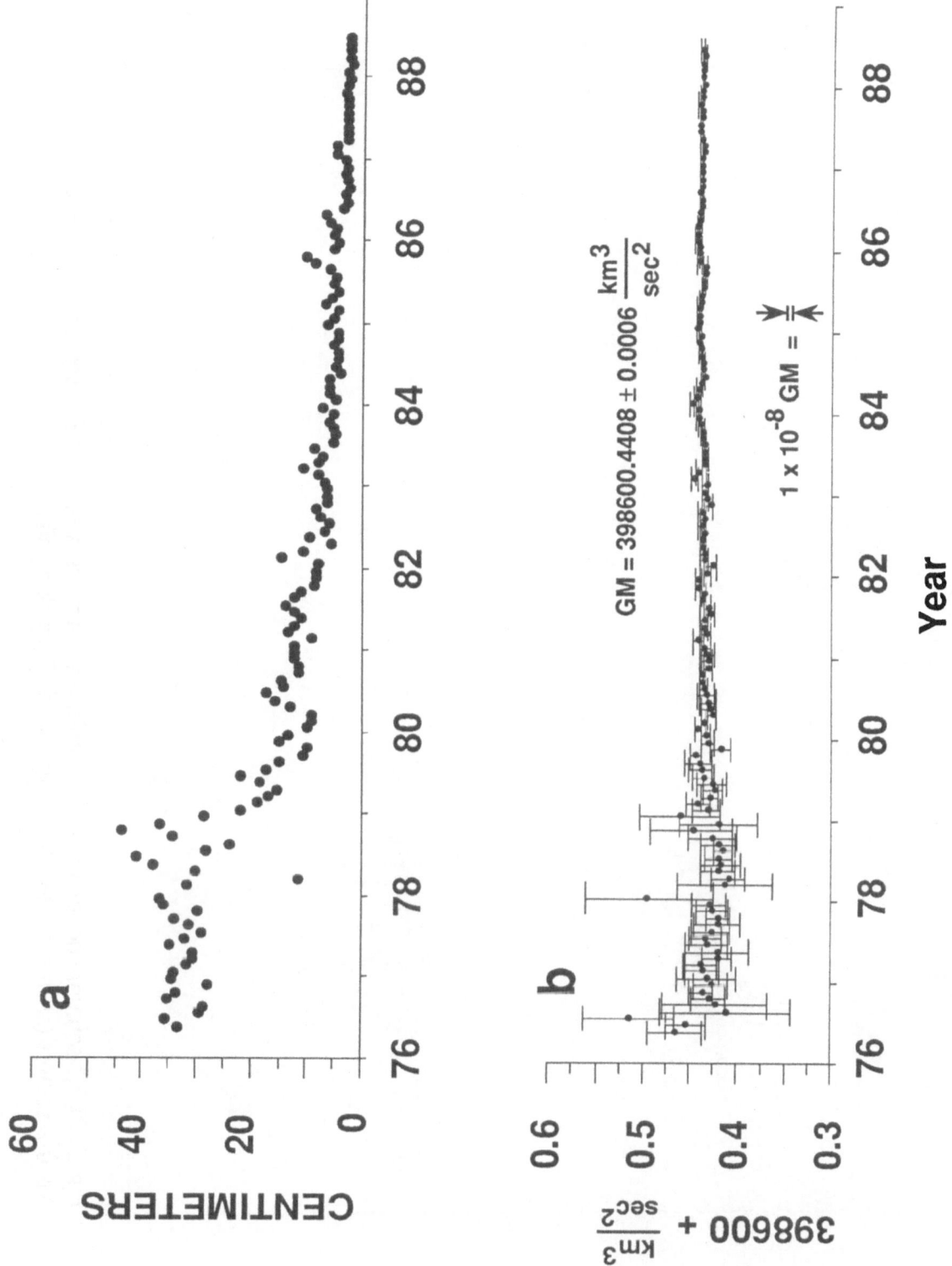

Fig. 1. (a) RMS of orbital fit for each monthly arc of LAGEOS data. (b) Values of the product of the mass of the Earth and the gravitational constant (GM) determined in each monthly arc.

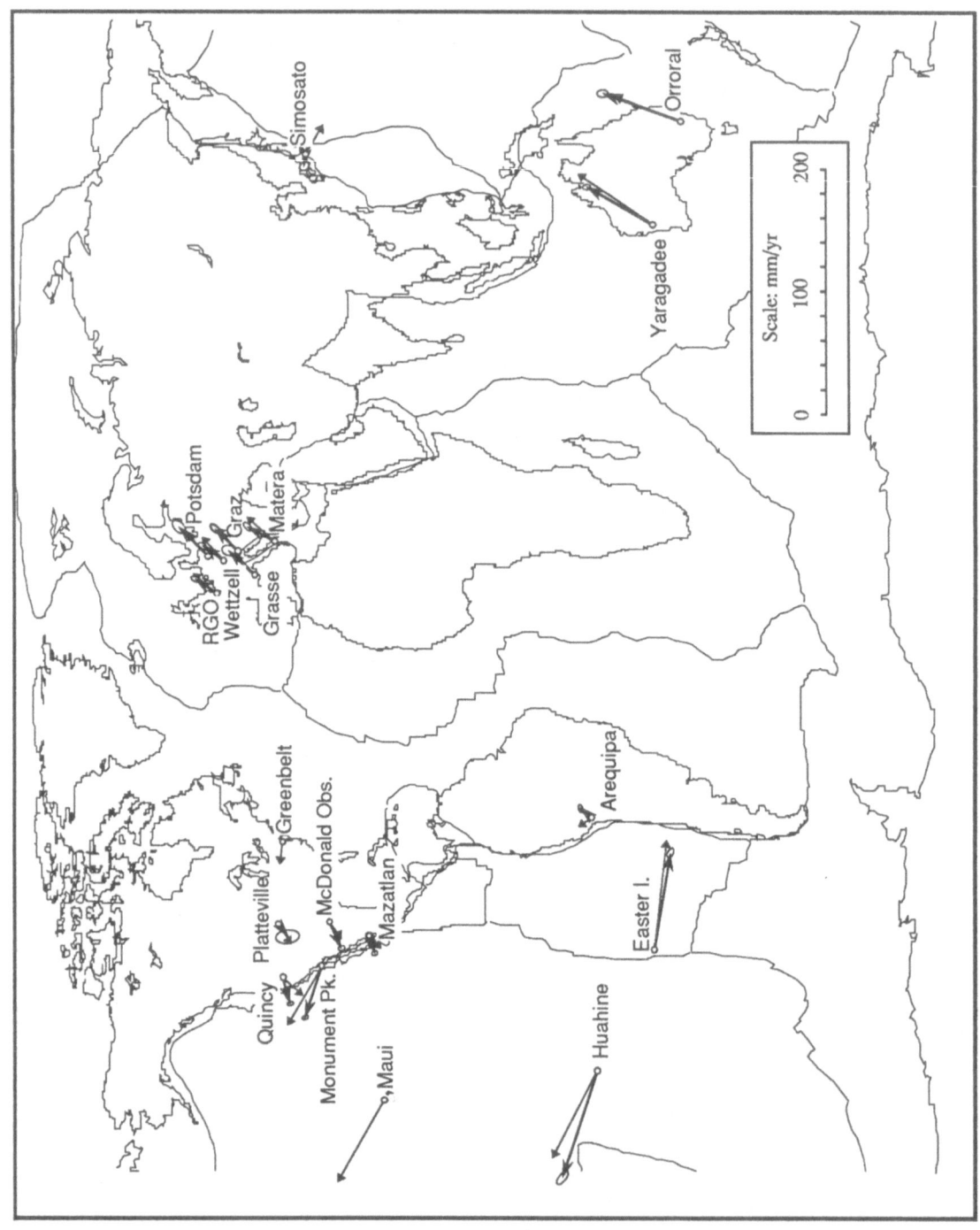

Fig. 2. SL7.1 solution motion vectors (as bold black arrows with error ellipses) and Minster & Jordan AM0-2 vectors (as thin arrows) for the global SLR tracking sites.

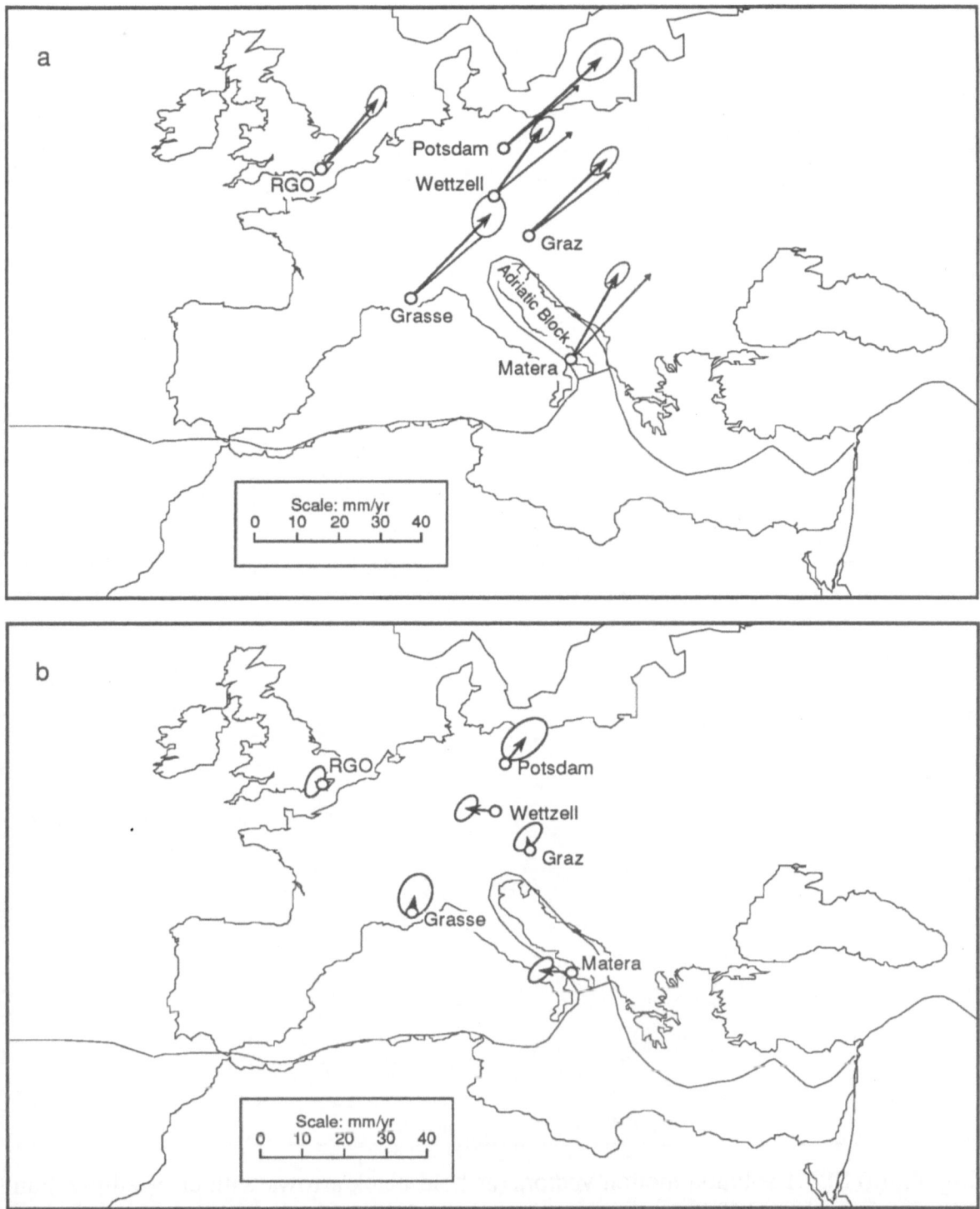

Fig. 3. (a) SL7.1 solution motion vectors (as bold black arrows with error ellipses) and Minster and Jordan AM0-2 vectors (as thin arrows) for the European laser tracking sites. The AM0-2 vector shown for Matera is based upon African motion as described in the text. (b) The same sites showing residual motion vectors between SL7.1 and AM0-2. Figure is taken from Smith et al. (1989a).

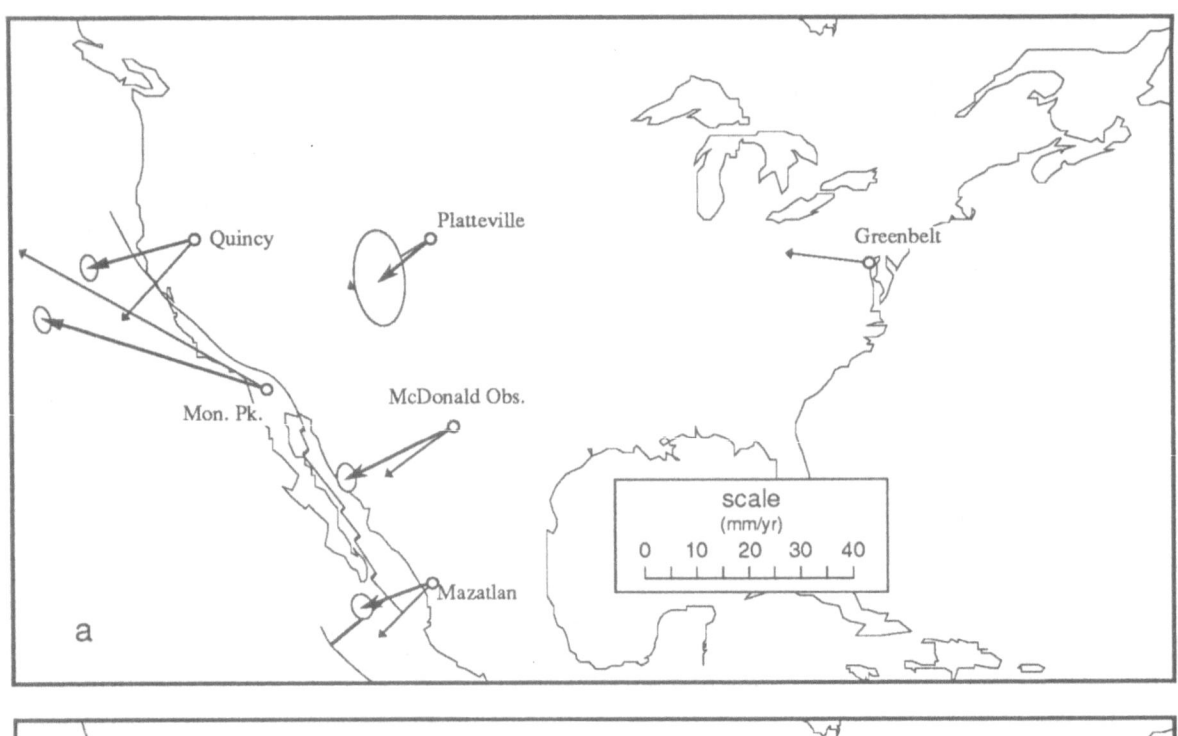

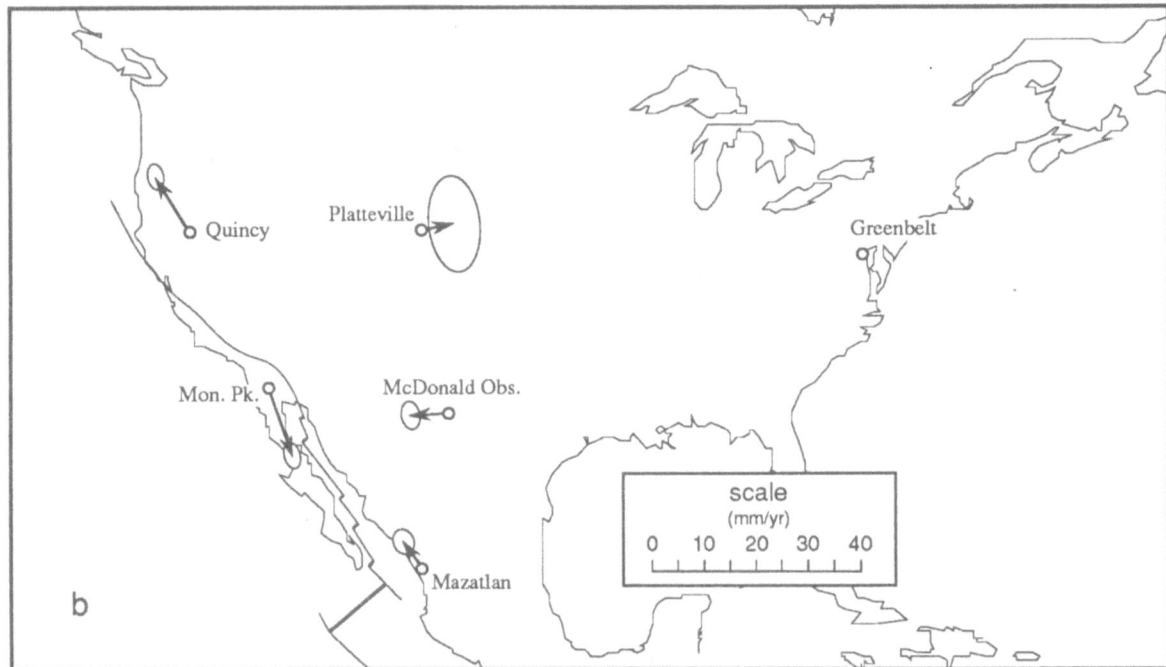

Fig. 4. (a) SL7.1 solution motion vectors (as bold black arrows with error ellipses) and Minster & Jordan AM0-2 vectors (as thin arrows) for the North American laser tracking sites. (b) The same sites showing residual motion vectors between SL7.1 and AM0-2. Figure is taken from Smith et al. (1989a).

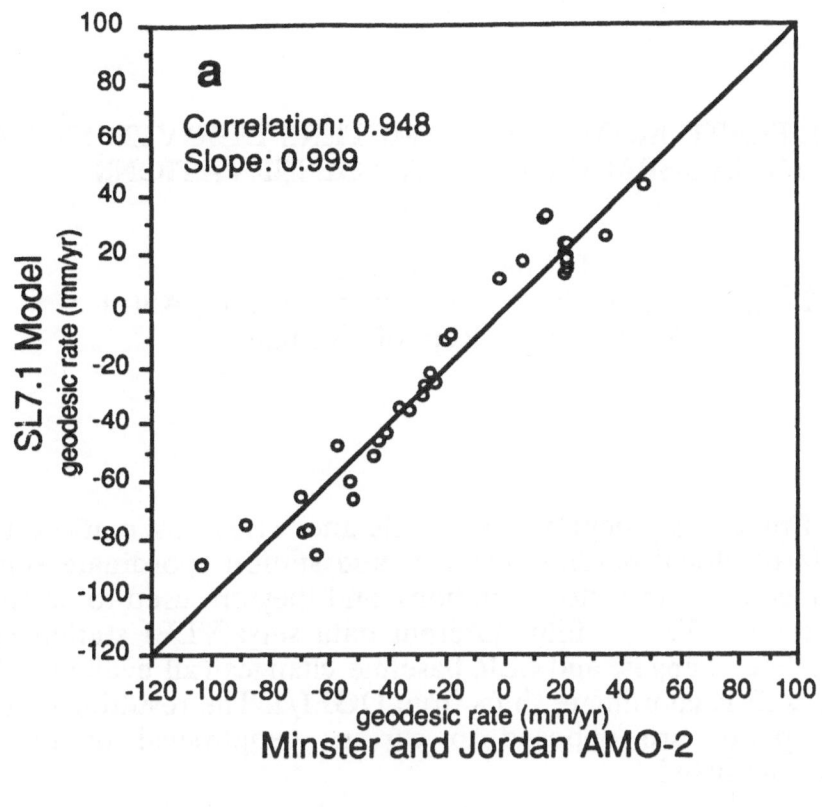

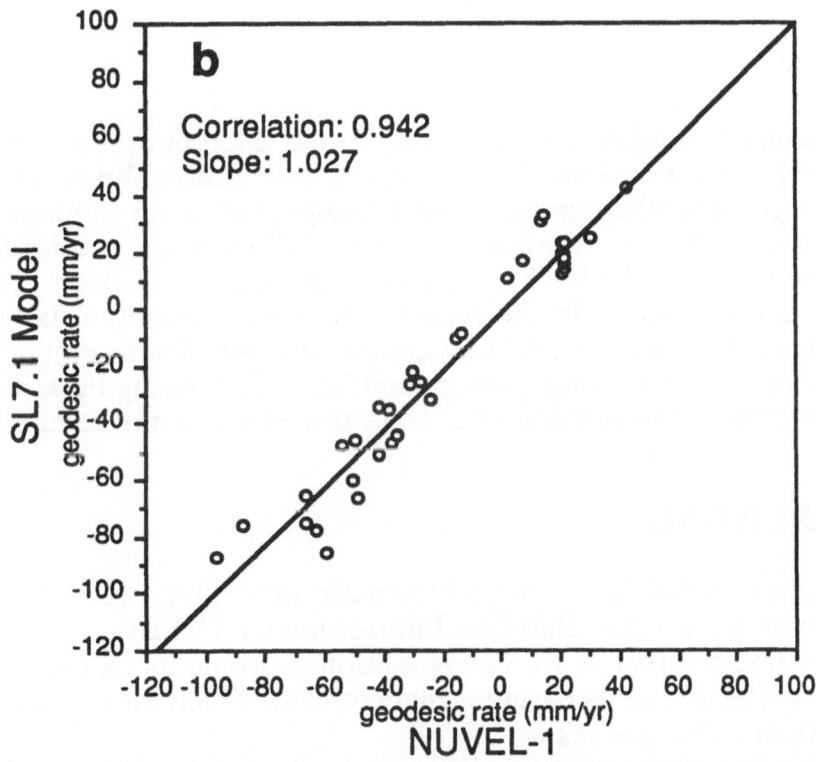

Fig. 5. SL7.1 geodesic rates for SLR stations centrally located within a plate:
(a) Correlation with rates predicted by AMO-2. (b) Correlation with rates
predicted by NUVEL-1. Figure taken from Smith et al. (1989a).

GLOBAL PLATE MOTION PARAMETERS DERIVED FROM ACTUAL SPACE GEODETIC OBSERVATIONS

Hermann Drewes
Deutsches Geodätisches Forschungsinstitut/Abt. I
Munich, Fed. Rep. of Germany

ABSTRACT

Recent results of time-dependent geodetic SLR and VLBI observations provide us with precise data of global baseline changes and station coordinate shifts. These data are interpreted as plate tectonic motions, and they are used to derive uniform plate rotation vectors. We use four different data sets: VLBI station coordinate shifts, VLBI baseline changes, and SLR baseline changes (all evaluated by NASA GSFC), as well as SLR coordinate shifts from DGFI/I. The resulting plate rotation vectors of five plates are opposed to current geophysical parameters. The discrepancies are discussed.

INTRODUCTION

Current plate kinematic models are exclusively based on geophysical observations (sea floor spreading rates, transform fault azimuths, earthquake slip vectors). These data reflect *recent geologic* phenomena as an average over some millions of years. Frequently cited models are the "absolute" motions AM2 (Minster and Jordan, 1978) and the relative motion model NUVEL-1 (Gordon et al. 1988, DeMets et al, 1989).

In geodesy we are interested in the *present-day* plate motions to be estimated from observations over a time interval of a couple of years. These motions are not necessarily identical with the recent geologic motions. Comparing both data sets is equivalent to checking the inclination of a 1-km trace by a 1-mm section.

GEODETIC MODELLING

Geodetic observations suitable for plate kinematic modelling are Satellite Laser Ranging (SLR) and Very Long Baseline Interferometry (VLBI). From repeated observations in a time interval Δt we derive station movements $\overline{\Delta x}$ either in terms of coordinate shifts $(\Delta\varphi, \Delta\lambda)$ or as baseline changes (Δb), which may be transformed into spherical distance changes (Δs).

The kinematic plate parameters to be estimated are the individual rotation vectors $\overline{\Omega}$. Any instantaneous motion of a spherical cap can be uniquely and completely described by a geocentric rotation vector. We represent $\vec{\Omega}$ by the geographical position of the pole of rotation (Φ, Λ) and the rotational velocity (ω). The geometric relations between $\vec{\Omega}$ (Φ, Λ, ω) and $\vec{\Delta x}$ ($\Delta\varphi$, $\Delta\lambda$) or Δs, respectively, are shown in Figure 1.

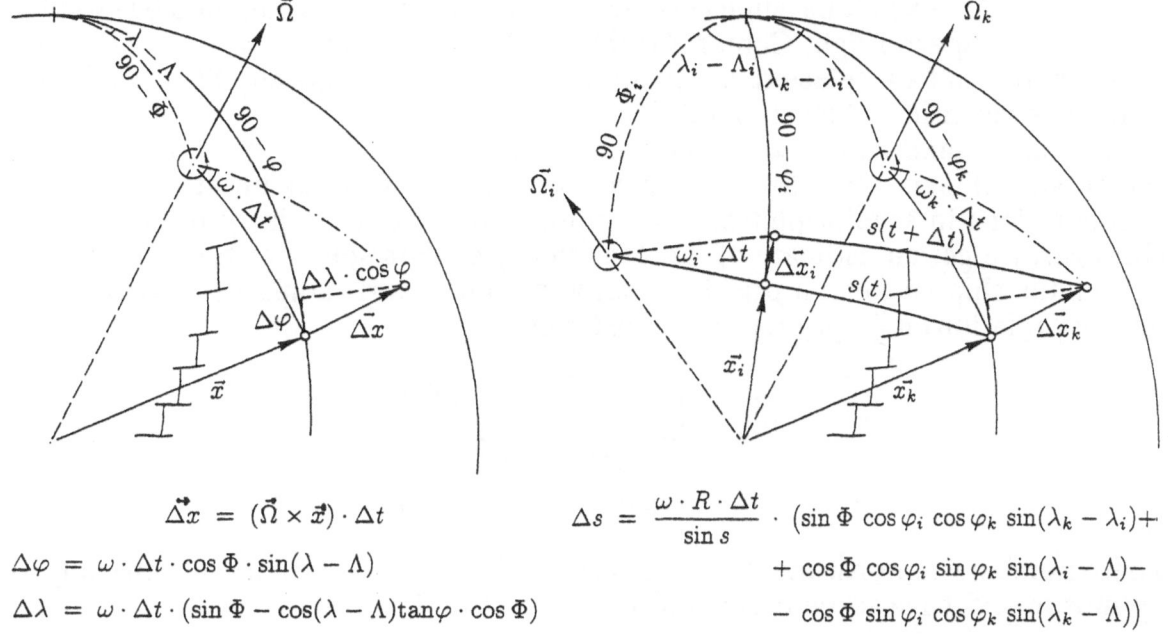

$$\vec{\Delta x} = (\vec{\Omega} \times \vec{x}) \cdot \Delta t$$

$$\Delta\varphi = \omega \cdot \Delta t \cdot \cos\Phi \cdot \sin(\lambda - \Lambda)$$

$$\Delta\lambda = \omega \cdot \Delta t \cdot (\sin\Phi - \cos(\lambda - \Lambda)\tan\varphi \cdot \cos\Phi)$$

$$\Delta s = \frac{\omega \cdot R \cdot \Delta t}{\sin s} \cdot (\sin\Phi \cos\varphi_i \cos\varphi_k \sin(\lambda_k - \lambda_i) +$$
$$+ \cos\Phi \cos\varphi_i \sin\varphi_k \sin(\lambda_i - \Lambda) -$$
$$- \cos\Phi \sin\varphi_i \cos\varphi_k \sin(\lambda_k - \Lambda))$$

Fig. 1.   Geometric relations between geodetic observables and plate kinematic parameters.

DATA ADJUSTMENT

For estimating each three parameters of n plates we need at least 3 · n observations $\Delta\varphi$, $\Delta\lambda$, or Δs. In case of redundant observations we perform a least squares adjustment (Drewes, 1982). The kinematic datum is realized by fixing the rotation parameters of one plate ($\vec{\Omega}$ = const.) or by introducing at least three coordinate shifts ($\Delta\varphi$ and/or $\Delta\lambda$). In the latter case the plate kinematic datum is given with respect to the underlying reference frame of the station coordinate estimation (SLR or VLBI reference system).

The accuracy of estimated plate rotation parameters depends on the accuracy of geodetic observations and the configuration of observation stations. The configuration is an optimum, if the great circles from two stations intersect rectangularly in the pole, and if the geocentric distance to the pole is 90°. A first order network design is given by Drewes (1982).

USED DATA SETS

Four different data sets were used for the present study
-        44 SLR station coordinate shifts computed by the German Geodetic Research Institute, DGFI L03 (Reigber et al., 1989),

- 104 SLR spherical distance changes (geodesic rates) between 15 stations from Goddard Space Flight Center, GSFC SL7.1 (Smith et al., 1989),
- 55 VLBI derived coordinate shifts computed by Goddard Space Flight Center, GLB401 (Ma et al., 1989),
- 124 VLBI baseline changes between 30 stations computed by Goddard Space Flight Center, GLB405 (Ma et al., 1989).

A comparison of selected observations referring to adjacent SLR and VLBI stations is given in Tables 2 and 3.

The individual data sets were analysed and processed in particular plate rotation vector adjustments after formulae given in Fig. 1. Due to unsatisfying station configurations in several plates, the solutions are very weak in some parameters. However, we get an estimate of the consistency of data sets, which is expressed by the relationship between a priori (given) and a posteriori (adjusted) variances. We find the relations σ (a priori) : σ (a posteriori)

L03	1 : 6.8
SL7.1	1 : 1.7
GLB401	1 : 7.1
GLB405	1 : 5.0

These numbers demonstrate that the space geodetic error estimates all generally too optimistic and have to be corrected by factors 2 to 7.

A GEODETIC PLATE KINEMATIC MODEL

For the final result we computed two independent combined adjustments of SLR and VLBI observations of the different data types:
- a coordinate shift solution including L03 and GLB401 data sets
- a baseline solution including SL7.1 and GLB405 data sets.

We did not combine the two SLR solutions or the two VLBI solutions, respectively, because they are highly correlated by using the same observation material. The group weights were introduced according to the a posteriori σ of single adjustments (see above) and controlled by an a posteriori variance estimation. The datum was realized by the GLB401 coordinate shifts, which refer to the motion of station Westford/MA and the azimuth change from Westford to Gilcreek/Alaska as derived from the geophysical model AM0-2 of Minster and Jordan (1978). Table 1 presents the results of the two adjustments in terms of plate rotation vectors. They are opposed to the geophysical models (see next chapter).

In general, we find a very good agreement between the two geodetic solutions. There never occurs a significant difference in the parameters. A notable phenomenon is the large error in the longitude of rotation pole of the European plate, which is due to the small extension of the European SLR/VLBI networks.

As a demonstration of the residuals after adjustment we present in tables 2 and 3 the modelled observables (computed from adjusted parameters) of selected stations. They are compared with the original observations and opposed to the geophysical predictions.

32

Table 1.   Geodetically determined plate rotation vectors and comparison with
geophysical models (Φ, Λ in degrees, ω in degrees per M.yrs.)

PLATE PARAMETER		GEODETIC GLB401/DGFI L03	GLB405/SL7.1	GEOPHYSICAL AM0-2	NUVEL-1
PCFC	Φ	-59.8 ± 2.1	-60.7 ± 1.2	-62.9	-62.9
	Λ	81.4 ± 5.4	76.6 ± 1.7	111.5	111.5
	ω	0.78 ± 0.07	0.84 ± 0.03	0.74	0.74
NOAM	Φ	-33.1 ± 8.5	-36.0 ⎫	-3.5	-19.0
	Λ	-89.4 ± 3.2	-89.8 ⎬ fixed	-81.8	-95.0
	ω	0.18 ± 0.01	0.18 ⎭	0.23	0.21
EURO	Φ	64.2 ± 3.3	55.2 ± 7.3	50.5	40.7
	Λ	-45.2 ± 37.2	-71.3 ± 22.5	-101.0	-126.8
	ω	0.36 ± 0.18	0.34 ± 0.08	0.25	0.20
ASIA	Φ	-36.8 ± 3.9		50.5	40.7
	Λ	-45.2 ± 5.3		-101.0	-126.8
	ω	1.40 ± 1.40		0.25	0.20
AUST	Φ		15.9 ± 4.5	38.4	30.2
	Λ		69.3 ± 4.9	27.3	36.9
	ω		0.59 ± 0.03	0.69	0.63

COMPARISON WITH GEOPHYSICAL MODELS

Tables 1, 2, and 3 present the opposition of geodetically and geophysically determined parameters. The plate rotation vectors (Table 1) can directly be compared with the geophysical AM0-2 numbers (Minster and Jordan, 1978) because the geodetic datum (VLBI model GLB401) refers to AM0-2 by fixing one station motion (Westford) and one azimuth change (Westford-Gilcreek) with these model values. The recent NUVEL-1 model (DeMets et al., 1989), which originally refers to a fixed Pacific plate, was transformed to the same datum by introducing identical Pacific plate parameters with AM0-2.

It can be noted that the geodetically determined parameters of the North American and Australian plates fit better to the NUVEL-1 model, while the European plate parameters are in better agreement with AM0-2. It is evident that the Asian plate (Japanese and Chinese stations) does not move like the European plate as postulated by the geophysical models.

More obvious than the rotation parameters are the derived station displacements from the different models. These are presented in Tables 2 and 3. In Table 2 we have to keep in mind that the given sigmas of GLB401 and DGFI L03 observations have to be multiplied by factor 6.8 after the a-posteriori variance estimation (see above). The factor in Table 3 for GLB405 and SL7.1 is 1.5.

Table 2.   Comparison of geodetically observed station coordinate shifts, modelled (from adjusted plate rotation parameters) observables, and geophysical predictions (cm/yr)
(DGFI L03 shifts are transformed to the VLBI GLB401 datum)

STATION		GLB401	DGFI L03	Model	AM0-2	NUVEL-1
EUROPE						
Wettzell	$\Delta\varphi$	1.6 ± 0.0	0.4 ± 0.1	1.5 ± 0.1	1.6	1.1
	$\Delta\lambda$	1.6 ± 0.1	1.6 ± 0.1	1.7 ± 0.3	2.0	1.9
Effelsbg.	$\Delta\lambda$	1.3 ± 0.1	1.0 ± 0.1	1.3 ± 0.2	1.8	1.4
RGO, resp.	$\Delta\lambda$	1.7 ± 0.4	2.2 ± 0.1	1.4 ± 0.3	1.7	1.7
JAPAN						
Simosato	$\Delta\varphi$	–	-0.2 ± 0.2	-0.2 ± 1.0	-1.5	-1.7
	$\Delta\lambda$	–	-0.6 ± 0.2	-0.9 ± 1.0	2.4	1.3
Kashima	$\Delta\varphi$	-1.1 ± 0.2	–	-1.3 ± 1.0	-1.6	-1.7
	$\Delta\lambda$	-0.6 ± 0.2	–	-0.3 ± 1.0	2.3	1.2
PACIFIC						
Hawaii	$\Delta\varphi$	3.3 ± 0.2	3.8 ± 0.1	3.7 ± 0.5	3.7	3.7
	$\Delta\lambda$	-6.1 ± 0.1	-6.3 ± 0.1	-6.1 ± 0.3	-6.8	-6.8
Monument	$\Delta\varphi$	1.6 ± 0.1	1.5 ± 0.1	1.3 ± 0.3	2.8	2.8
Peak	$\Delta\lambda$	-4.1 ± 0.0	-3.1 ± 0.1	-4.0 ± 0.1	-4.8	-4.8
NORTH AMERICA						
Quincy	$\Delta\varphi$	-0.7 ± 0.2	-0.9 ± 0.1	-0.9 ± 0.2	-1.6	-1.0
	$\Delta\lambda$	-2.1 ± 0.1	-2.2 ± 0.1	-1.8 ± 0.1	-1.4	-1.8
Ft. Davis	$\Delta\varphi$	-0.6 ± 0.1	-0.1 ± 0.1	-0.4 ± 0.1	-1.0	-0.3
	$\Delta\lambda$	-1.4 ± 0.0	0.6 ± 0.2	-1.8 ± 0.1	-1.4	-1.7
Green Bank	$\Delta\varphi$	0.5 ± 0.1	0.3 ± 0.1	0.4 ± 0.1	0.2	0.7
Greenbelt	$\Delta\lambda$	-2.2 ± 0.1	0.1 ± 0.1	-1.9 ± 0.1	-1.7	-1.9

In Table 2 we find an excellent agreement of the numbers within the Pacific. The Pacific plate stations in North America, however, differ by the factor two in latitude motions. This is the well-known San Andreas discrepancy, which is not necessarily a local effect, but appears also in this global solution. That means, all the Pacific plate motion can be modelled in the way of a slower rotation, i.e. the motions along the San Andreas fault become smaller. In North America we find again a better agreement of geodetically derived motions with the NUVEL-1 model than with AMO-2. In Europe, the geodetic longitude variation is about 20 % smaller than predicted from the geophysical models.

For better review we summarize the motions graphically in Figure 2, where shifts relative to a fixed North American plate are shown.

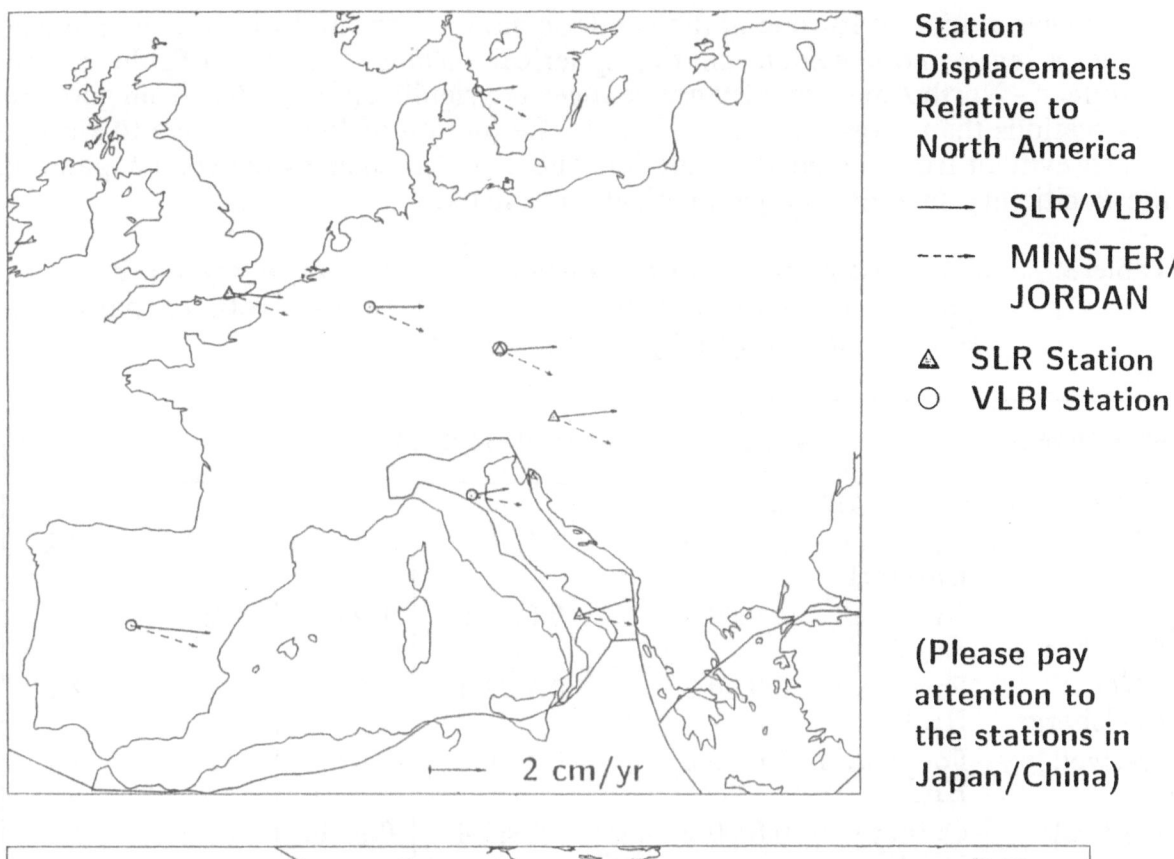

Station
Displacements
Relative to
North America

—— SLR/VLBI

---- MINSTER/
      JORDAN

△ SLR Station
○ VLBI Station

(Please pay
attention to
the stations in
Japan/China)

⊢—— 2 cm/yr

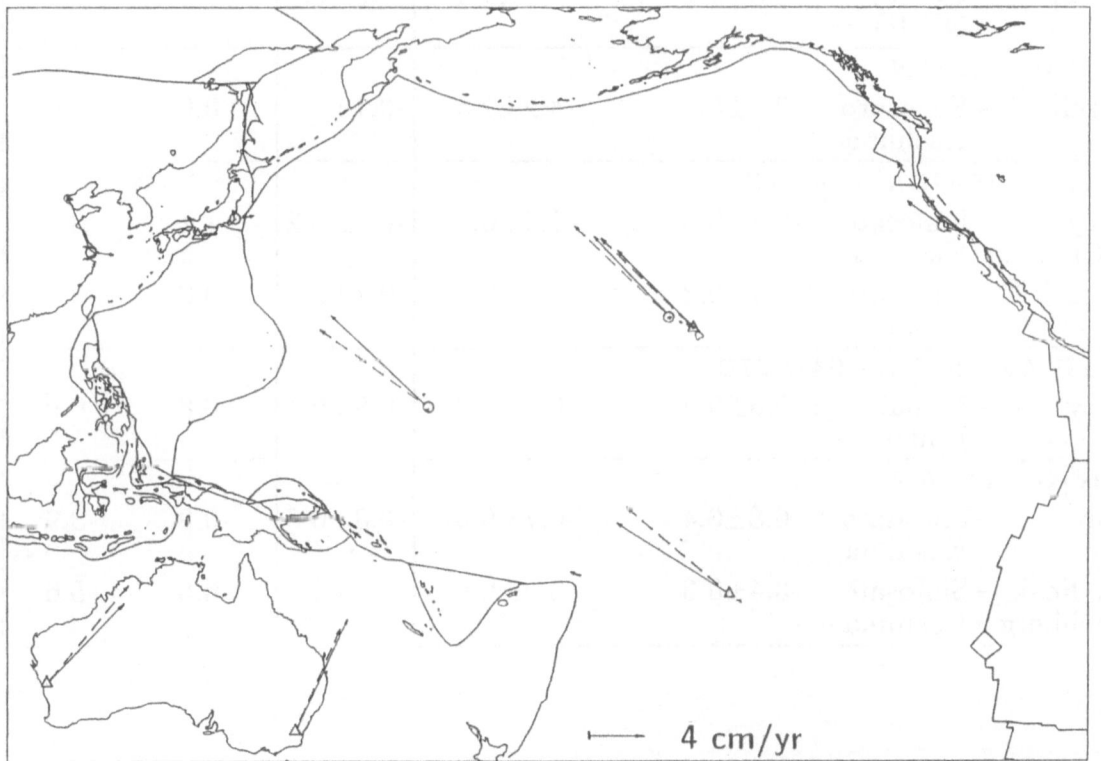

⊢—— 4 cm/yr

Fig.2   Station displacements from geodetic and geophysical models

The slower motion of Europe in the geodetic model compared with the geophysical model is even more evident in the spherical distance changes (Table 3). All European - North American distance changes are significantly smaller from geodetic observations than geophysically predicted. The motion of Japan relative to Europe is also evident from the geodetic model. The distance changes between Japan and the Pacific are smaller than geophysically predicted.

Table 3.   Comparison of spherical distance changes as observed by space geodetic methods, modelled from common plate rotation parameter adjustment, and geophysical models (cm/yr).

STATIONS		SLR(SL7.1)	VLBI(GLB405)	MODEL	AMO-2	NUVEL-1
EUROPE – NORTH AMERICA						
Wettzell	– Greenbelt Haystack	1.4±0.2	1.6±0.1	1.8±0.1	2.1	2.2
RGO Effelsberg	– Greenbelt Haystack	1.8±0.2	3.0±0.9	1.9±0.1	2.2	2.2
RGO Effelsberg	– Fort Davis HRAS	2.3±0.3	5.1±1.1	1.6±0.1	2.1	2.1
Wettzell	– Fort Davis HRAS	2.0±0.3	0.1±0.2	1.5±0.1	2.1	2.1
Wettzell	– Quincy OVRO	0.8±0.2	0.3±1.4	0.9±0.1	1.8	1.9
EUROPE – JAPAN						
Wettzell	– Simosato Kashima	-2.8±0.4	-4.2±2.3	-3.1±1.0	0.0	0.0
NORTH AMERICA – JAPAN						
Quincy Hat Creek	– Simosato Kashima	-0.7±0.3	-1.3±0.7	-0.3±0.2	-1.1	-0.9
Greenbelt Haystack	– Simosato Kashima	-0.7±0.3	-4.4±2.7	-0.7±0.6	-0.6	-0.5
NORTH AMERICA – PACIFIC						
Quincy Hat Creek	– Mauai Kauai	0.6±0.2	0.4±0.2	0.8±0.1	0.8	0.9
PACIFIC – JAPAN						
Mauai Kauai	– Simosato Kashima	-6.8±0.4	-6.7±0.3	-6.8±0.6	-9.9	-8.9
Mon. Peak Vandenberg	– Simosato Kashima	-3.4±0.3	-3.5±0.8	-3.4±0.2	-6.5	-5.6

CONCLUSION

Space geodetic observations are now capable of modelling global plate kinematics. These models are not identical with geophysical models. With future observations, especially in deformation belts like the Mediterranean, geodesy will provide the opportunity of including new kinematic units ("micro-plates") in the puzzle of plates.

REFERENCES

DeMets, Ch., Gordon, R.G., Argus, D.F., Stein, S. (1989). Current Plate Motions. *Pers. Comm., Publ. in preparation.*

Drewes, H. (1982). A geodetic approach for the recovery of global kinematic plate parameters. *Bull. Géod.*, 56, 70-79.

Gordon, R.G., DeMets, Ch., Argus, D.F., Stein, S. (1988). Current Plate Motions. AGU Fall Meeting, *EOS*, 69, 1416 (abstr.).

Ma, C., Ryan, J., Caprette, D. (1989). Crustal Dynamics Project data analysis - 1988 VLBI geodetic results, 1979-1987. *NASA TM-100723.*

Minster, J.B., Jordan, T.H. (1978). Present-day plate motions. *J. Geophys. Res.*, 83, 5331-5354.

Reigber, Ch., Ellmer, W., Müller, H., Geiß, E., Schwintzer, P. (1989). Plate motions derived from the DGFI-L03 solution. *IAG Symp. 101, Edinburgh* (this volume).

Smith, D.E., Kolenkiewicz, R., Dunn, Pl, Torrence, M.H., Klosko, S.M., Robbins, J.W., Williamson, R.G., Pavlis, E.C., Douglas, N.B., Fricke, S.K. (1989). The determination of present-day tectonic motions from Laser ranging to Lageos. *Ron Mather Symp.*, Kensington/Australia.

A QUALITATIVE AND QUANTITATIVE COMPARISON OF GEODETIC RESULTS FROM SLR AND VLBI

Ronald Kolenkiewicz, Chopo Ma, and James W. Ryan
NASA, Goddard Space Flight Center, Laboratory for Terrestrial Physics,
Code 920, Greenbelt, MD 20771, U.S.A.

Mark H. Torrence
ST Systems Corp., Lanham, MD 20706, U.S.A.

ABSTRACT

During the past decade, both Satellite Laser Ranging (SLR) and Very Long Baseline Interferometry (VLBI) systems have occupied the same, or nearly the same, locations on the surface of the Earth. Two of the commonly used geologic models which describe the plate geodesic rates between locations on the Earth's surface are the AM0-2 and the NUVEL-1.

Qualitative results indicate that both the SLR and VLBI measure motion between Monument Peak and Quincy to be about − 26 mm/yr, which is much less than predicted by either the AM0-2 (- 53.3 mm/yr) or the NUVEL-1 (– 45.4 mm/yr) geologic models. SLR and VLBI measure extension within the North American plate in the Basin and Range province of the Western United States. Both technologies measure the widening of the North Atlantic to be about 20% less than predicted by either the AM0-2 or the NUVEL-1 models.

Quantative results have been obtained for six locations which have been occupied by both the SLR and VLBI systems for an adequate time period to enable the calculation and comparison of the intersite geodesic rates for the two technologies. These six locations are: Platteville, Co, Quincy, CA, and McDonald, TX, on the North American plate; Monument Peak, CA, and Maui, HW on the Pacific plate; and Wettzell, FRG, on the Eurasian plate. For the 15-geodesic rates between these sites, the mean difference between SLR and VLBI (SLR minus VLBI) is − 0.2 mm/yr with a standard deviation of 5.2 mm/yr (− 0.2 ± 5.2 mm/yr). Similar values between SLR and AM0-2 are 0.6 ± 10.9 mm/yr and between VLBI and AM0-2 are 0.4 ± 8.8 mm/yr. For the SLR and NUVEL-1, the value is 0.6 ± 9.1 mm/yr; and for VLBI and NUVEL-1, the value is 0.4 ± 7.0 mm/yr. For the above data set, the SLR and VLBI derived-tectonic rates show better agreement with each other than with either the AM0-2 or the NUVEL-1 geologic models.

INTRODUCTION

A goal of the National Aeronautics and Space Adiministration's (NASA) Crustal Dynamics Project (CDP) is to improve the understanding of geodynamics by measuring crustal deformation, tectonic motion, polar motion and earth rotation. Two of the technologies the

CDP uses for these investigations are satellite laser ranging (SLR) and very long baseline interferometry (VLBI). An important part of the CDP's activities is to periodically compare the results from the two different technologies. SLR and VLBI have occupied nineteen different locations on the Earth. In this paper we briefly discuss the qualitative agreement of observations of tectonic processes as observed by the two technologies in two regions on Earth, though not necessarily from collocated locations. In this brief paper the rate of change of the geodesic distance for fifteen geodesic distances between six of the locations will be compared.

THE SLR AND VLBI SOLUTIONS

The VLBI solution was produced by NASA's VLBI analysis group and is called GLB484 (Ma, 1989 a). The solution derived an epoch position and tectonic velocity for both fixed and mobile VLBI sites scattered around the world. The position of the Westford, MA and Chilbolton, England antennas were fixed, and the velocities of the two sites were fixed to be that predicted by the Minster/Jordan AM0-2 tectonic model. (Minster and Jordan, 1978) Further details about the parameterization for this solution can be found in a discussion of a previous similar solution, GLB401, (Ma, 1989 b).

The SLR solution is called the SL7.1 tectonic motion solution, and was produced by NASA's SLR analysis group. Independently determined station positions from three months of laser ranging data to LAGEOS were used to determine station tectonic velocities. This was done in a system in which the latitude and longitude of Greenbelt, MD, and the latitude of Maui, HW were fixed and constrained to move with the AM0-2 predicted motion. Further details of this solution can be found in Smith, 1989.

QUALITATIVE COMPARISON

SLR systems have measured geodesic rates in many of the same general regions that VLBI has. In particular, we will look at the region in Western United States covering the San Andreas Fault and the Basin and Range Province, and across the North Atlantic.

Figure 1 shows locations in the United States occupied by the SLR and VLBI; these sites are all located on the North American Plate except for Monument Peak which is located on the Pacific Plate. Figure 1 also shows the geodesic rates as measured by each technology. The Monument Peak-Quincy rate shows excellent agreement between the SLR and VLBI. The VLBI geodesic rate for Monument Peak-Hat Creek further confirms the extension rate for this direction along the San Andreas Region as measured by both technologies is less that geologically predicted rate. The measurements from Monument Peak across the San Andreas to Platteville, McDonald and the East Coast show very good agreement, though the Monument Peak-McDonald rate is somewhat different.

In the Western United States an indication of overall spreading is detected across the Basin and Range region (Figure 2). The slight compression predicted by AM0-2 and NUVEL-1 for the Monument Peak-Platteville line (of -1.5 and -0.4 mm/yr respectively) disagrees with the extension detected by SLR. The rates from Platteville to both Quincy and Monument Peak are larger than that expected from geological models. Results from VLBI group indicate rates ranging from 3 mm/yr between Platteville and Hat Creek (~100 km NNW of Quincy) to 6 mm/yr between Platteville and Quincy. The SLR determined rate from Quincy to Greenbelt of 4 ± 2 mm/yr is consistent with these slower rates if one assumes no deformation is occurring east of the Rockies. While earlier geologic studies could only

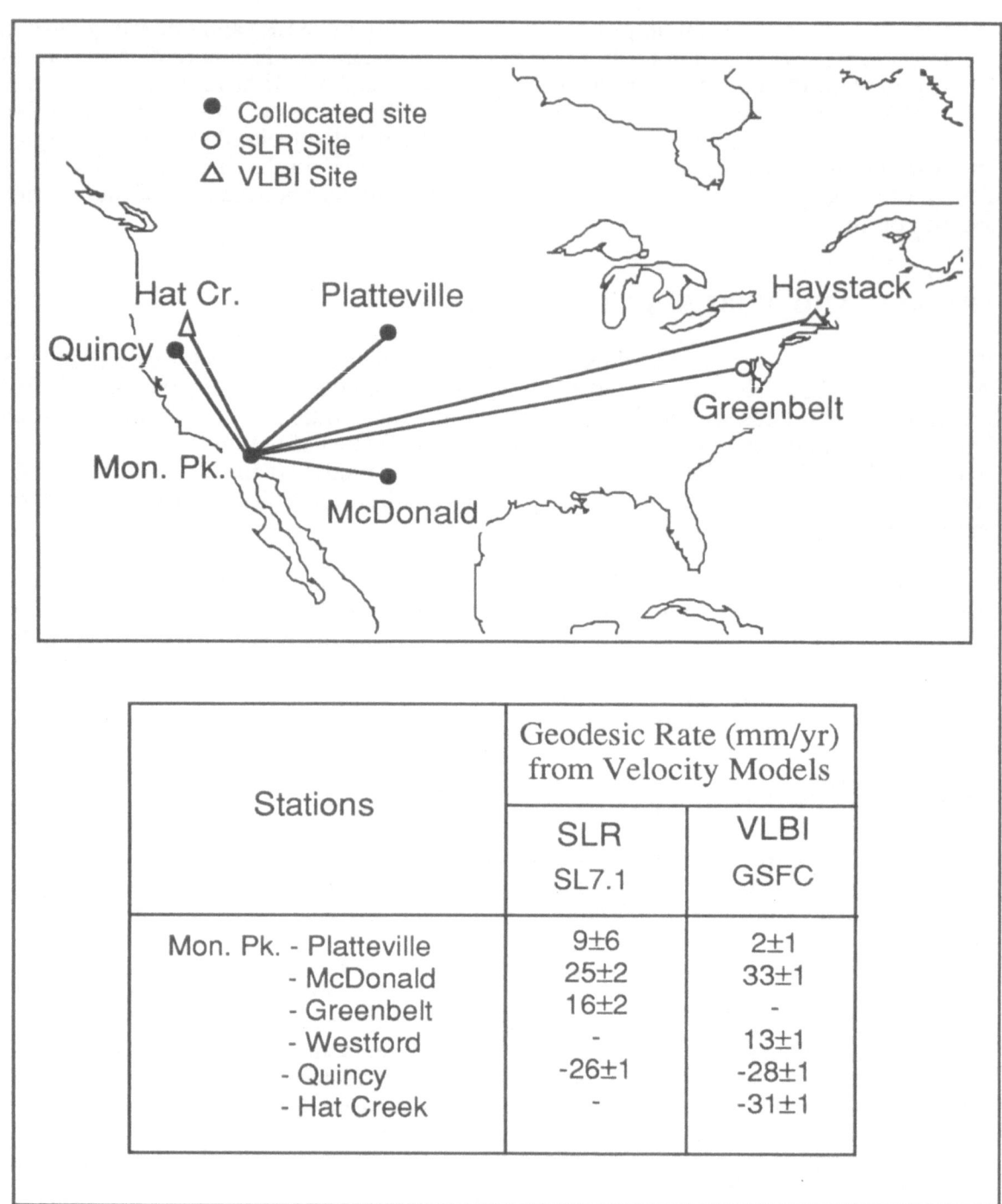

Stations	Geodesic Rate (mm/yr) from Velocity Models	
	SLR SL7.1	VLBI GSFC
Mon. Pk. - Platteville - McDonald - Greenbelt - Westford - Quincy - Hat Creek	9±6 25±2 16±2 - -26±1 -	2±1 33±1 - 13±1 -28±1 -31±1

Figure 1. Motions across the San Andreas Fault

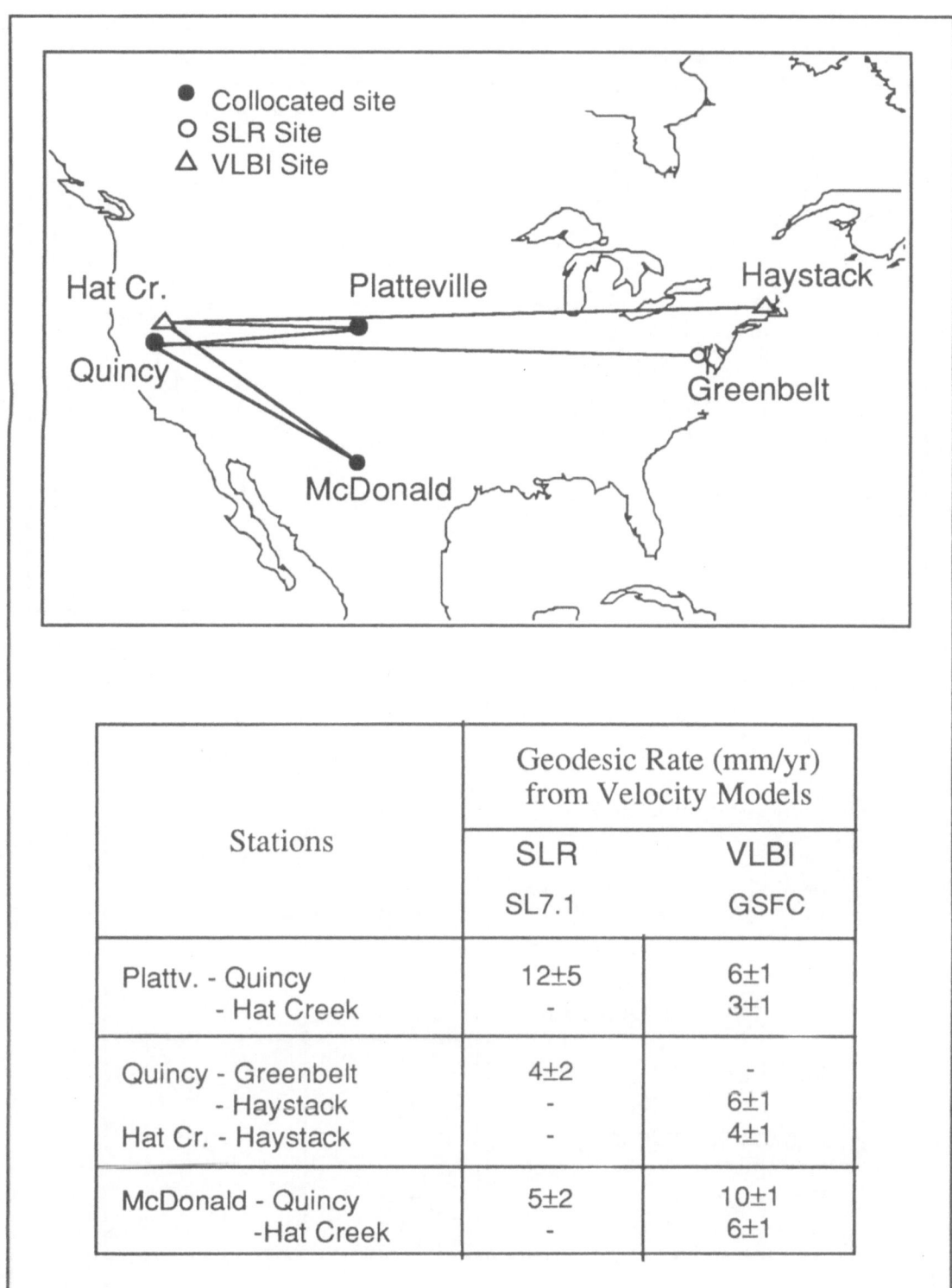

Stations	Geodesic Rate (mm/yr) from Velocity Models	
	SLR SL7.1	VLBI GSFC
Plattv. - Quincy - Hat Creek	12±5 -	6±1 3±1
Quincy - Greenbelt - Haystack Hat Cr. - Haystack	4±2 - -	- 6±1 4±1
McDonald - Quincy -Hat Creek	5±2 -	10±1 6±1

Figure 2. Extension in the Basin and Range Province

41

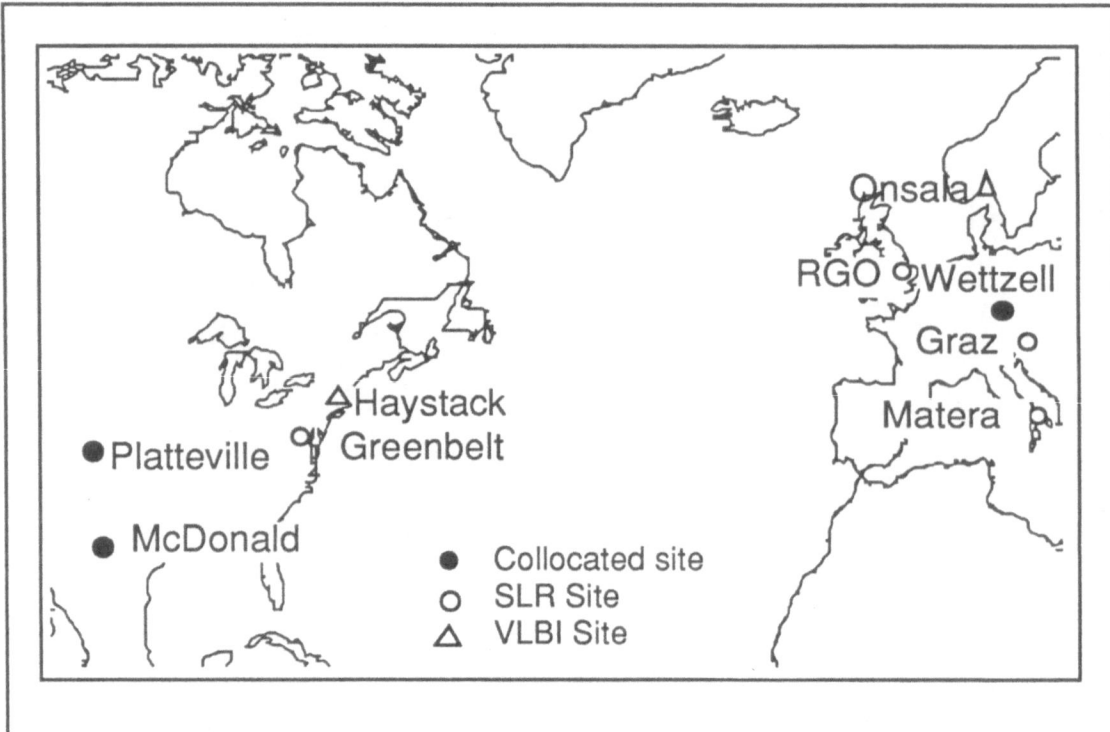

Stations	Geodesic Rate (mm/yr) from Velocity Models	
	SLR SL7.1	VLBI GSFC
Wettzell - Greenbelt	14±2	-
- Haystack	-	17±1
- Platteville	9±8	16±1
- McDonald	20±3	15±1
Greenbelt - Graz	19±3	-
- RGO	18±2	-
Haystack - Onsala	-	15±1
Platteville - Graz	12±8	-
- RGO	12±8	-
- Onsala	-	15±1

Figure 3. North Atlantic Spreading

predict a spreading rate of between 1 to 20 mm/yr (Minster and Jordan, 1987), these current measurements show that this extension is in the 5-10 mm/yr range based on directly measured intersite kinematics.

A comparison of the SLR and VLBI geodesic rates measured across the North Atlantic are shown in Figure 3. Of special note is the slower rate of spreading of approximately 4 mm/yr for the SLR North American to Eurasian sites versus that predicted by either of the geological models. The VLBI rates from the GSFC group for similar lines, such as Wettzell to McDonald and Wettzell to Platteville of 15 mm/yr and 13 mm/yr respectively, further indicate slower relative motion across the Atlantic. It appears that the results from both space geodetic techniques disagree with the magnitude of spreading across the Atlantic as compared with rates determined by the geologic models by about 20%.

QUANTITATIVE COMPARISON

The rates-of-change and the associated standard deviations for the 15 geodesic distances for the six locations as derived by the SLR, and VLBI are shown in Table 1. Also shown in Table 1 are the predicted rates from two geologic models: AM0-2, and a more recent model, NUVEL-1 (DeMets, 1989). The differences between the results for the two technologies and the two geological models are shown in Table 2.

TABLE 1. Comparison of SLR, VLBI, AM0-2, and NUVEL-1 Geodesic Rates

Stations		Geodesic Rate, mm/yr			
From	To	GSFC SLR	GSFC VLBI	AM0-2	NUVEL-1
McDonald	Quincy	4.5 ± 2.3	10.1 ± 1.1	0.0	0.0
McDonald	Monument Pk.	24.6 ± 2.2	32.8 ± 0.7	40.8	35.6
McDonald	Platteville	1.1 ± 9.6	-0.6 ± 1.0	0.0	0.0
McDonald	Maui	25.9 ± 0.2	31.9 ± 1.3	36.1	30.5
McDonald	Wettzell	19.5 ± 2.7	15.2 ± 1.1	20.7	21.0
Quincy	Monument Pk.	-25.9 ± 1.4	-27.6 ± 1.4	-53.3	-45.5
Quincy	Platteville	11.6 ± 5.4	5.7 ± 1.2	0.0	0.0
Quincy	Maui	5.7 ± 1.7	4.9 ± 1.6	13.0	6.9
Quincy	Wettzell	7.6 ± 2.4	11.2 ± 1.6	18.3	18.8
Monument Pk.	Platteville	8.9 ± 6.3	2.4 ± 1.1	1.1	-0.2
Monument Pk.	Maui	4.0 ± 1.6	0.8 ± 1.4	0.0	0.0
Monument Pk.	Wettzell	4.0 ± 2.4	5.5 ± 1.4	0.7	4.8
Platteville	Maui	21.4 ± 5.1	13.3 ± 1.4	17.5	13.3
Platteville	Wettzell	9.2 ± 7.8	16.0 ± 1.4	20.0	20.3
Maui	Wettzell	-42.9 ± 3.2	-45.1 ± 1.8	-44.9	-35.0

The overall agreement between the SLR and VLBI geodesic rates is good, being on average - 0.2 ± 5.2 mm/yr. Both the SLR and VLBI results agree better with each other than with the predicted results of the either geologic model. The largest discrepancy between the space-based measurements and the geologic models is for the Quincy, CA to Monument Peak, CA geodesic line. Even though the prediction of the NUVEL-1 model is 8 mm/yr less than the prediction of AM0-2 for that intersite geodesic rate, the agreement of the SLR and VLBI results is much better.

TABLE 2. Differences Between SLR, VLBI, AM0-2, and NUVEL-1 Geodesic Rates

Stations		Difference in Geodesic Rate, mm/yr				
From	To	VLBI minus SLR	SLR minus AM02	VLBI minus AM0-2	SLR minus NUVEL-1	VLBI minus NUVEL-1
McDonald	Quincy	5.6	4.5	10.1	4.5	10.1
McDonald	Monument Pk.	8.2	-16.2	-8.0	-11.0	-2.8
McDonald	Platteville	-1.7	1.1	-0.6	1.1	-0.6
McDonald	Maui	6.0	-10.2	-4.2	-4.6	1.4
McDonald	Wettzell	-4.0	-1.2	-5.5	-1.5	-5.8
Quincy	Monument Pk.	-1.7	27.4	25.7	19.6	17.9
Quincy	Platteville	-5.9	11.6	5.7	11.6	5.7
Quincy	Maui	-0.8	-7.3	-8.1	-1.2	-2.0
Quincy	Wettzell	3.6	-10.7	-7.1	-11.2	-7.6
Monument Pk.	Platteville	-6.5	7.8	1.3	9.1	2.6
Monument Pk.	Maui	-3.2	4.0	0.8	4.0	0.8
Monument Pk.	Wettzell	1.5	3.3	4.8	-0.8	0.7
Platteville	Maui	-8.1	3.9	-4.2	8.1	0.0
Platteville	Wettzell	6.8	-10.8	-4.0	-11.1	-4.3
Maui	Wettzell	-2.2	2.0	-0.2	-7.9	-10.1
	mean	-0.2	0.6	0.4	0.6	0.4
	std. dev.	5.2	10.9	8.8	9.1	7.0

SUMMARY

Qualitatively the SLR and VLBI systems are measuring similar geodesic rates for locations in the same tectonic regions but not collocated. The so-called "San Andreas discrepancy" (Minster and Jordan, 1987) for motion "missing" along the San Andreas is observed by both SLR and VLBI. The SLR and VLBI systems both measure similar motions across the San Andreas as measured from Monument Peak. The spreading thought to be ongoing within the Basin and Range Province is confirmed by both SLR and VLBI. And spreading across the North Atlantic between the North American and Eurasian tectonic plates as measured by both SLR and VLBI systems is about 20% less than that predicted by geological models.

To summarize the quantitative results from the directly observed locations, for 15 geodesic rates-of-chage between six locations on the Earth, measured by SLR and VLBI technologies there is a -0.2 ± 5.2 mm/yr agreement. The agreement between the rates predicted by the NUVEL-1 geologic model and SLR results 0.6 ± 9.1 mm/yr, and is 0.4 ± 7.0 mm/yr for the comparison of the NUVEL-1 model and the VLBI results. For the above data set the SLR and VLBI derived-tectonic rates show better agreement with each other than with either the AM0-2 or the NUVEL-1 geologic models.

REFERENCES

DeMets, C., R. G. Gordon, D. F. Argus, and S. Stein, Current Plate Motions, Geophys. J. of the RAS, DGG and EGS, in press, 1989.

Ma, C., and J. W. Ryan, private communication, 1989 a

Ma, C., J. W. Ryan, and D Caprette, Cristal Dyanamics Project Data Analysis, NASA TM 100723, February 1989

Minster, J. B., and T. H. Jordan, Present-day Plate Motions, J. Geophys. Res., 83, 5331-5254, 1978.

Minster, J. B., and T. H. Jordan, Vector constraints on western U. S. deformation from space geodesy, neotectonics, and plate motions, J. Geophys. Res., 92, 4798-4804, 1987.

Smith, D. E. R. Kolenkiewicz, P. J. Dunn, M. H. Torrence, J. W. Robbins, S. M. Klosko, R. G. Williamson, E. C. Pavlis, N. B. Douglas, and S. K. Fricke, Tecotnic Motion and Deformation from Satellite Laser Ranging to Lageos, J. Geophys. Res, in press, 1989.

GEODYNAMIC RESEARCH USING LAGEOS LASER RANGING DATA AT THE CENTRAL INSTITUTE FOR PHYSICS OF THE EARTH POTSDAM (GDR)

R. Dietrich, G. Gendt and F. Barthelmes
Academy of Sciences of the GDR
Central Institute for Physics of the Earth
Telegrafenberg A17, Potsdam, 1561, (GDR)

Abstract: LAGEOS laser ranging data from 1983 to 1987 and the Potsdam-5 software complex were used to investigate the following phenomena: station coordinates, tectonic deformations, tidal deformation (LOVE numbers h_2 and l_2), ocean tides resp. frequency-dependent LOVE number k_2, tailored gravity field modelling based on point masses, post-glacial rebound modelling by point masses changing with time.

1. Introduction

During the last decade theory and software developments for satellite data analysis have been advanced at our institute. Especially laser ranging data of LAGEOS were analyzed beginning with MERIT short campaign 1980, continued in MERIT main campaign and post MERIT activities up to the new International Earth Rotation Service (IERS). For these purposes a software complex called POTSDAM-5 was developed. Its main sections are a data selection part, a part for orbital integration and differential orbit improvement, one for the integration of variational equations and a separate adjustment part based on pass by pass stored observational equations.

2. Standards and Data

All parameters in our software are in accordance with the MERIT STANDARDS (Melbourne et al. 1983), except the gravity field model. Here GEM-L2 was replaced in 1988 by GEM-T1. The nominal value for the geocentric gravitational constant is $GM = 398600.440 \ km^3 \ s^{-2}$.
The results presented here are based on LAGEOS data gained from September 1983 up to December 1987 with a gap from June 1985 up to December 1986. The data consist of 126292 normal points, containing 10583 passes from 39 stations.
The standard arc length in the computations was 6 days for coordinate and tectonic motion determinations and 15 days for tidal research. For gravity field determination a combination of 6-days-arcs and 30-days-arcs was used.

3. Investigations

3.1. Station coordinates and tectonic motions

Station coordinate evaluation can be performed by comparison of two pairs of station coordinate sets by a Helmert-transformation (7-parameter similarity transformation). In this way the accuracy of the datum (especially origin and scale) and of the inner geometry (via the resulting rms value) can be estimated. We analysed various station coordinate sets (see Tab. 1). Before similarity transformation the coordinate sets were reduced to common epoch by AMO-2 (Minster and Jordan 1978). The results in Tab. 2 show an accuracy of the origin of $\pm 1...4$ cm in x and y and of $\pm 10...20$ cm in z. The scale has an accuracy of $\pm 1...2 \cdot 10^{-8}$, the inner geometry can be determined with $\pm 2...4$ cm.

For tectonic motion investigations we adjusted in addition to the station coordinates R, λ and φ linear changes of longitude and latitude with time ($\dot{\lambda}$ and $\dot{\varphi}$). The station coordinate set together with the adjusted motion values are presented in Tab. 3. Here all stations were included which have covered the time interval 1983-1987 regularly with data. The resulting global deformations are shown in Fig. 1 and for the regions of North America and Europa in Fig. 2 and 3. The accuracy of the adjusted values is about ± 10 mm/year. This should be taken into account when looking for significant differences between the LAGEOS solution and AMO-2.

3.2. Tidal research

The station position displacements due to tidal forces can be used to adjust the LOVE numbers h_2 and l_2. We performed a determination of these parameters in one variant as globally adjusted parameters and in another variant as one pair per station, and from these station-dependent parameters the mean and rms values of Tab. 4 were computed. One can state no significant difference to the WAHR model; the accuracy measures of h_2 and l_2 correspond to relative accuracies of 2 % resp. 10 %.

The data of September 83 – May 85 (21 months) were used to adjust the second and third degree harmonics of the M_2, N_2, S_2, K_2, O_1, Q_1 and P_1 tides (see also Gendt and Dietrich 1988). Here the solid earth tidal response was introduced by using the WAHR model. The results are presented and compared to the Schwiderski model in Tab. 5. Fixing the ocean tide model (Schwiderski 1980) we used the same investigations to compute corrections to the frequency-dependent LOVE numbers k_2 (see Fig. 4). In these tidal investigations 15-days-arcs were used as a standard. Solutions for 30-days-arcs and 7.5-days-arcs are presented for comparison – they give an insight into the stability of the solution.

3.3. Gravity field determination

In 1988 a gravity field model based on point masses and tailored for LAGEOS was computed at our institute (for details see Dietrich and Gendt 1988). It consists in 74 point masses on optimized positions (see Fig. 5). The position optimization strategy was worked out by Barthelmes (1986). With the POTSDAM EARTH MODEL for LAGEOS No. 1 (POEM-L1) from 1988 as an initial model now an improved model named POEM-L2 was computed. The positions of the masses were fixed and LAGEOS data were used to improve their magnitudes. In opposite to POEM-L1, in which we used 15-days-arcs from LAGEOS and no additional information, in the POEM-L2 determination a combination of 6-days-arcs and 30-days-arcs of extended LAGEOS data was used, and additional information on zonal harmonics up to degree 11 from GEM-T1 (Marsh et al. 1987) was introduced in form of low-weighted observations.

The new model POEM-L2 exists in its original point mass representation and also transformed into spherical harmonics. The coefficients can be made available upon request. The geocentric gravitational constant GM in POEM-L2 has a value of $GM = 398600.4383 \pm 0.0010 \ km^3 s^{-2}$.

One possibility for model evaluation is the orbital fit to LAGEOS. In Tab. 6 the results of two months are represented and compared to GEM-T1. One can conclude that the models are equivalent for LAGEOS. As external information on zonal harmonics is included in POEM-L2 it should (in opposite to POEM-L1) also work well for other satellites at high altitudes.

Another way for model evaluation is the computation of geoid undulation differences for different pairs of gravity field models. The undulation differences between POEM-L2 and GEM-T1 (rss values) are ± 8.4 cm up to degree 4 and ± 13.8 cm up to degree 5. One can conclude that POEM-L2 contains an accurate and quite independent estimate of the long-wavelength part of the geopotential.

3.4. Postglacial rebound modelling by point masses

The secular change of J_2 was determined by analysis of long satellite orbits (Yoder et al. 1983, Rubincam 1984, Cheng et al. 1989) and explained by geophysical models (Peltier 1985, Sabadini et al. 1988) predominantly as a result of postglacial rebound. Mainly three regions contribute here: Laurentia, Fennoscandia and Antarctica (Antarctica with a combined effect of isostatic rebound and possible present ice mass budget imbalance). It was the question to find a simple model for which the observed J_2 value could be split up and separately related to these 3 regions.
We decided to put three point masses on these regions (see stars in Fig. 5), and additionally to the POEM-L2 determination the linear changes of their magnitudes with time were adjusted. The constraint of $J_2 = -3 \cdot 10^{-11}$/year (after Yoder et al. 1983) was added. A compensating mass changing with time was put into the geocenter. Expressing the point masses in terms of spherical harmonics one can say that by fixing the information on J_2 the change of the

nonzonal part is used for the subdivision of this information into the three regions.

Test computations with simulated data as well as the computations with real data showed that it was not possible to derive results with high significance for the given time interval. The adjusted values (see Tab. 7) still have quite large errors. At present we would therefore emphasize the methodological aspect of this investigation. An extension of the time interval of LAGEOS data should lead to more stable results.

In principle it is possible to convert every set of three $\dot{m}$ - values into time derivatives of spherical harmonics, (where J_2 will be equal to the introduced condition and) where the long-wavelength part (relevant to LAGEOS orbit) can be compared for instance with calculations from geophysical models.

4. Concluding remark

It was the aim of this paper to give an impression on the different fields of actual geodynamic research, to which LAGEOS laser ranging data analysis can contribute, and to present some recent results achieved at our institute.

References

Barthelmes, F. (1986): Untersuchungen zur Approximation des äußeren Schwerefeldes der Erde durch Punktmassen mit optimierten Positionen. Veröff. Zentralinstitut Physik der Erde Nr. 92, Potsdam 1986.

Cheng, M. K.; Eanes, R. J.; Shum, C. K.; Schutz, B. E.; Tapley, B. D. (1989): Temporal variations in low degree zonal harmonics from STARLETTE orbit analysis. Geoph. Res. letters 16 (1989) 5, 393-396.

Dietrich, R.;Gendt, G.(1988): A gravity field model from LAGEOS based on point masses (POEM-L1). Paper 6th Int. Symp. "Geodesy and Physics of the Earth", GDR, Potsdam, August 1988. Proceedings Part II, 180-197, Potsdam 1989.

Gendt, G.;Dietrich, R.(1988): On the determination of tidal parameters using LAGEOS laser ranging data. Paper 6th Int. Symp. "Geodesy and Physics of the Earth", GDR, Potsdam, August 1988. Proceedings Part I, 275-283, Potsdam 1989.

Marsh, J.G.; Lerch, F.J.; Putney, B.H.; Christodoulidis, D.C.; Smith, D.E.; Felsentreger, T.L.; Sanchez, B.V.; Klosko, S.M.; Pavlis, E.C.; Martin, T.V.; Robbins, J.W.; Williamson, R.G.; Colombo, O.L.; Rowlands, D.D.; Eddy, W. F.; Chandler, N.L.; Rachlin, K.E.; Patel, G.B.; Bhati, S.; Chinn, D.S.(1988): A new gravitational model for the earth from satellite tracking data: GEM-T1. J. Geophys. Res., 93, 1988, B6, 6169-6215

Melbourne, W.; Anderle, R.; Feissel, M.; King, R.; McCarthy, D.; Smith, D.; Tapley, B.; Vicente, R. (1983): Project MERIT Standards. U.S. Nav. Obs., Circ. No. 167, Washington 1983

Minster, J. B.; Jordan, T. H. (1978): Present-day plate motions. J. Geophys. Res. 83 (1978), 5331-5353.

Peltier, W. R. (1985): The LAGEOS constraints on deep mantle viscosity: Results from a new normal mode method for the inversion of viscoelastic relaxation spectra. J. Geophys. Res. 90 (1985), 9411-9421.

Rubincam, D. P. (1984): Postglacial rebound observed by LAGEOS and the effective viscosity of the lower mantle. J. Geophys. Res. 89 (1984), 1077-1087.

Sabadini, R.; Yuen, D. A.; Gasperni, P. (1988): Mantle rheology and satellite signatures from present-day glacial forcings. J. Geophys. Res. 93 (1988), 437-447.

Schwiderski, E.W. (1980): On charting global ocean tides. Rev. Geophys. Space Phys. 18 (1980) 243-268.

Yoder, C. F.; Williams, J. G.; Dickey, J. O.; Schutz, B. E.; Eanes, R. J.; Tapley, B. D. (1983): Secular variation of Earth's gravitational harmonic J_2 coefficient from LAGEOS and nontidal acceleration of Earth rotation. Nature 303 (1983), 757-762.

Contribution of the Central Institute for Physics of the Earth, Potsdam, GDR, No. 1837

Table 1. Characteristics of various station coordinate sets.
(sets 2 – 6 from BIH Annual Report for 1987)

No.	SSC	Epoch	Station motion	Used LAGEOS-data
1	ZIPE-83/87ADJ	45800	adjusted	Sept.1983-Dec.1987
2	SSC(DGFI)87L03	45700	trend	1980 - Oct. 1986
3	SSC(SHA)87L01	45700	AM1-2	Nov.1984-Jan.1986
4	SSC(DUT)87L04	46612	–	1986
5	SSC(CSR)88L01	45335	AM1-2	May 1976-Jan. 1988
6	SSC(GSFC)87L14	45335	AM0-2	May 1976-June 1987

Table 2. Results of a HELMERT transformation (7 parameters) for coordinate solutions of different institutions (about 15 stations common to all SSC were used; rms for coordinate differences in cm, scale difference μ in 10^{-8}, difference in geocentre in cm)

		ZIPE-83/87 ADJ	DGFI	SHA	DUT	CSR	GSFC
ZIPE-83/87 ADJ			3.6	2.8	3.2	2.3	2.8
DGFI	μ	-1.71					
	Δx	3.9		3.8	3.7	3.6	3.4
	Δy	-1.5					
	Δz	-21.1					
SHA	μ	-1.33	.33				
	Δx	0.4	-3.0		3.9	3.3	3.6
	Δy	0.6	2.6				
	Δz	-17.0	4.3				
DUT	μ	-1.20	.34	.01			
	Δx	-1.7	-4.7	-1.8		2.6	1.5
	Δy	1.0	3.0	-0.1			
	Δz	-16.2	7.0	1.6			
CSR	μ	-.18	1.46	1.12	1.11		
	Δx	-2.1	-6.1	-2.8	-1.1		3.3
	Δy	-0.4	1.1	-0.9	-0.4		
	Δz	-9.2	12.8	8.0	6.7		
GSFC	μ	+.31	1.97	1.66	1.55	.42	
	Δx	-1.4	-5.1	-2.5	0.2	1.3	
	Δy	2.0	3.1	1.3	0.5	1.4	
	Δz	-9.9	11.5	7.3	4.8	-1.3	

Table 3. Solution ZIPE-83/87ADJ using the LAGEOS laser ranging data in 1983/85 and 1987 with motion in longitude and latitude (Epoch 1984.3)

Station	x [m]	y [m]	z [m]	$\dot{\lambda}$	$\dot{\varphi}$ [mm/year]
1181 POTSDAM	3800621.070	882007.241	5028859.355	-13	-32
7086 F.DAVIS	-1330118.043	-5328532.977	3236146.694	-15	-34
7090 YARRAGA.	-2389010.241	5043330.536	-3078527.051	27	48
7105 GREENBE.	1130723.011	-4831352.393	3994108.317	-28	-21
7109 QUINCY	-2517233.569	-4198559.650	4076571.655	-24	-28
7110 M.PEAK	-2386276.431	-4802358.118	3444883.043	-48	- 3
7122 MAZATLAN	-1660087.124	-5619104.077	2511639.025	-16	-11
7210 MAUI	-5466005.595	-2404431.172	2242188.408	-63	23
7834 WETTZELL	4075529.829	931783.355	4801617.984	-28	-33
7835 GRASSE	4581691.709	556161.703	4389359.075	-19	-16
7838 SIMOSATO	-3822390.106	3699361.456	3507573.109	-28	6
7839 GRAZ	4194426.333	1162696.014	4647246.260	-22	- 7
7840 HERSTMO.	4033464.056	23664.353	4924304.743	-14	2
7907 AREQUIPA	1942794.882	-5804076.658	-1796919.513	7	-22
7939 MATERA	4641964.573	1393072.267	4133261.929	- 8	1

Table 4. Results for ocean tidal constituents

Tide deg.	POSTDAM-5 results with arc-length of						Model of SCHWIDERSKI (1980)	
	15-days		for comparison 7.5-days		30-days			
	ampl [cm]	phase [deg]	ampl [cm]	phase [deg]	ampl [cm]	phase [deg]	ampl [cm]	phase [deg]
M_2 2	3.26	320.1	3.22	320.9	3.30	317.6	2.96	310.6
3	.39	171.5	.26	182.3	.47	166.6	.36	168.6
N_2 2	.62	321.0	.73	327.8	.63	323.9	.65	321.8
3	.06	130.1	.06	56.0	.12	73.5	.11	171.9
S_2 2	1.07	308.1	1.04	309.5	1.10	308.9	.93	314.0
3	.35	261.0	.30	265.5	.34	259.5	.26	202.0
K_2 2	.36	296.9	.36	300.6	.31	297.2	.26	315.1
3	.24	200.1	.22	198.9	.24	200.0	.09	195.0
O_1 2	3.06	312.1	2.93	314.8	3.32	317.0	2.42	313.7
3	1.40	81.0	1.69	87.6	1.47	75.8	1.32	83.6
Q_1 2	0.98	310.0	.83	323.8	1.05	311.3	.54	313.7
3	.41	123.3	.42	137.4	.66	167.2	.31	107.3
P_1 2	1.03	330.4	.71	321.3	.90	337.9	.90	313.9
3	.43	34.7	.43	47.6	.46	44.5	.30	40.0

Table 5. Different variants of global determination of LOVE numbers h_2 and l_2 from laser ranging data of LAGEOS

h_2	l_2	Characteristics of determination
0.605 ±0.004	0.097 ±0.002	direct global adjustment
0.610 ±0.004	0.098 ±0.002	direct global adjustment (ocean tides simultaneausly adjusted)
0.598 ±0.012	0.092 ±0.009	mean of local determinations, Sept.83 – May 85
0.583 ±0.012	0.083 ±0.008	mean of local determinations, 1987
0.609	0.085	WAHR model

Table 6. Orbital fit rms value for 2 months LAGEOS data in dependence of the arc length (3rd generation stations). Time interval: March 13th – May 12th 1987 (MJD: 46867–46927) Additionally adjusted: station coordinates, earth rotation parameters (3-days-intervals), empirical acceleration (15-days-intervals)

Model	6-days-arcs	15-days-arcs	30-days-arcs
POEM-L2	4.0 cm	5.2 cm	6.5 cm
GEM-T1	4.0 cm	5.2 cm	6.4 cm

Table 7. Changes of point masses with time (in units of the earth mass $\cdot 10^{11}$/year)

Mass	λ [deg]	φ [deg]	$\dot{m}$
Laurentia	−85°	+55°	+5.0 ±2
Fennoscandia	+23°	+65°	+1.2 ±2
Antarctica	+54°	−81°	−0.4 ±2
Geocenter (compensating mass)			−5.8

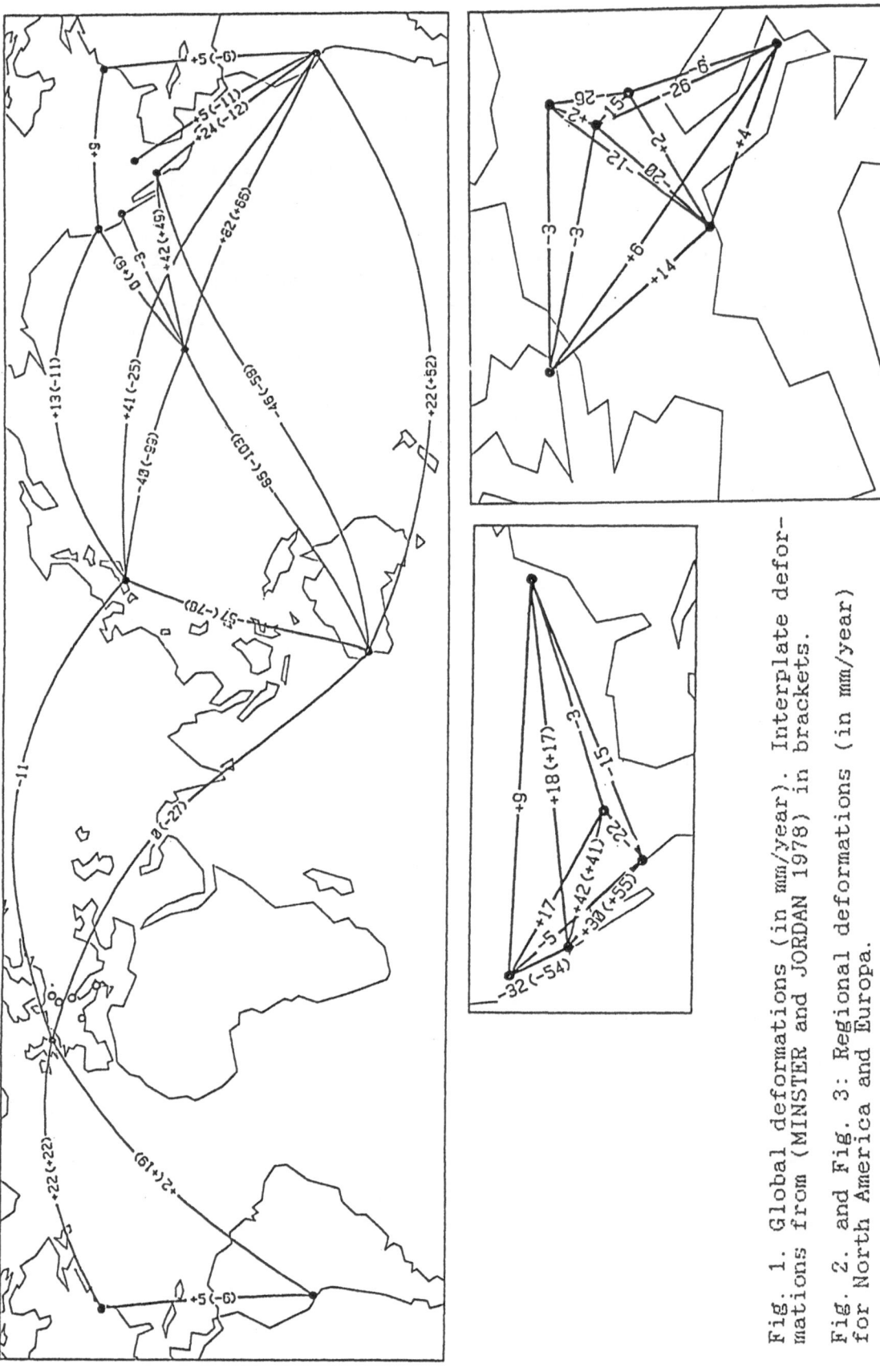

Fig. 1. Global deformations (in mm/year). Interplate deformations from (MINSTER and JORDAN 1978) in brackets.

Fig. 2. and Fig. 3: Regional deformations (in mm/year)
for North America and Europa.

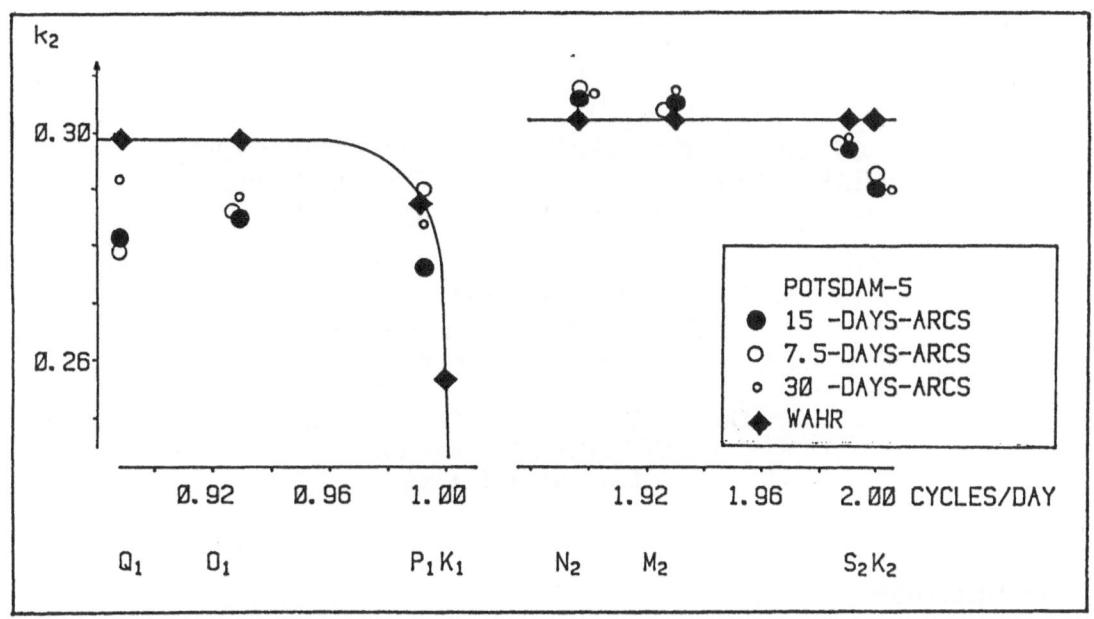

Fig. 4. Frequency-dependent LOVE number k_2

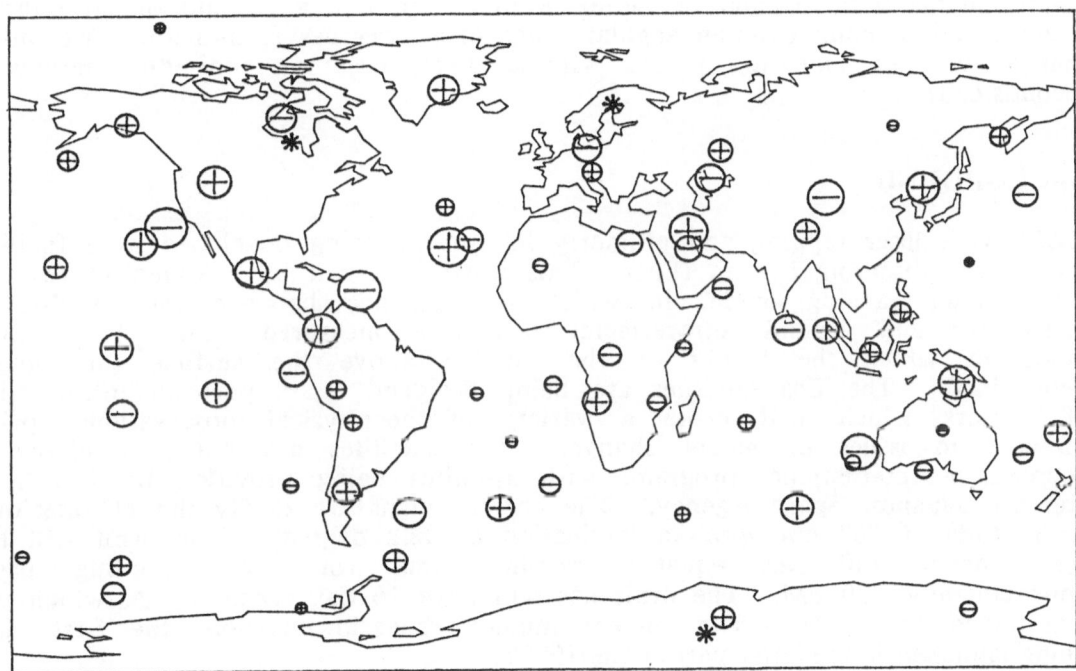

Fig. 5. Gravity field model POEM-L2: distribution of the point masses

GEOSCIENCE LASER RANGING SYSTEM (GLRS): CHARACTERISTICS AND EXPECTED PERFORMANCE IN GEODYNAMIC APPLICATIONS

Steven C. Cohen
Goddard Space Flight Center
Greenbelt, MD USA

Douglas S. Chinn and Peter J. Dunn
ST Systems Corporation
Lanham, MD USA

INTRODUCTION

Covariance analysis of the performance of the Geoscience Laser Ranging System (GLRS) indicates that three dimensional relative positioning can be recovered to an accuracy of several millimeters over spatial scales from a few kilometers to several hundred kilometers and over temporal scales as short as several days. Key factors influencing the accuracy are the range noise, target number and location, system pointing capability, dwell time on the targets, orbital geometry, and gravity field uncertainties. GLRS is being designed to provide range measurements with 10 mm or better accuracy, to fire at a rate of 40 pulses-per-second, and to point over an angular range of 50 degrees from nadir. We anticipate that it will be able to operate with a dwell time on individual targets of 2 seconds or less.

BACKGROUND

GLRS is a laser ranging and altimetry instrument being developed as a facility for the Earth Observing System (Eos). This instrument can be operated in two modes: (1) a laser ranging mode, in which the distance between the satellite borne instrument and ground retroreflector targets is measured, and (2) an altimeter mode in which the height of the satellite above the surface (or clouds) is determined. The Eos satellites are being designed to support an integrated suite of sensors which will make a variety of geophysical observations, primarily directed to issues of global change. Eos satellites are being developed as a cooperative international program, with satellites being provided by NASA, ESA, and the Japanese Space Agency. The current plans are to fly the US platforms at an altitude of 700 km with an inclination of 98.2 degrees. The orbit will be sun synchronous, with an equator crossing time for the ascending pass of approximately 1:30 PM. The orbit has will have 16 day repeat cycle, which will be maintained to 1 km. The current mission scenario envisions the first platform being launched in the latter part of the 1990's.

The scientific requirements for the laser ranging measurements include horizontal length accuracies of 5 to 20 mm and relative height accuracies of 5 to 30 mm over baselines from a few km to 1000 km. The period between surveys is variable, typically being a month for most geodynamic applications, but being a week or less during active periods. The requirement for laser altimetry over relatively flat ice sheets is a vertical height precision of 10 cm while for the determination of cloud-top heights an accuracy of 100 m is required. The ground spot size will be 75 m and the spot location will be known to 20 m. The spacing between successive pulses will be approximately 150 m alongtrack.

INSTRUMENT DESCRIPTION AND SPECIFICATIONS

GLRS builds on a rich heritage of scientific and engineering analysis of laser ranging and altimetry (Fitzmaurice et al, 1975; Cohen and Cook, 1979; Kahn, et al, 1980; Bufton, et al, 1982; Degnan, 1984; Thomas, et al, 1985) The basic features of GLRS were described in (Cohen, et al, 1987). The conceptual block diagram is shown in Fig. 1; the instrument specifications are given in Table I. Here we summarize the salient features. The Nd:YAG laser transmitter is a laser diode pumped rod, whose 1064 nm output pulse is frequency doubled and tripled to produce pulses at 535 and 355 nm. The choice of the pulse width at the fundamental frequency has not been finalized, but will be 50 ps or less. The ultraviolet and visible pulses are directed to a precision pointing mount which in turn directs the pulses to the ground retroreflector targets. In addition, a small portion of the green pulse is diverted from the transmitter optical train to the receiver in order to start the range timer. The transmitted pulses are reflected from the targets back toward the satellite. Return pulses are collected by an 18-cm diameter telescope. A portion of the collected green pulse is used to generate a stop pulse for the timing subsystem. The residual green energy along with the entire ultraviolet pulse are directed toward a streak camera. The streak camera is used to record the differential arrival time of these two pulses and thus provide a measure of the atmospheric propagation correction which must be applied to the range timing data. The infrared pulse is transmitted at nadir and reflected off the earth's surface or clouds. In either case, the reflected energy is collected by a 50-cm diameter telescope. The arriving signal generates a stop pulse for the altimeter time of flight measurement. In addition the returned pulse is digitized in order to provide information on height distributions, pulse spreading, and surface slopes. The navigation subsystem includes a GPS receiver, 3-axis gyro, and dual star-trackers.

Among the more important parameters that impact GLRS performance are the pointing capability of the instrument of 50 degrees both along and across-track, the pulse rate of 40 pulses/s, the energy distribution of 120 mj at 1064 nm, 60 mj at 532 nm, and 40 mj at 355 nm, the ranger pointing precision of 0.01 milliradians, the range receiver resolution of 10 ps (equivalent to a range precision of 1.5 mm), and the streak camera timing resolution of 2 ps (which corresponds to an uncertainty in the atmospheric propagation range correction of about 4 mm). For surface altimetry, the timing resolution is 300 ps with a sampling interval of 750 ps. For cloud altimetry the resolution and sampling interval are both about 500 ns. The laser has a life time of 1 billion shots; however, a redundant laser head will also be available. Since GLRS development is still in the conceptual design stage, some of the specifications may be modified due to scientific or engineering considerations.

COVARIANCE ANALYSIS FOR LASER RANGING

The rest of this paper will focus on the retro-ranging applications of GLRS. For geodynamic research, there are several important features of GLRS that make it an attractive instrument for regional scale crustal movement studies. The first of these is its high spatial resolution. GLRS targets can be placed in a network with intersite distances ranging between a few kilometers to hundreds of kilometers. The quasi-simultaneous motion of all the sites within a grid can be captured in

GEOSCIENCE LASER RANGING SYSTEM

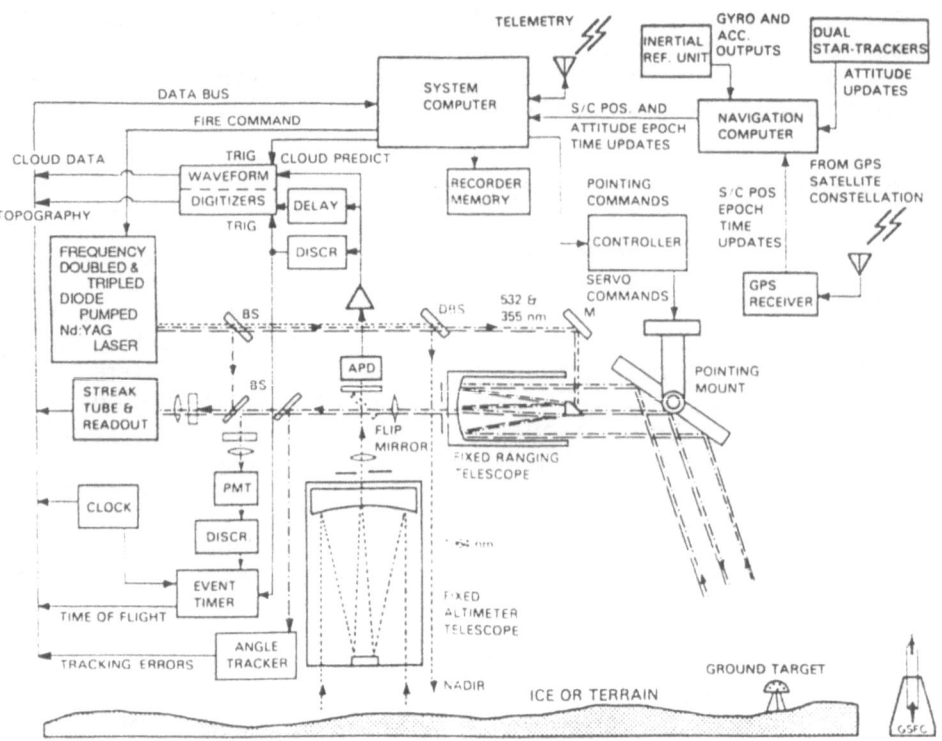

Fig. 1. GLRS conceptual block diagram.

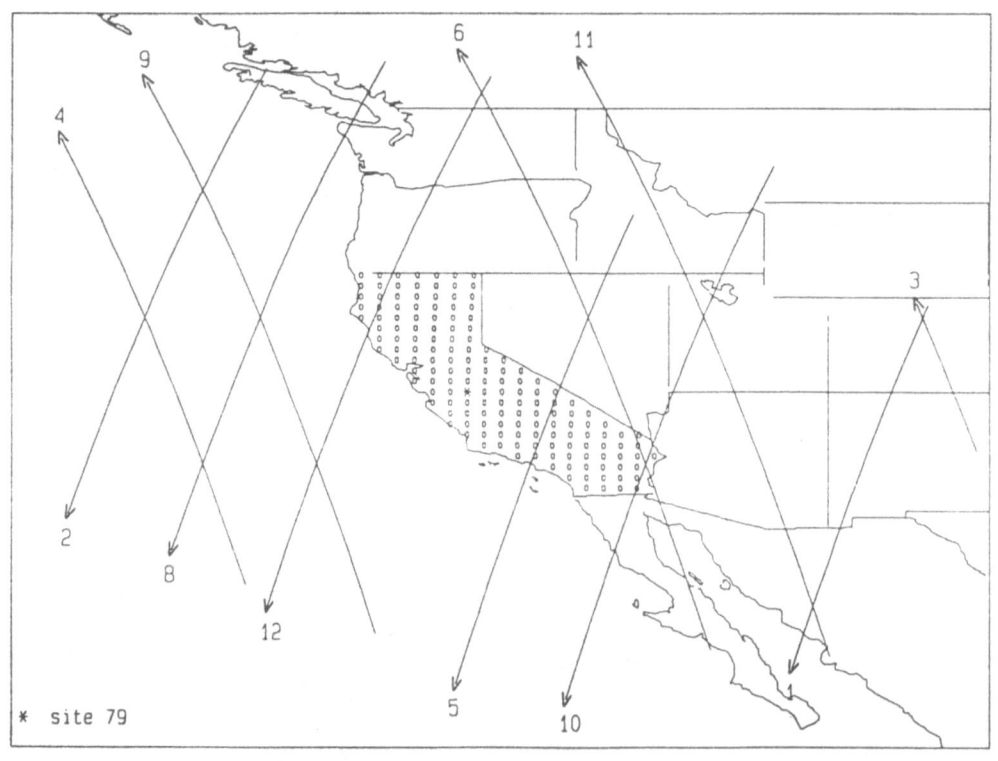

Fig. 2. Eos ground tracks over California grid

Table I. GLRS INSTRUMENT SPECIFICATIONS

Laser:	Nd:YAG, diode pumped, frequency doubled & tripled
Pointing Capability:	50 degrees
Pulse Rate:	40 pulses/s
Beam Divergence:	0.125 millirad (532 nm); 0.1 millirad (355 nm)
	0.1 millirad (1064 nm); 80 m footprint
Telescope Diameter:	18 cm (ranging function)
	50 cm (altimetry function)
Weight:	325 kg
Power:	550 W (peak)
Laser Lifetime:	> 1 billion shots (5 yrs at approx. 20% duty cycle)
Pulsewidth:	50 ps (FWHM @ 1064 nm)
Energy:	120 millijoules (1064 nm)
	60 millijoules (532 nm)
	40 millijoules (355 nm)
Ranger Pointing Prec.:	0.01 milliradians
Alt. Pointing Knowledge:	0.025 milliradians
Timing Resolution:	10 ps (range receiver)
	2 ps (streak camera receiver)
	300 ps (surface alt./750 ps sampling interval)
	500 ns (cloud altimeter receiver)
Data Rate:	800 kb/s

Table II. NOMINAL SIMULATION PARAMETERS

Range Noise:	10 mm
Pulse Rate:	20 pulses/s
Pointing:	50 degrees maximum; along and across track
Dwell Time on Targets:	1.5 s
Number of Targets:	157 with 50 km spacing in California grid
Orbital Parameters:	700 km altitude; 98.2 degrees inclination;
	0 eccentricity
A Priori Sat. Coordinate Error:	1 m
A Priori Sat. Velocity Error:	1 mm/s
A Priori Target Coordinate Error:	1 m
Gravity Field Covariances:	GEM T2 (complete to degree and order 36)

Drag and solar radiation pressure included in force model and covariances

---

Table III. ARC STATISTICS FOR PASSES OVER CALIFORNIA GRID

arc	duration (min sec)	number of acquisitions	number of targets	comments
1	4 00	147	43	day 1 am
2	4 23	161	70	day 1 am
3	1 58	60	5	day 1 pm; not used
4	5 05	187	94	day 1 pm
5	6 36	245	156	day 2 am
6	7 13	263	157	day 2 pm
8	5 23	198	134	day 3 am
9	6 18	233	150	day 3 pm
10	6 14	232	139	day 4 am
11	6 42	248	144	day 4 am
12	6 05	225	156	day 5 am

a single survey since solutions can be generated over a wide range of time scales as short as a few days or even less. Since GLRS will be deployed on a polar orbiting satellite it will provide a convenient on-call capability for monitoring episodic events such as volcanic eruptions and earthquakes. The current plans for GLRS envision lightweight, portable, and inexpensive ground targets which require neither electrical power nor telecommunications lines and can be installed by individual researchers and organizations.

The accuracy with which the three dimensional position of GLRS targets can be determined depends on a variety of factors (see e.g, Kumar and Mueller, 1978; Kahn, *et al*, 1980). The instrument factors include the range noise, the pointing capability, the time spent dwelling on a target, and the pulse rate. The geodetic factors include the number and location of targets, the orbit geometry, the number of passes used in a solution, the length of the orbital arcs used in the analysis, the *a priori* constraints on orbit and target locations, and the force model. Other factors include the details of the analysis technique, the acquisition logic, system biases, and unmodeled effects. We have conducted a number of covariance analyses and simulations to determine how accurately baseline lengths and relative heights can be determined and to study trade-offs between geodetic needs and engineering complexity. Here we present the results obtained with the nominal simulation parameters shown in Table II. More detailed trade-off studies are to be published elsewhere (Cohen, *et al*, 1989). The range noise of 10 mm includes not only the intrinsic precision of the instrument but also unmodeled random effects. The pulse rate of 20 pulses/s is less than the GLRS specifications, but when considered in combination with the dwell time of 1.5 s, gives performance similar to GLRS operational conditions, and was more economical to implement in our computer program. For example, the simulation parameters produce 30 shots on a target per acquisition. The actual instrument will pulse at 40 pulses/s and the combined dwell and slew time will be 2 s. Since the slew time will be a significant fraction of this two seconds, the rate of data acquisition will be comparable to that generated in the simulation. The *a priori* target coordinate accuracy of 1 m for each coordinate is easily achieved. We have verified that our assumptions on the *a priori* satellite state vector do not have a major effect on the results. Nevertheless, GPS tracking systems will be able to provide much tighter *a priori* constraints which may be important for dealing with the effects of errors in the gravity model as discussed below. The target grid, which consists of 157 equally spaced (50 km separation) targets in California is one which has been used in several previous analyses and provides a basis for comparing the present results with the earlier ones. We use the GEM T2 gravity field covariances (Marsh, *et al*, 1989) up to degree and order 36 and have included drag and solar radiation pressure uncertainty. Neither these latter two effects, nor uncertainties in GM affect our results. Unmodeled biases which remain constant over the limited time that the satellite is over a target grid, such as instrument biases, variations in the satellite center of mass, uncorrected atmospheric biases, etc., are also unlikely to significantly affect the results.

Fig. 2 shows the orbital tracks and the target grid that were used in the analysis and Table III shows some of the statistics concerning various passes or arcs over the grid. Most of the targets were observed in one pass every twelve hours. Arcs 1 and 2 occurred on successive revolutions, but non-overlapping portions of the grid were observed in these passes. Arc 7, which is not shown in Fig.2 or Table III, had no useful visibility over the target grid; similarly, arc 3 had such limited visibility that it too was excluded from the analysis. The column labeled "duration" on Table III is the time from the first observation to the last in a pass, and is typically several minutes. A simple sequential acquisition of the targets is assumed; at each stage the computer determines whether a target is visible with the then current observing geometry. If it is, observations are made during the specified dwell interval. Subsequently, the beam is slewed to the next target. If a target can not be observed, the computer cycles to the next target in the numerical sequence. The number of acquisitions refers to the number of times the system acquires a target in the pass. A target may be acquired more than once in a pass or not at all; the number of targets that are observed at least once in the pass is indicated in the labeled column. The analysis which we have conducted is a dynamic, short arc approach in which an independent adjustment of the satellite orbit is made for each pass over the target grid. This offers advantages over long arc techniques (McNamee, *et al*, 1988) in that force model errors have less of a corrupting influence on the determination of local site coordinates. The deduced station coordinates have a common bias in a geocentric coordinate system (unless fiducial reference points are used in the grid) which are largely removed by transforming to a local tangent plane coordinate system.

Fig. 3a shows the computed uncertainty in the baseline length and the relative height which are due to the assumed tracking system noise of 10 mm. The height is measured in a local tangent plane coordinate system with an origin at site number 79, which is near the middle of the California grid. The curves on the left and right side of the figure refer to sites lying to the north and east of this origin respectively (the stations are sequentially numbered from left to right and top to bottom). The length uncertainties are 1 to 2 mm and show little variation with distance or direction. The height uncertainties range from just over 1 mm to 6 mm at 500 km north of the origin. Since the analysis involves the simultaneous solution for the satellite coordinates and the target locations, errors in the force model used to model the satellite dynamics potentially affect the data reduction. Fig. 3b shows the baseline length and relative height uncertainties due to gravity model errors. The length uncertainty is a few millimeters for baselines of 100 km or less, but grows with increasing distance, particularly for sites to the east of the origin. This orientation influence was previously noticed by Cohen, *et al* (1987) and appears to be due to the dominant north-south orientation of the orbit. The height uncertainties are 3 to 5 mm at 50 km, but increase more rapidly with distance than the length errors.

When these results are compared to our earlier work (Cohen, *et al*, 1987) we find that the uncertainties due to laser noise are very comparable but the uncertainties due to gravity field errors are larger, particularly in height. This is due to the lower orbit (700 km vs 800 km) now planned for Eos. For the future, significant improvements in these uncertainties are envisioned. It is likely that GPS receivers on-board Eos will be used to measure the satellite position. The measured orbital positions may be sufficient to obviate the integration of the satellite equations of motion or at least provide important constraints on errors due to gravity model errors. In addition, general model improvements and the incorporation of ground and GLRS tracking data on Eos into the gravity field model will reduce the orbit errors and hence their effect on baseline length and heights. In this regard, it should be noted that GLRS can be used to provide a backup orbit determination capability for Eos by ranging to a globally distributed network of ground targets. We have made a preliminary assessment of the accuracy which can be achieved. As expected, with current models, the errors are dominated by gravity field effects. The accuracy in the radial component of the satellite position is 20-30 cm rms for an arc consisting of one entire revolution of the Earth and about 1 meter for a three day arc. These results have been confirmed by independent analysis (J. McNamee, University of Texas, unpublished memo, 1988).

CONCLUSIONS

The calculations we have presented here have verified that GLRS can provide data of high spatial and temporal density data at several millimeters accuracy for baselines up to several hundred kilometers long. Adequate observational geometry is achieved after several passes over the assumed fair-weather

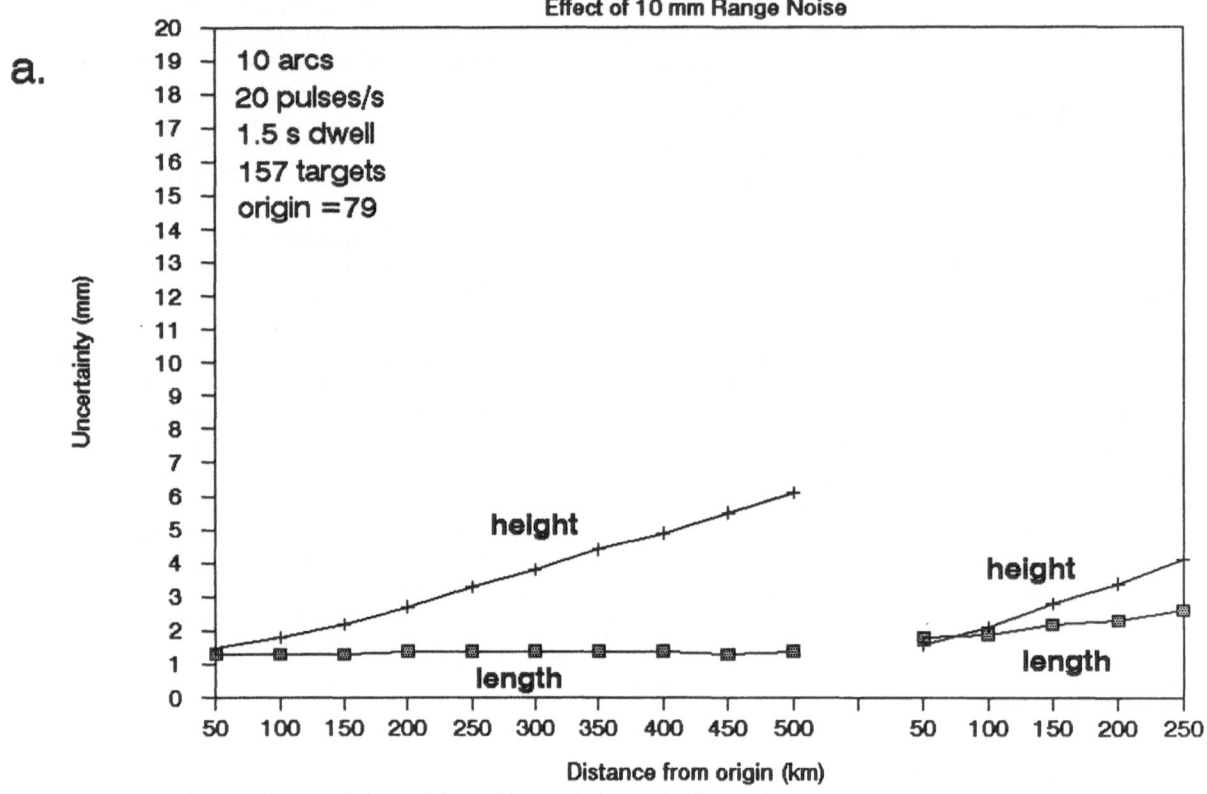

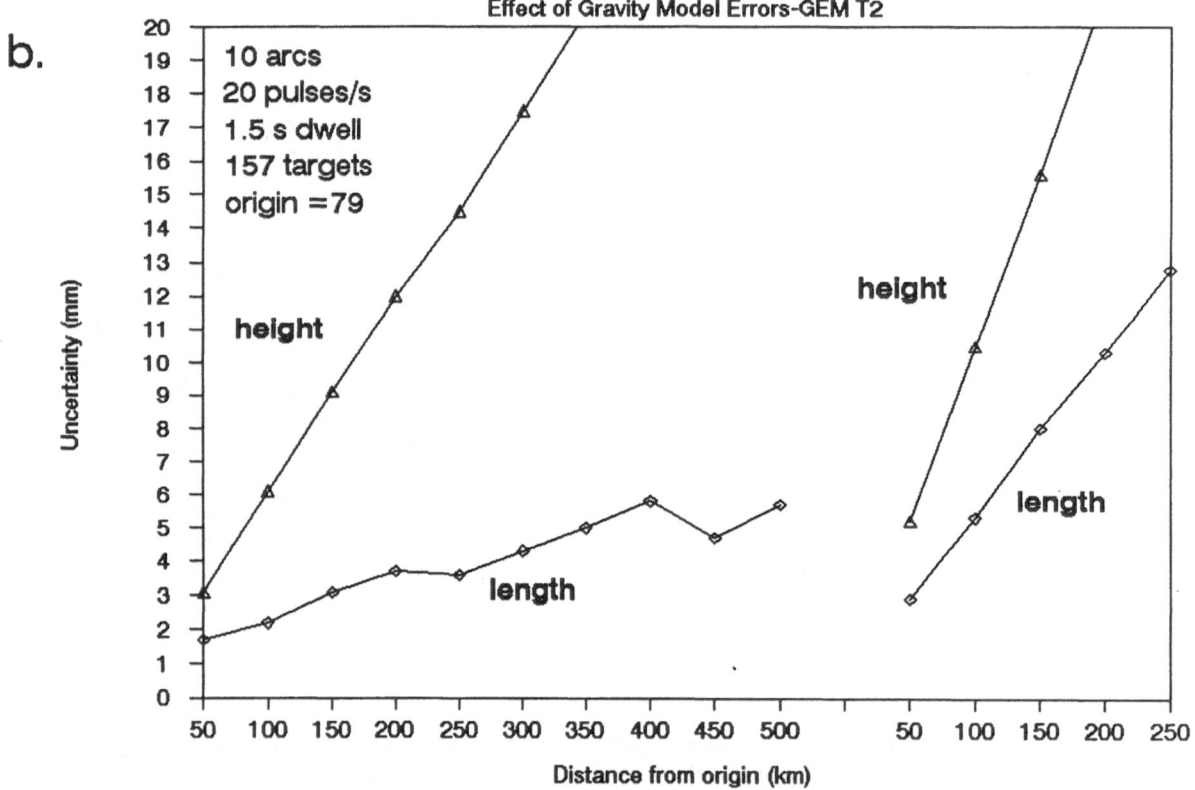

Fig. 3. Length and height uncertainty vs baseline distance. The curves to the left and right are for targets north and east of the origin respectively. a. Due to 10 mm range noise; b. Due to GEM T2 gravity error

California grid.

Acknowledgements: We thank our colleagues Chreston Martin for collaboration on the orbit determination analysis and George Wyatt for producing the ground track plots. We also appreciate the many contributions to GLRS development of the GLRS Science Team and the Goddard scientific and engineering teams including support contractors. The Science Team members are: Steven Cohen (Team Leader), Goddard Space Flight Center (GSFC), Charles Bentley, University of Wisconsin; Michael Bevis, North Carolina State University; Jack Bufton, GSFC; Thomas Herring, MIT; Kim Kastens, Lamont-Doherty Geological Observatory; Jean-Bernard Minster, Scripps Institution of Oceanography; William Prescott, US Geological Survey; Robert Reilinger, MIT; James Spinhirne, GSFC; Robert Thomas, Joint Oceanographic Institution; Jay Zwally, GSFC; and Scott Shipley, (Co-Investigator) ST Systems Corporation.

REFERENCES

Bufton, J., Robinson, J., Femiano, M. and Flatow, F. (1985). Satellite laser altimeter for measurement of ice-sheet topography, *IEEE Trans. Geosci. Remote Sensing*, GE-23, 414-425.

Cohen, S. and Cook, G., (1979). Determining Crustal Strain Rates with Spaceborne Geodynamics Ranging System Data, *Manuscripta Geodaetica*, 4, 245-260.

Cohen, S., Degnan, J., Bufton, J., Garvin, J., and Abshire, J. (1987). The Geoscience Laser Altimetry/Ranging System, *IEEE Trans. Geosci. Remote Sensing*, GE 26.

Cohen, S., Chinn, D., and Dunn, P., (1989). Geodetic Analysis for the Geoscience Laser Ranging System (manuscript in preparation).

Degnan, J. (1984). An overview of NASA Airborne and Spaceborne Laser Ranging Development, *Proc. 5th Int. Workshop on Laser Ranging Instrumentation (Royal Greenwich Observatory, East Sussex, England)*, 102-111.

Fitzmaurice, M., Minott, P., and Kahn, W. (1975). Development and Testing of a Spaceborne Laser Ranging System Engineering Model, *NASA X-723-75-307.*

Kahn, W., VonBun, F., Smith, D., Englar, T., and Gibbs, B. (1980). Performance Analysis of the Spaceborne Laser Ranging System, *Bull. Geod.*, 54, 165-180.

Kumar M., and Mueller, I. (1978). Detection of Crustal Motions Using Spaceborne Laser Ranging Systems, *Bull. Geod.*, 52, 115-130.

Marsh, J., Lerch, F., Putney, B., Felsentreger, T., Sanchez, B., Klosko, S., Patel, G., Robbins, J., Williamson, R., Engelis, T., Eddy, W., Chandler, N., Chinn, D., Kapoor, S., Rachlin, K. Braatz, L., and Pavlis, E. (1989). The GEM-T2 Gravitational Model, *NASA TM 100746.*

McNamee, J., Schutz, B., and Tapley, B. (1988). Recovery of Geophysical Parameters using a Spaceborne Laser Ranging System, *Eos*, 69, 1153.

Thomas, R., Bindschadler, R., Cameron, R., Carsey, F., Holt, B., Hughes, T., Swithinbank, C., Whillans, I., and Zwally, H.J. (1985). Satellite Remote Sensing for Ice Sheet Research, *NASA TM 86233.*

ORBIT DETERMINATION ACCURACY IMPROVEMENT BY SPACE-VLBI OBSERVABLES AS TRACKING DATA

I. Fejes, T. Borza, Sz. Mihály, L. Szánthó
FÖMI Satellite Geodetic Observatory
H-1373 Budapest, Pf. 546.

INTRODUCTION

Precise orbit determination (OD) has two basic ingradients:
 – precise orbit model and
 – precise tracking data.

Spacecraft tracking data exist in great variety of forms and the terms used to name different data types are often confusing. We hope we are not going to add to the confusion by calling attention to a new type of tracking data wich will be available in the near future during space-VLBI missions.

Traditional tracking data types can be ordered into three main categories:

a) Directional data type:

Telemetry and control (T/C) stations can provide rough 0.01-0.05 deg. directional information on the spacecraft in the local (antenna) reference frame. These data are neglected in most OD computations because of insufficient accuracy.

In the early days of satellite geodesy photographic observations were used to provide directional information with respect to the optical reference frame (e.g. FK 4 stars). The accuracy of photographic observations was in the order of a few tenth of an arc second which is insufficient for present day requirements of space geodesy.

High precision directional information in the inertial reference frame can be obtained by ΔVLBI, where the angular distance (and/or distance rate) between the spacecraft and a compact natural radio source can be measured with milli arc second (mas) precision. This method assumes that the artificial (radio transmission) radiation of the satellite and the natural radiation of the radio source is observed simultaneously or in short period directional switching mode with the same telescope - elements of the ground based very long baseline interferometer.

b) Range and range-rate data type:

This is the most widely used data type produced by a variety of instruments in variety of forms (one way, two ways, multiple ways - satellite to ground, satellite to satellite etc.).

T/C stations provide routinely range and range-rate data with limited accuracy (typically 5-10 m, 3-10 mm/s). This is sufficient for most OD requirements.

For high precision OD several dedicated microwave tracking instruments appeared or are under development e.g. the PRARE (Precise Range and Range-rate Equipment) developed at University of Stuttgart by a group led by Ph. Hartl (1984), a GPS based tracking instrument developed by NASA (Younck et al. 1985) or the DORIS system developed by the French CNES (Dorrer and Lefebre 1985). Measurement precisions are tipically in the order of 5-15 cm in range and 0.01-3 mm/s in range-rate.

Perhaps the most accurate range data is obtained by the fixed 3-rd generation laser tracking stations where the precisions are in the order of 1-3 cm.

c) Difference of Range (DOR) data:

Is obtained by simultaneous observations of the spacecraft transmissions by 2 or more tracking stations. Although DOR type data could be readily obtained by 2 synchronously working ranging equipment, we consider it as a separate data type because DOR can be the results of interferometric processing. It is sometimes combined with ΔVLBI which is called ΔDOR in JPL terminology (Jacobi et al. 1985).

We must be aware of the fact that precision is only one parameter in the evaluation process when we try to figure out which technique should be considered for the precise OD of a particular spacecraft. Spacecraft characteristiques, orbit coverage, data rate, tracking efficiency, data availability, operational aspects and cost should also be considered accordingly.

Space VLBI satellites will provide a new data type for orbit improvement (Tang 1984) which has not only high precision but requires no additional equipment, has good orbit coverage, high data rate, high tracking efficiency - all this because it comes with the spacecraft science data inherently.

In the following paragraphs we give a short description of the concept and present the results of our orbit determination simulation computations.

PLANNED SPACE-VLBI MISSIONS

The feasibility of space VLBI has already been demonstrated experimentally (Levy et al. 1986) by using the 5 m diameter antenna of a TDRS satellite.

Two dedicated space-VLBI projects are in preparation for the years 1993-96. One in the Soviet Union called RADIOASTRON (Kardashev and Slysh 1988) with a 10 m diameter antenna the other one in Japan at ISAS a space VLBI project called VSOP (VLBI Space Observatory Program) (Morimoto et al. 1988.).

The Soviet RADIOASTRON project at IKI has already been approved and funded. The operational frequencies are 0.327, 1.6, 5 and 22 GHz. The anticipated launch data is 1993. The 10 m diameter RADIOASTRON antenna will fly on a highly elliptical 24 h orbit with apogee height 76000 km, perigee height 2000 km and inclination 65°. Expected lifetime is 2-5 years.

The 10 m antenna diameter VSOP is recently informally approved by the Japanese government (Hirabayashi 1989.). The operational frequencies are 1.6, 5, and 22 GHz. The orbit can be characterised with 20 000 km apogee height, 1000 km perigee height, and inclination of 46° or 31°. Expected lifetime is longer than 3 years.

Both project are primarily designed for astronomical investigations. The main objectives of these missions are to improve imaging quality and/or angular resolution of compact galactic and extragalactic radio sources. Wether these missions can be used for astrometric and geodetic projects depends on the orbit determination accuracy. Requirements are at the submeter level. However routine astrophysical data processing will also require rather precise orbit knowledge in the range of 5-50 m.

THE SPACE VLBI OBSERVABLES

Although future space VLBI satellites will be tracked by a combination of conventional methods as well, there will be an additional data type which contains information on the mutual geometry of the ground stations - satellite - extragalactic radio sources. This tracking data is the space-VLBI delay and delay rate which is obtained after regular scientific observing sessions by standard VLBI data processing.

Consequently the special charachteristics of the delay and delay rate tracking data is twofold:

– contain information on the ground station - satellite - EGRS geometric configuration simultaneously

– they are obtained by the VLBI observing network instead of a "tracking network" (although at least one telemetry station should be included in order to record the space data on the ground).

Both points, have some interesting consequences

1. The orbit of the space VLBI satellite will be tied directly to two reference frames simultaneously (Fejes et al.1986):
   – to the geocentric terrestrial reference frame (CTS) defined by a set of reference VLBI ground station positions
   – to the inertial reference frame (CIS) defined by the positions of the EGRS.

2. The direct line of sight visibility between the spacecraft and the ground VLBI telescopes is not required in order to obtain useful tracking data.

It should be noted however that in reality space VLBI data is always combined with range and range-rate type data because of at least one telemetry station should receive the observed raw data on the ground from the satellite, uplink stable reference frequency, control satellite clock etc. This means that a two way coherent link between the satellite and the T/C station exists which provides precise range and range-rate tracking data anyway.

THE SIMULATION COMPUTATIONS

In order to demonstrate the contribution of space-VLBI delay and delay rate data to orbit improvement of space-VLBI satellites, we have developed our own software called SGOFAKE. This is a deterministic orbit improvement software package capable to simulate and use range, range-rate, DOR and space VLBI delay and delay rate type tracking data.

Table 1. shows the SGOFAKE program in the environment of some internationally known programs and compare the capabilities in receiving different tracking data types. The comparison demonstrates why we had to develop our own package.

Table 1.

Type of observations / Programs	Direction α / δ	Range r	Range rate $\dot{r}$	DOR dr	DeltaVLBI $d\alpha / d\delta$	Space VLBI $\tau / \dot{\tau}$	Sat-sat
GEODYN 2	yes	yes	yes	yes	yes	no	yes
ORAN	yes	yes	yes	yes	yes	no	yes
POTSDAM5	yes	yes	yes	yes	no	no	no
GEORAN	no	yes	no	no	no	no	no
SGOFAKE	no	yes	yes	yes	no	yes	no

66

The error model of SGOFAKE is rather simple and contain terms which were strictly necessary for the above purpose (i.e. measurement errors and geometric modell errors). Using the SGOFAKE program we have simulated a RADIOASTRON observation scenario. The nominal parameters are given in Table 2.

Table 2. Space-VLBI observing scenario for RADIOASTRON

ORBIT:	epoch (MJD)	:	48988	d	(1993.01.01)
	semi-major axis	:	42350	km	
	excentricity	:	0.82		
	inclination	:	70	degr	
	arg. of perigee	:	290	degr	
	long. of asc.node	:	220	degr	
	mean anomaly	:	0	degr	

VLBI STATIONS	Lat			Long			X (m)	Y (m)	Z (m)
	°	′	″	°	′	″			
Jevpatoria	45	13	19	-33	10	28;	3756200,	-2455600,	4522200
Suffa	40	0	0	-70	0	0;	1669200,	-4586100,	4095200
Ussurijsk	44	0	0	-131	45	36;	3052300,	-3418600,	4425700

TELEMETRY STATION (T/C)									
Jevpatoria	45	13	19	-33	10	28;	3756200,	-2455600,	4522200

MINIMUM NETWORK
Jevpatoria-Radioastron
Data-rate   1/20 m
Cut off elevation   10 degr
Arc length   12 h
        (e.g.) 1993. 01. 02.   10 h 00 m – 22 h 00 m

SOURCE
        3C147 RA: 5 h 38 m 43.5 s   DECL: 49° 49′ 42″

In order to assess the orbit improvement by space-VLBI data we have used the most simplified strategy we could. The initial 6 orbit parameters were offset according to Table 3. which produced 6.4 km average error in satellite position as compared with the "true" orbit.

Table 3. Errors in orbital elements to be adjusted

	semi-major axis	:	10	m
	excentricity	:	0.0001	
	inclination	:	0.0001	degr
	arg. of perigee	:	0.001	degr
	long. of asc. node	:	0.001	degr
	mean anomaly	:	0.001	degr

Two basic tracking data type were used in different combinations in the orbit parameter adjustment process:

1. delay, delay-rate which are produced by
   − telemetry and control station (T/C data)
   − Precise Range and Range-rate Equipment (PRARE data) and

2. Space-VLBI delay and delay-rate which were produced by one ground to space baseline (Jevpatoria-RADIOASTRON).

The adjustment iterations were carried out using 4 combination of the above datatypes.

1. Assuming only T/C tracking.

2. T/C tracking combined with space-VLBI data.

3. Only PRARE tracking.

4. PRARE combined with Space-VLBI data.

The assumed measurement errors, geometric errors   are given in Table 4. We have used a short 12h orbit arc in order to minimize the nongravitational dynamic effects.

Table 4.

MEASUREMENT ERRORS:		
	VLBI/PRARE	T/C
Delay	0.5 ns	
Delay-Rate	1 .0 ps/s	
Range	0.1 m	2.0 m
Range-Rate	0.01 mm/s	3.0 mm/s
tropospheric ref.	2.0%	
ionospheris ref.	2.0%	
GEOMETRIC ERRORS		
Station coordinates	0.1 m	5.0 m
Station timing	1.0 microsec	
Source coordinates	1.0 mas	
ERP Pole	2.0 mas (6 cm)	
UT1	0.1 ms	

THE RESULTS

The orbit parameter adjustments showed no or slow convergence after the 2nd iteration using TC or combined T/C + Space-VLBI data.

The average orbit accuracy using only T/C tracking data was in the order of 1 km. The combination of T/C data with Space-VLBI delay-delay rate data gave more than one order of magnitude higher orbit accuracy; 40-50 m.

Using PRARE or combined PRARE + Space-VLBI data the meaningful number of iterations were 4.

The average orbit accuracy using only PRARE data was slightly worse than 10 m while the combined PRARE + Space-VLBI delay, delay rate data produced 1 m level orbit accuracy. The summary of results are shown graphically on Fig. 1.

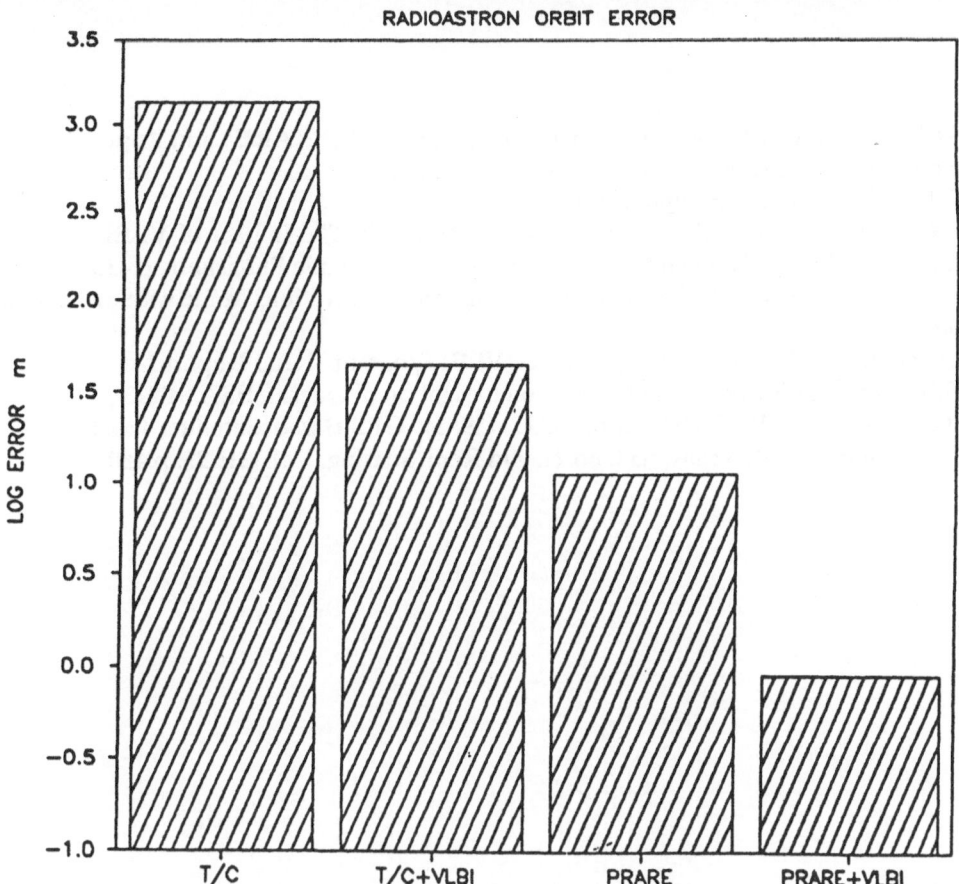

Fig. 1. Average orbit errors as function of applied tracking data types.

CONCLUSIONS

The space-VLBI observables (delay, delay-rate) may be considered as a new type of tracking data for the space-VLBI satellites. The place and role of this new data type was reviewed with respect of the different other tracking data types. Since the astrometric and geodetic applications of space-VLBI require high precision orbit we have investigated the contribution of this new data type to the orbit determination accuracy. Simulation computations were presented for the RADIOASTRON satellite. Results indicate that the combination of space VLBI delay and delay rate with traditional (range range-rate) tracking data yield to significant orbit accuracy improvement. For the scenario simulated, one order of magnitud average orbit accuracy improvement has been demonstrated.

REFERENCES

Dorrer, M and Lefebre, M. (1985): Doppler Orbitography and Radiopositioning Integrated by Satellite: DORIS, CSTG Bulletin 8

Fejes, I., Almár, I., Ádám, J., Mihály, Sz. (1986): Space VLBI potential applications in geodynamics. Adv. Space Res. Vol. 6. No 9. p.205.

Hartl, P. (1984): The Precise Range and Range Rate Equipment (PRARE) and its possible support to the radar altimeter measurements for ERS-1, ESA SP221

Hirabayashi, H. (1989): Private communication

Kardashev, N.S., Slysh, V.I. (1988): The Radioastron Project. in The Impact of VLBI on Astrophysics and Geophysics. Ed. M.J.Reid and J.M.Moran. Kluwer Academic Publishers. p.433.

Levy, G.S. et al. (1986): Very Long Baseline Interferometric Observations Made with an Orbiting Radio Telescope. Science, 234. 187.

Morimoto M., et al. (1988): VSOP - Japan's Space VLBI Program. NRO Report No 183.

Tang, G. (1984): A short note on the high-precision navigation of Quasat, ESA SP213 p. 185.

Younck, T.P., Melbourne, W.G., Thornton, C.L. (1985): GPS-Based Satellite Tracking System for Precise Positioning, IEEE Trans. on Geosci. and Rem. Sensing. Vol. GE-23. p.450.

THE EFFECT OF MODEL ERRORS ON LASER RANGING RESIDUALS

M. Carpino
Oss. Astronomico di Brera
Milano (Italy)

A. de Haan, F. Migliaccio, F. Sansò
Ist. di Topografia, Fotogrammetria e Geofisica
Politecnico di Milano - Milano (Italy)

ABSTRACT

The effect of the imperfect modelling of the Earth gravitational field on the residuals of a least squares orbital fitting was investigated. The residuals of an orbital adjustment of Lageos laser ranging data were analyzed, and covariance and crosscovariance functions estimated and compared with the results of the preceding theoretical analysis, in order to get some insight about the physical mechanism responsible for the signal.

INTRODUCTION

In previous years, an analysis of laser ranging data treatment was already carried on by the Italian study groups to which the authors belong (Betti et al., 1987 and Carpino et al., 1988). The conclusions achieved after that research were that the normal reduction procedures for S.L.R. data leave a significant signal in the residuals. That was obtained by means of a time correlation analysis, and was interpreted as the effect of the unmodelled or erroneously modelled forces.
As an intermediate result, a higher quality procedure for outliers rejection was set up, based on the empirical analysis of the residuals. But it was clear that the real goal was to interpret the empirically revealed signal in terms of the dynamics of the satellite.
So the questions that the authors asked themselves were: are there any errors in the gravity field model (due to truncation of the coefficients or to wrong coefficients)? Are there any other errors of which one must take care?
A way of providing answers to these questions was to look again at the correlation among the residuals, this time estimating not only the timelike but also the spacelike covariance functions, and comparing them.

THE TIMELIKE CORRELATION ANALYSIS

First of all, it was decided to analyze S.L.R. residuals again by estimating the timelike covariance function. While previously the research had dealt both with Lageos and Starlette data, now it was focused only on Lageos data.

The data used were normal points residuals sampled every 2 minutes. They were provided by the Italian firm Telespazio, which obtained it as a part of the TPZ-88.1 Lageos solution, achieved by using a unique kinematical and dynamical model to represent high precision satellite data.
The main features of the residuals can be found in Table 1.

Period covered	Sept. 1983 - June 1986
Gravity field model	GEM-T1 (complete to degree and order 36)
Tidal model	MERIT subset of the earth and ocean tides
Tropospheric correction	Marini and Murray model (with 20° elevation cutoff)
Arcs	short (15 days), combined in multi-arc solutions covering 6 months each

Table 1 - Main features of the set of normal points residuals used for the correlation analysis.

The covariance function for the complete set of residuals was therefore estimated, and it is shown in Fig. 1. It is evident that a signal was left, and with a simple calculation it is found to be of the same order of the noise (see Table 2).

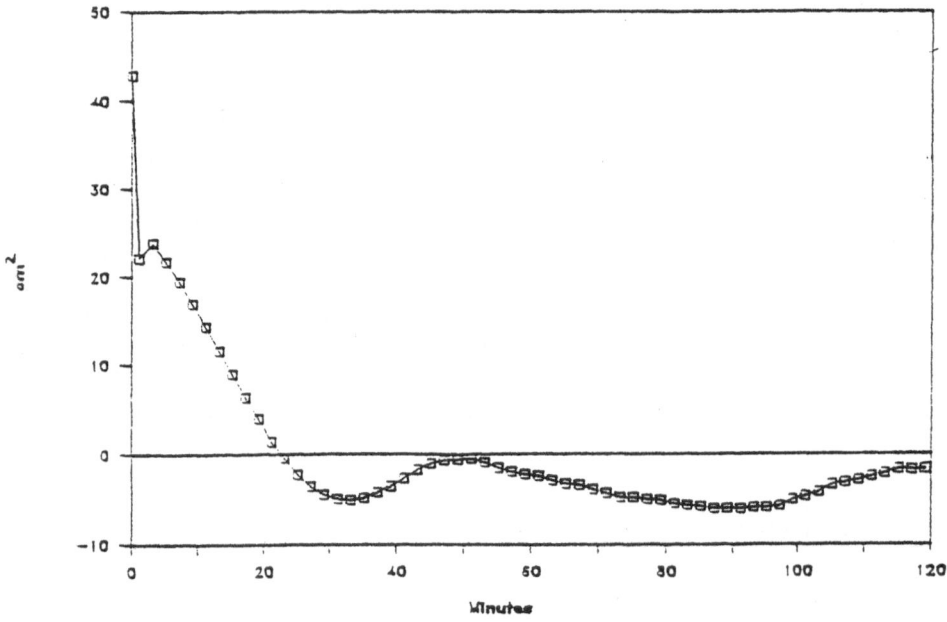

Fig. 1 - Timelike correlation: covariance function, all residuals, Sept. 1983 - June 1986.

Number of residuals	n = 144539
Bin size	$\delta t = 2$ min
Signal	$\sigma_s = 4.7$ cm
Noise	$\sigma_n = 4.6$ cm
Ratio $\sigma^2_s/(\sigma^2_s+\sigma^2_n)$	R = 51%
Correlation length (for $C(t)=1/2\sigma^2_s$)	$\tau_c = 13$ min

Table 2 - Results of timelike correlation analysis using all residuals.

This confirms the results of the previous works (Betti et al., 1987), and one must also notice that it is quite a reassuring confirm as the number of residuals analyzed is very high, reaching nearly 150000.
Another experiment was done, estimating the covariance function by using only the residuals of data collected over Europe and superimposing an elevation cutoff at 60°, that is cutting off all residuals coming from observations lower than a 60° angle on the horizon. The reasons for these choices will be explained in next paragraph.
The results are shown in Fig. 2 and Table 3: they are in very good agreement with those obtained using the whole set of residuals.

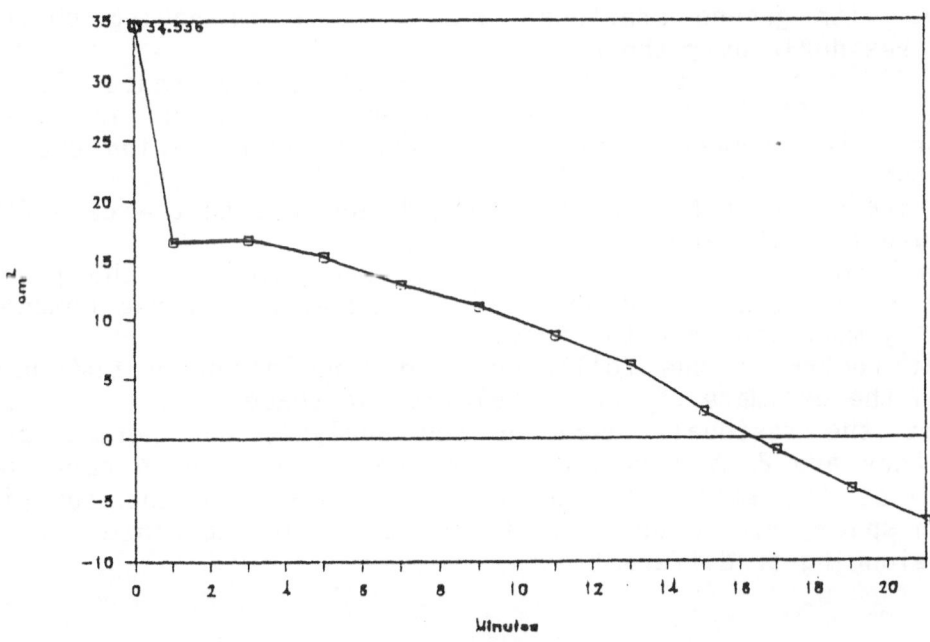

Fig. 2 - Timelike correlation: covariance function, Europe, Sept. 1983 - June 1986.

Number of residuals	n = 9170
Bin size	$\delta t = 2$ min
Signal	$\sigma_s = 4.1$ cm
Noise	$\sigma_n = 4.2$ cm
Ratio $\sigma^2_s/(\sigma^2_s + \sigma^2_n)$	R = 48%
Correlation length (for $C(t)=1/2\sigma^2_s$)	$\tau_c = 11$ min

Table 3 - Results of timelike correlation analysis using residuals over Europe.

THE SPACELIKE CORRELATION ANALYSIS

As written in the introduction, the really new part of this work was to look at the spacelike correlation among the residuals.
Also in this case, the covariance functions were built by the standard procedure for the isotropic processes, that is by taking the products of the residuals with angular distance less than or equal to a fixed bin θ, which was then doubled, tripled, etc. as to enlarge the area progressively by adding an annulus at a time.
It was decided to work on the residuals over Europe, as this region is well covered with ground stations which results in having uniformly distributed residuals over the area.
Besides, as it is clear that a Lageos residual is a distance along the sight line, i.e. it depends on the tracking station, and in order to evidence the radial component of the residuals, an elevation cutoff at $60°$ was imposed.
The results are shown in Fig. 3 and Table 4 (where also the case for an elevation cutoff at $70°$ appears).
What one must notice is the σ_s which has been revealed: though quite weak, the 1.5 cm signal could possibly be the effect of the unmodelled or erroneously modelled gravity field.
An important remark is the following: the correlation in time has no influence on the estimate of the correlation in space.
In fact, as the residuals used in the analysis are normal points residuals, they are 2 minutes apart from each other, which corresponds to about $3°$ at the speed of Lageos. This means that the covariance estimates in space, within an area of radius $3°$, are obtained only with residuals belonging to different orbits.

74

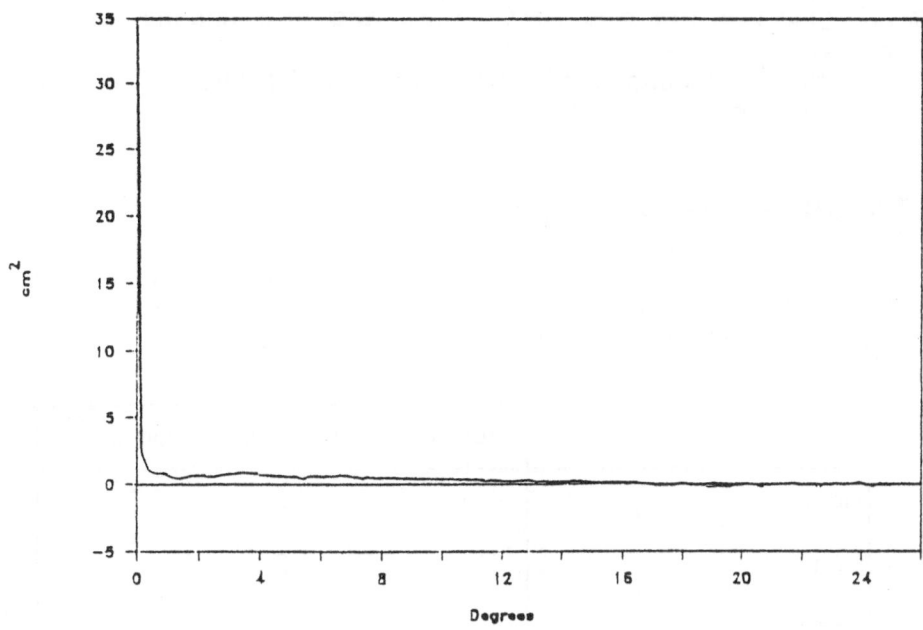

Fig. 3 - Spacelike correlation: covariance function, Europe, elevation cutoff 60°.

	Cutoff = 60°	Cutoff = 70°
Number of residuals	n = 9170	n = 3820
Bin size	θ = $0^{\circ}.25$	θ = $0^{\circ}.25$
Signal	σ_s = 1.5 cm	σ_s = 1.7 cm
Noise	σ_n = 5.7 cm	σ_n = 5.2 cm
Ratio $\sigma^2_s/(\sigma^2_s+\sigma^2_n)$	R = 6.5%	R = 6.5%

Table 4 - Results of spacelike correlation analysis using residuals over Europe.

PRELIMINARY CONCLUSIONS

The experiments so far performed lead to the following conclusions.
A signal has been definitely evidenced in the residuals, with a σ_s of the order of 4 ÷ 5 cm, a correlation length of 11 ÷ 13 min and the first zero at 20 minutes.
The signal seems to consist of two parts. One part (though quite little) is spatially correlated ($\sigma \sim 1.5$ cm), the other part is timelike correlated.
While the part which is spatially correlated could possibly represent

75

the effect of gravity field errors, what are the causes of the signal depending on time? .
There were a number of causes of which one could think, of course. The authors looked at some.

CAUSES OF THE SIGNAL DEPENDING ON TIME

At first, the uncertainties of the dynamic model were taken into account. The estimated accuracy in the forces acting on Lageos (according to Milani et al., 1987) are shown in Table 5.

CAUSE	ACCELERATION (10^{-10} cm s^{-2})
moon	20
sun	3
drag	3
solar radiation pressure	60
Earth albedo radiation pressure	80
thermal emission	2

Table 5 - Estimates of the uncertainties of the accelerations on Lageos due to perturbative forces other than the geopotential.

As it can be easily computed, the uncertainty of the position of the satellite after 20 minutes (which correspond to the first zero of the covariance function) is of the order of 0.01 cm.
The conclusion is that these forces cannot explain a signal of about 5 cm on a time length of 20 minutes, as our case is.
So, as the signal doesn't seem to be generated by an error in the force model, other possible sources of error were investigated, which may have an orbit like pattern, without being correlated area like.
Local errors (that is errors depending on the station) were checked: they could come from not good enough accuration in the calibration of ground instruments tracking Lageos. In order to prove this, only the residuals coming alternately from different stations, along the same arc, were correlated. The result of this experiment appears in Fig. 4 where, for the sake of comparison, also the correlation among all residuals is shown.
The σ_s has only slightly (about 0.5 cm) lowered, that is: only a little part of the signal has been lost, which could be ascribed to instrumental behaviour.

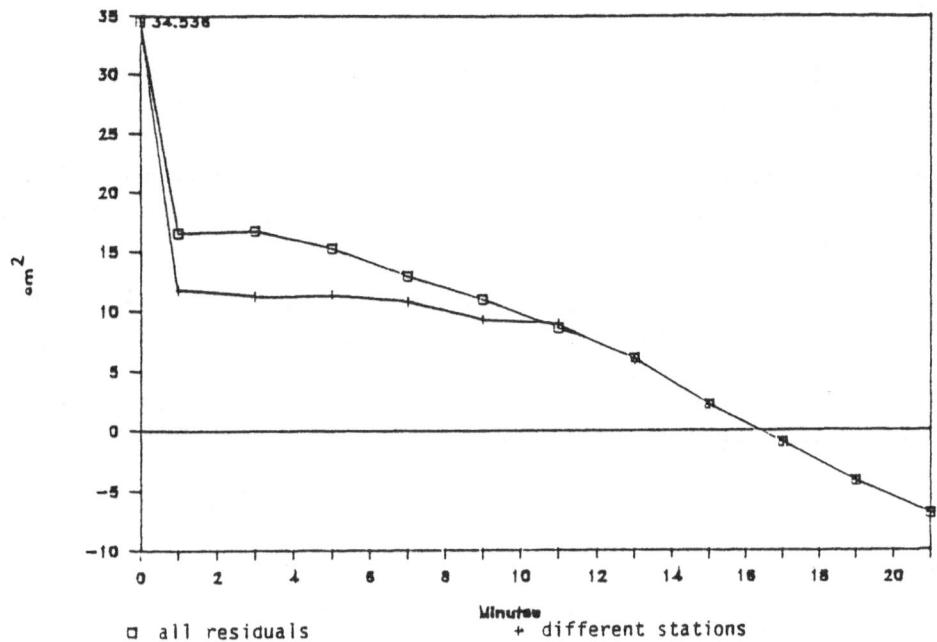

□ all residuals   + different stations

Fig. 4 - Local errors: timelike correlation, residuals over Europe, elevation cutoff 60°.

As a last attempt, the question was investigated if there could be errors in the atmospheric refraction model. In particular they could be:
1) an error in the model of the refraction index along the path;
2) an error in the layered assumption (in particular the shape of the upper limit of the atmosphere could be not adequate).
A more refined atmospheric model could explain the signal which is still left in the residuals. Of course the answer to this question should come from real experts of these problems.

CONCLUSIONS

Analysing once again the Lageos normal points residuals, but treating a much larger number of points than in previous works, the timelike correlation among them confirmed the existence of a signal with a σ_s of the order of 4÷5 cm.
The spacelike correlation analysis proved the existence of a σ_s of about 1.5 cm, which is not influenced by the correlation in time, and could represent the effect of the gravity field model.
Possible causes of the signal depending on time were investigated and partially rejected, still leaving doubts in particular about the atmospheric refraction model, and leaving the door open to a further improvement in this research.

ACKNOWLEDGMENT

The authors thank A. Cenci and M. Fermi of Telespazio S.p.A., Rome, for providing the Lageos normal points residuals used for the correlation analysis.

77

REFERENCES

Betti, B., Carpino, M., Migliaccio, F. and Sansò, F. (1987): Signal and Noise in SRL Data. Bull. Geod., n. 61, pp. 235-260.

Carpino, M. and Migliaccio, F. (1988): On the Preprocessing of Satellite Laser Ranging Data. Proc. Third International Conference on the Wegener/Medlas Project, Univ. Bologna.

Cenci, A., Fermi, M. and Caporali, A. (1989): European Baselines Determined with Lageos: the TPZ-88.1 Solution. Telespazio S.p.A., Roma.

Milani, A., Nobili, A.M. and Farinella, P. (1987): Non-gravitational Perturbations and Satellite Geodesy, Adam Hilger Ltd, Bristol.

DESIGN OF AFRICAN PLATE SLR/VLBI NETWORK FOR GEODYNAMICS AND EARTHQUAKE RESEARCH

by

Professor Dagogo M J Fubara
Dean, Faculty of Environmental Sciences
University of Science & Technology
Port Harcourt, Nigeria

ABSTRACT

Laser ranging to geodetic artificial satellites is now a proven technique for centimetre accuracy level determination of three-dimensional geodetic coordinates of widely spaced network points. Repeated determinations of the network point coordinates enable the determination of spatial crustal movements as a tool for geodynamics and earthquake research. The Mediterranean forms the boundary zone where the African, Arabian and Eurasian tectonic plates are said to be causing deformation due to collision. The WEGENER-MEDLAS project has for years since the early part of this decade set up a net work of several points on the Eurasian and Arabian plates for Satellite Laser Ranging (SLR) observations and analyses. The island of LAMPEDUSA near Malta is the only station on the African plate. As long as Africa remains the "dark continent" in the world-wide Crustal Dynamics programme, full understanding of global geodynamics cannot be achieved; earthquake research in Africa will remain deficient and other attendant scientific knowledge from SLR will be lost.

A design of the African plate SLR network for observing recent crustal movements on global and regional scales as an aid to earthquake research is herein proposed. Applicable tectonic platemodels are discussed. As an initial approximation, the rigid plate hypothesis is adopted. Network design criteria included not only geodetic, geophysical, geologic and meteorologic considerations but also reliable recovery of plate rotations when combined with the WEGENER-MEDLAS project.

1. INTRODUCTION

The International Lithosphere Programme launched in 1981 for interdisciplinary research in the solid earth sciences is focused on the "Dynamics and Evolution of the Lithosphere as the framework of Earth Resources and the Reduction of Hazards". Among its prime goals, besides mineral and energy resources exploration and development are the mitigation of geological hazards, the protection of the environment and the strengthening of the earth sciences and their effective application in developing countries [Price, 1987].

Most earthquakes occur within tectonic plate boundaries that are typically less than a few hundred kilometers wide and subject to internal stress and strain arising from plate motions. For this reason, international geoscientists of the developed nations have in this decade focused   on programmes and projects with objectives including the following:

(i)   to measure the motion and deformation of plate;

(ii)   to establish the distribution of stress and strain in the lithosphere and to models for the way these are concentrated;

(iii)   to provide the basic data required for modelling of the internal mechanisms responsible for driving the plates;

(iv)   to study polar motion and variations in the earth's rotation to determine possible relationships between these quantities and geophysical phenomena such as earthquakes and volcanic eruptions.

1.1 *The Role of Space Geodetic Techniques*

In order to achieve the above objectives the measurement activities must determine (a) gross motion of the plate system; (b) micro-plates versus plastic deformations within continental collision zones; (c) stress and strain in the lithosphere, in relation to an elasticity of the asthenosphere; (d) earthquakes in relation to plate motion, energetics and driving mechanism of plate motion; and (e) vertical movements and deformation of the lithosphere.

Established space geodetic methods such as Satellite Laser Ranging (SLR) to artificial satellites and independent-clock stellar radio interferometry called Very Long Baseline Interferometry (VLBI) offer 3-dimensional relative point positions determinations or baseline measurements with accuracies of the order of 1cm for widely separated points or baselines up to several thousand kilometres.   Vertical accuracies are generally less than horizontal accuracies but by less than a factor of 2 [Bilham, et al, 1989] A new space geodetic technique - an artificial satellite-based radio-interferometry-the Global Positioning System (GPS) can attain accuracies similar to VLBI.   GPS offers the possibility of cost-effective densification of geodetic networks with spatial separation of 1km to 1000km [King, et al, 1985].   Indeed, plate motions which constitute a 1 to 10cm/year displacement impute signal to the earthquake process have been measured directly with SLR and VLBI methods [Carter and Robinson, 1986; Christodoulidis, et al, 1985].

If seismologists were in possession of 2 to 3 centuries of this type of geodetic data from earthquake belts or all the plates of the world, we would be in much stronger position to understand the mechanisms and hence more effectively predict the occurrence and mitigate the adverse impacts of earthquakes.   We currently have the tools to implement a world-wide geodetic network with SLR, VLBI, GPS that will benefit future generations.   Such studies are particularly important in developing nations where geodetic measurements are currently of low accuracy or non-existent even where seismic risks exist. Recognising this, the Inter-Union Commission (ICL) on the Lithosphere adopted a resolution in 1988 as follows:

"In view of the unprecedented increase in population in the developing nations ... we recommend the implementation of geodetic networks with a scale and distribution appropriate to characterising earthquake risk in these nations."

FIG. 1.—CONTINENTAL AND OCEANIC FRACTURE SYSTEMS

FIG 1   PLATE DISTRIBUTION OF CRUSTAL DYNAMICS PROJECT STATIONS

Table 1:   CRUSTAL DYNAMICS PROJECT
WORLD-WIDE SITES DISTRIBUTION ON PLATES

1	North American Plate	36	Stations
2	Eurasian Plate	29	"
3	Pacific Plate	17	"
4	Nazca Plate	1	"
5	South American Plate	5	"
6	Arabian Plate	4	"
7	Indo-Australian Plate	3	"
8	African Plate	Nil	"

In spite of all this, the international community of planetary scientists has developed necessary programmes and projects for global geodynamics and crustal motion in which only one of the 7 major world plates - the African plate, is not actually involved. See Figure 1. Some salient features of these international programmes, exclusive of Africa are discussed below. This is followed by the pertinent geologic features of the African plate to justify that it qualifies to be included in the global network of SLR/VLBI/GPS geodetic network for geodynamics and earthquake research. This is the focus of this paper. The motion of the lithospheric plates is the most fundamental aspect of plate tectonic theory. An effective monitoring programme should strive for international coordination and global in extent with a dedicated global network configuration since the problem of monitoring the gross motion of the plate is essentially one of global scope. Yet, Africa which is used to being called "the Dark Continent" remains dark and void dark and void in currently existing geodynamic programmes as the startling discussions that follow will show.

2. CURRENT GEODYNAMICS PROGRAMMES

In recognition of the potential practical applications of SLR and VLBI, the USA established in 1979 a coordinated federal programme for the application of this space geodetic technology to crustal dynamics and earthquake research - the NASA Geodynamics Programme. Recognising the global nature of the studies and need for international collaboration, the programme started with investigators from 14 countries. To date, 22 countries are or have been involved in the NASA programme. Outside the USA, the European scientific community and some Middle East countries formed a consortium of collaborating investigators known as the "Working Group of European Geoscientists for the Establishment of Networks for Earthquake Research" - WEGENER, to study crustal deformations in the Mediterranean region.

2.1 NASA Geodynamics Programme

According to [NASA-TM 4065, 1988], the two-fold programe objectives are:

(i)   to contribute to the understanding of the solid Earth, in particular, the processes that result in movement and deformation of the tectonic plates;

(ii)   to obtain measurements of the Earth's rotational dynamics and its gravity and magnetic fields in order to better understand the internal dynamics of the Earth.

The major areas of research of the programme are (a) Earth Dynamics, (b) Crustal Motion and (c) Geopotential research.

The Earth Dynamics research objectives are to develop models of polar motion and earth rotation and to relate studies of global plate motion to the dynamics of the earth's interior. This programme is expected to lead to increased understanding of the global structure of the Earth and the evolution of the crust and lithosphere. The research conducted includes studies of the dynamic interaction between different regions of the Earth's tectonic features. How all these can be accomplished without involving the African plate is difficult to imagine.

For the Crustal Motion case styled the Crustal Dynamics Project, the primary focus is conduct of field measurements and modelling studies of crustal deformation in various tectonic plate settings except the African plate. From the measurements and analyses, models which describe the accumulation and release of crustal strain (earthquake precursors) and the crustal motion between and within the tectonic plates, particularly the North American, Pacific, Eurasian, South American, and Indo-Australian plates will be developed as an aid to earthquake prediction and mitigation. Surely, these phenomena occur on the African plate which, though dorminant in size, is nevertheless omitted.

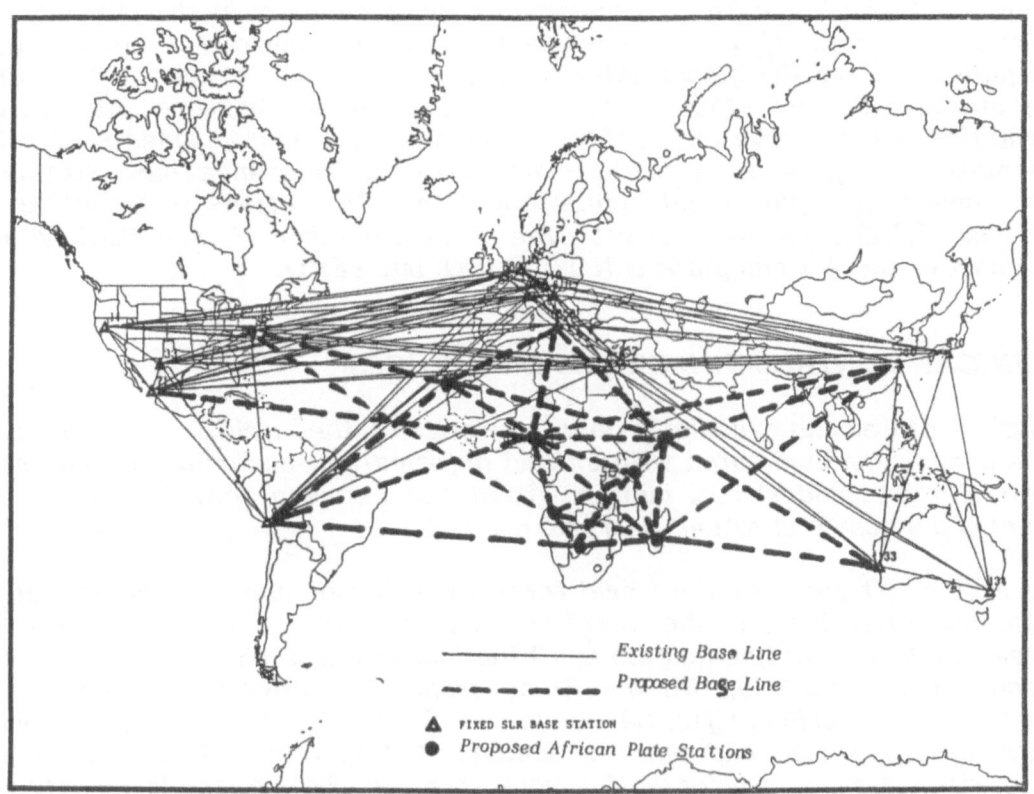

Figure 2 Crustal Dynamics Project - SLR baselines: *Existing and Proposed*

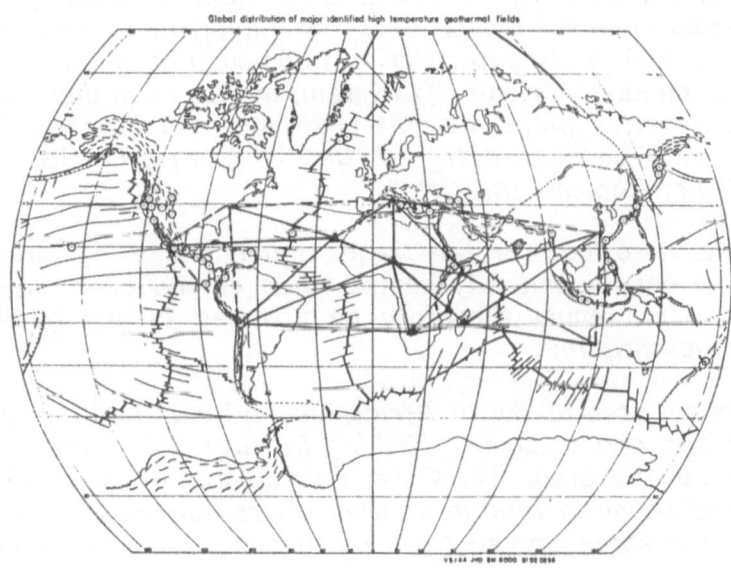

Fig. 3 Global distribution of major identified high temperature geothermal fields.

*Proposed African Plate Network
in relation to Existing Baselines
and Major Tectonic Plates*

KEY

Existing Baselines	– – – – –
Proposed Baselines	———————
Tectonic Plate boundaries	⊤⊢⊤⊢⊤⊢⊢
Proposed African SLR Stations	– – – – ▲
High Temperature	– – – – – – – – ⊙

83

Figures 1, 2 and 3 and Table 1 underscore gross anomaly in the current world-wide (except Africa) efforts in geodynamics and earthquake research. Yet, the basic philosophy for the global network required has been to establish "...at least three stations on the stable part of each plate to measure internal plate deformation and to serve as a reference network for measurements of relative plate motions. These same stations also serve as reference base stations for measurements of regional deformation with mobile stations near active plate boundaries" [NASA-TM 4065, 1988]. The current number of such stations on the stable part of the African plate is NOT THREE but ZERO.

3. AFRICAN CONTINENTAL CRUSTAL FEATURES

Geologic features and seismicity characteristics of the African continent exhibit all the phenomena associated with crustal deformation, accumulation and release of crustal stress and strain, motion within and between tectonic plates, earth tremors and volcanic eruptions past and present.

The active rift zones of the world form a continuous chain of volcanically and seismically active belts in the world oceans, the mid ocean ridges, with a few extensions into the continents such as (i) the East-African rift zone; (ii) the Baikal rift zone in Central Asia; (iii) the Rhine graben in Southern Germany; (iv) the Western North American rifts; (v) the Jordan Valley rift; (vi) the Rio Grande rift in New Mexico, U.S.A. and (vii) the Iceland Rift system. Of these, the East-African rift system is considered by many to be the best example of extensional tectonics relating to an incipient plate boundary with a clear structural link with the World-wide Oceanic Rift System via the Afar depression of Ethiopia and the Gulf of Aden. Consequently, the East African Rift system probably reflects the initial processes leading to future continental disruption and provide a unique area to investigate the interaction of crustal and upper mantle rifting processes at an incipient plate boundary and the transitional tectonics from oceanic to continental rifting. The other active continental rift systems such as Baikal, Rhine and Rio Grande rifts, unlike the East African rift system, do not have any clear structural connections to constructive plate margins [Fairhead and Stuart, 1982] which is thorough treatise on "The Seismicity of the East African Rift System and Comparison With Other Continental Rifts".

Africa may not be notorious for the frequent occurrence of large destructive earthquakes but it is very seismically active while earthquakes have occurred in several locations. Africa must therefore be involved in any geodynamics and earthquake research programme.

The Kaapvaal Craton area of South Africa has had long history of earthquakes since 1620 [Fernandez and Guzman, 1979]. In July 1976, the Koffienfontein earthquake occurred in this area. The Cares Earthquakes are located about 100 km NE of Cape Town and began in 1969 in an area where earthquakes were previously uncommon. Since 1950 three Gabon-Zaire Earthquakes have occurred to the west of the rift system, the most recent being on September 23, 1974 in Gabon and May 15, 1976 in Zaire. Several West African countries including Ghana and Nigeria have in the last few years experienced sizeable earth tremors. The Cameroons have had history of volcanic eruptions while the recent cause(s) of the Lake Nyos gas disaster is (are) still being investigated.

Fig 4 A ; General geological map showing main sedimentary basins of Nigeria and general structural data. The tentative location of the proposed SLR Stations

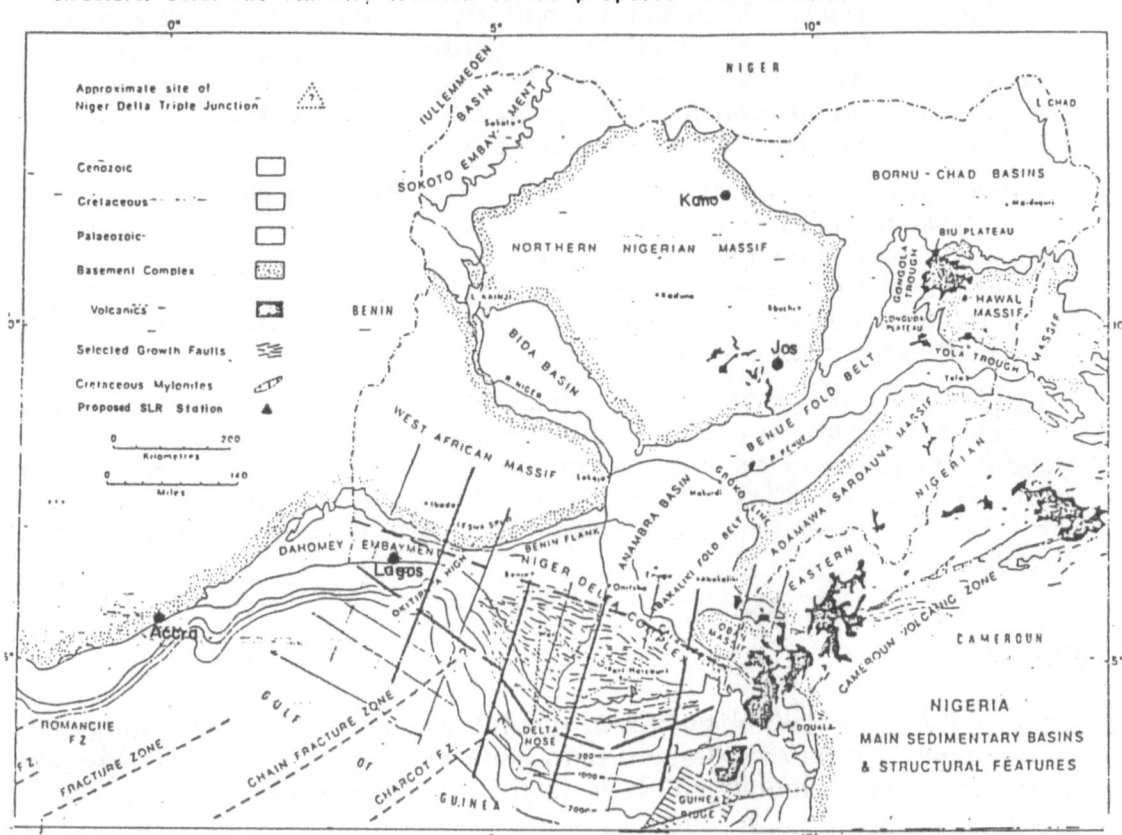

4. CONCLUSIONS AND RECOMMENDATIONS

From the discussions herein,

(i)   *It is obvious that the current international activities and area coverage excluding the African plate cannot in fact effectively accomplish the objectives of the NASA and associated Geodynamics and earthquake research programme;*

(ii)   *Application of space geodetic techniques involving the setting up of SLR, VLBI and GPS geodetic networks is the key strategy for geodynamics and earthquake forecast research. This demands international coordination and a dedicated global network configuration including at least 3 stations on the main African plate. Such fundamental network stations should be chosen away from deforming boundary zones, so that relative station motion may be interpreted in terms of plate motion or broad scale deformation of plate interiors;*

(iii)   *Due to non-availability of permanent SLR and VLBI facilities in Africa, the USA National Aeronautics and Space Administration (NASA), the Commission for Coordination of Space Techniques for Geodesy and Geodynamics (CSTG) and the Working Group of European Geo-Scientists for*

85

the Establishment of Networks for Earthquake Research (WEGENER) should set up a programme to use mobile SLR and VLBI, and GPS for observation at African stations. Given the geologic and seismic characteristics of the African continent and the availability of trained geodesists and government disposition, it is recommended that Jos, Nigeria (see Fig. 4) be chosen as the experimental station for setting up an African network of SLR, VLBI stations for geodynamics and earthquake research;

(iv) *Ultimately, a 7-station African Plate SLR/VLBI or GPS Network (see Figs. 2 and 3) should be established as part of the Global Crustal Dynamics Project. This will complete global coverage to permit the world-wide within and between tectonic plates motion studies and enable African plate regional deformation studies indicated by the African Rift Systems and the continents seismicity characteristics;*

(v) *Given the current and future economic outlook in Africa which is struggling to meet basic 20th century needs, international cooperation backed up with funds and equipment are required for the execution of the proposed African plate SLR/VLBI or GPS network for geodynamics and earthquake research. This approach enabled the execution of the African Doppler Survey (ADOS) project.*

REFERENCE

Aardoom, L. (1987). *Current activities in the measurement of lithospheric plate motion and deformation, Recent plate morement and deformation, Recent plate movements and deformation, Geodyn Series Vol. 20, K. Kasahara (Ed.) Amer. Geophys. Un., Washington D.C.*

Bilham, R., Yeats, R. and Zerbini, S. (1989). *Space geodesy and the global forecast of earthquakes, EOS, Trans. Amer. Geophys. Un., 70, 65.*

Carter, W.E. and Robinson, D.S. (1986). *Studing the earth by very long baseline interferometry, Sci. Amer., 46.*

Christodoulidis, D.C., Smith, D.E., Kolenkiewicz, R., Klosko, S.M., Torrence, M.H., and Dunn, P.J., (1985). *Observing tectonic plate motions and deformation from satellite laser ranging, J. Geophys. Res., 90, 9249.*

Coats, R.J., Frey, H., Boxworth, J., and Mead, G., (1985). *Space-Age geodesy: NASA Crustal dynamics project, IEEE Trans. Geosci. Remote Sensing, Vol. GE-23, 358.*

Fairhead, J.E. and Stuart, G.W. (1985). *The seismicity of the East African rift system and comparason with continental and oceanic rifts, Geodyn. Series Vol. 8, G. Palmason (Ed). Amer. Geophys. Un., Washington, D.C.*

Frey, H.V. and Bosworth, J.M., (1988). *Measuring contemporary crustal motions: NASA's crustal dynamics project, Earthquake and Volcanoes, 20, 96.*

Fubara, D.M.J., (1988). *Coastal zone erosion, land degradation and conservation in Nigeria, Uni. of Sci. & Tech., Port Harcourt.*

King, R.W., Masters, E.G., Rizos, C., Stolz, A., and Collins, J., (1985). Surveying with GPS, Mono., 9, Sch. of Surv., Univ. of New S. Wales.

NASA - Tech. Memo. (1988). NASA Geodynamics Program Summary report: 1979-1987, Office of Space Sci. Appl., Washington, D.C.

Price, R.A. (1987). Foreword, Recent plate Movements and deformation, Geodyn. Ser. 20, K. Kasahara (Ed.), Amer. Geophys. Un., Washington, D.C.

Scrutton, R.A. (Ed.), (1982). Dynamics of passive margins, Geodyn. Ser. 6, Amer. Geophys. Un., Washington, D.C.

DISPLACEMENT OF A JAPANESE VLBI STATION AS AN INDICATOR OF THE ISLAND ARC CONTRACTION

Kosuke Heki
Kashima Space Research Center,
Communications Research Laboratory
893-1, Hirai, Kashima-machi, Ibaraki 314, Japan

INTRODUCTION

Northeast Japan is a typical subduction zone characterized by active seismicity, volcanism and tectonics. Kashima very long baseline interferometry (VLBI) station is located on the "outer arc", that is, between the volcanic front and the Japan trench. The existence of many active faults in intraplate Japan suggests complicated crustal deformation making Kashima station somewhat unsuitable for the measurement of the tectonic plate motion itself. From another viewpoint, however, the measurment of the displacement of such an island arc station with respect to the stable plate interior will provide information on the on-going deformation of the island arc (Fig.1).

Lithospheres are considered to be quite thin under volcanic and back arcs due to the high temperature gradient while there is a cold and rigid fore-arc wedge under the outer arc. Also subduction zones are usually under some tectonic stress fields. These two factors result in the deformation of the island arc, that is, the displacement of the rigid outer arc with respect to the landward plate.

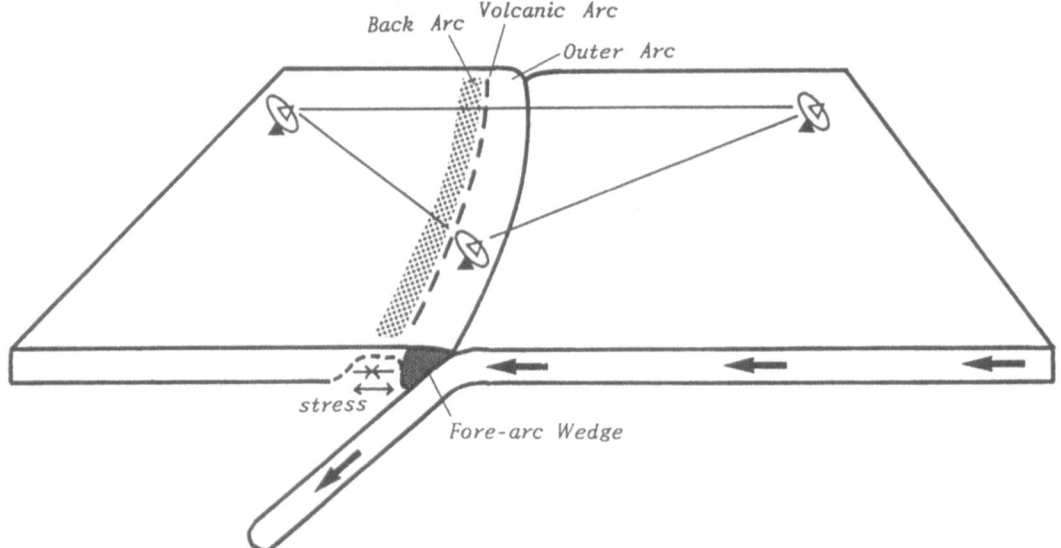

Fig.1. Displacement of an VLBI station on an outer arc with respect to the plate interior stations indicates the deformation of an island arc.

Island arcs are divided into two groups by the tectonic stress field status (Nakamura and Uyeda, 1980): either compressional (Chilean-type) or tensional (Mariana-type) in the direction of the plate convergence. It is a typical way of the compressional island arc deformation that compressional force from the subducting oceanic plate is once received by the fore-arc wedge and then deforms the thin-skinned back arc through this rigid fore-arc. Active folding and reverse (or strike-slip) faulting are expected to be observed as the results of the deformation (Fig.2). In this case, the movement of an outer arc with respect to the plate interior should be landward. In the tensional case, the direction should be oceanward. Stress fields are compressional in the Northeast Japan and Kashima is expected to move in the same direction as the subducting Pacific plate.

The displacement rate of the outer arc in a compressional subduction zone is equal to the contraction rate of the arc and can be estimated by integrating the strain rate in the direction of the stress field. In Japan, repeated geodetic surveys give values of a few times 10E-7/year as the recent crustal strain rates. If such deformation occurs over a zone of a few hundreds of kirometers, the movement of Kashima will be a few cm/year, which is large enough to be detected with a standard geodetic VLBI technique.

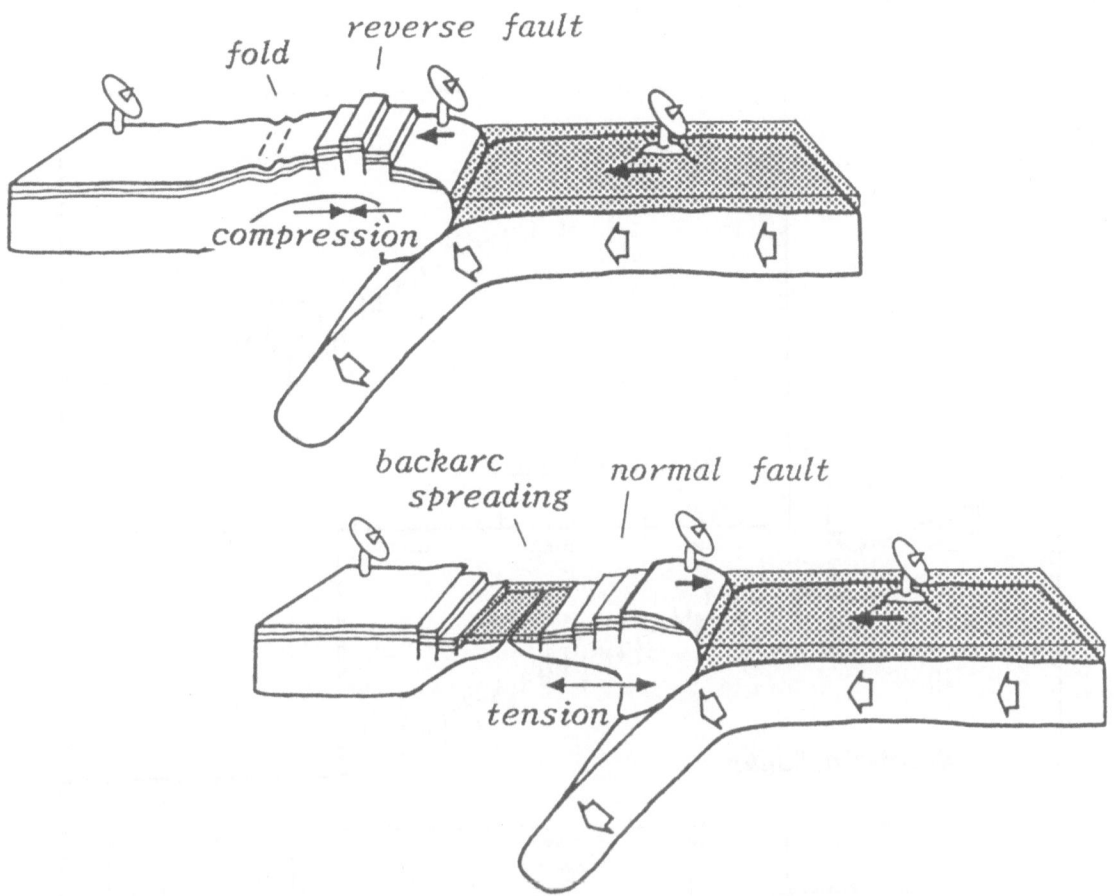

Fig.2. Two types of subduction zones. If tectonic stress fields are compressional/tensional (upper/lower), a VLBI station on the outer arc will move landward/oceanward with respect to the landward plate interior. We will recognize the deformation also with various tectonic activities such as active faulting and folding.

DATA ANALYSIS PROCEDURES

We used the data of 82 Crustal Dynamics Project (CDP) and International Radio Interferometric Surveying-Pacific Network (IRIS-P) VLBI observing sessions from 1984 January to 1988 October. Each observing session gives a set of relative station positions in terms of geocentric XYZ coordinate. This is converted into baseline lengths which are free from the uncertainty of the earth orientation. Then, changing rates of the baseline lengths are obtained by linear regression (Fig.3). Finally, horizontal site velocity vectors are adjusted with a least-squares procedure so that they best explain the observed baseline length changes.

Generally speaking, in order to convert distance changes into position changes, velocity vectors of some stations have to be known and fixed. In this study, those of four North American stations (Haystack, Fairbanks, Mojave, Hatcreek) are fixed and those of the stations on other plates (Kashima[NOAM?], Vandenberg[PCFC], Kauai[PCFC], Kwajalein[PCFC], Onsala[EURA], Wettzell[EURA]) are estimated. Kashima is considered to be either on Eurasian plate (Chapman and Solomon, 1976) or North American plate (Nakamura, 1983; Kobayashi, 1983) although Kashima is not expected to be completely fixed on a certain plate. In this study, we provisionally assume Kashima is on the North American plate and will discuss this point later.

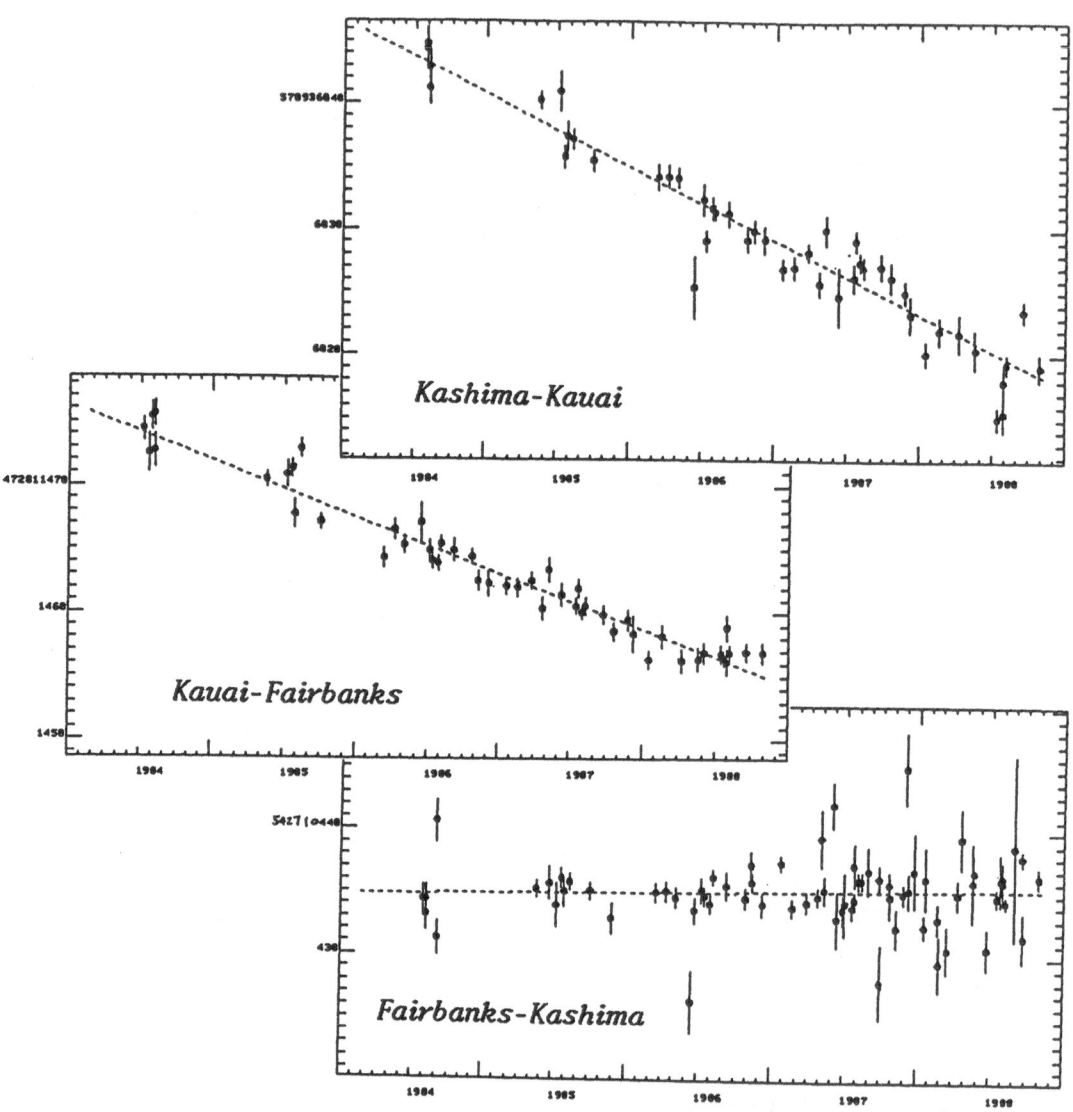

Fig.3. Three examples of the baseline length changes (Kashima-Kauai, Kauai-Fairbanks, Fairbanks-Kashima). Vertical and horizontal axes denote baseline lengths and times respectively. Each data point corresponds to a VLBI observing session and the attached error bar is the one-sigma formal error. Broken lines are the best-fit lines obtained by a weighted least-squares method.

Table 1. Movements of VLBI stations with respect to the North American plate.

Station(Plate)	North (cm/yr)	East (cm/yr)	Horizontal (cm/yr)	Azimuth (degree)
(Velocity Fixed)				
Fairbanks(NOAM)	0.0	0.0	0.0	---
Haystack(NOAM)	0.0	0.0	0.0	---
Mojave(NOAM)	1.0	-0.3	1.1	17
Hatcreek(NOAM)	0.8	-0.6	1.0	36
(Velocity Estimated)				
Kashima(NOAM?)	0.7+0.3	-1.2+0.3	1.4+0.4	-60+8
Vandenberg(PCFC)	3.8+0.1	-3.1+0.1	5.0+0.1	-39+1
Kauai(PCFC)	5.5+0.2	-5.9+0.2	8.1+0.2	-47+1
Kwajalein(PCFC)	5.2+0.4	-7.6+0.4	9.2+0.5	-56+2
Wettzell(EURA)	-0.5+0.7	1.2+1.2	1.3+1.2	112+34
Onsala(EURA)	-0.7+0.7	2.3+0.8	2.4+0.7	108+18

North,East,Horizontal: components of the estimated vectors, Azimuth: direction of the estimated vectors (from north, clockwise), NOAM: North America, PCFC: Pacific, EURA: Eurasia.
Errors are one-sigma formal errors and a-priori velocity vectors of Mojave and Hatcreek are calculated after Ma et al. (1988).

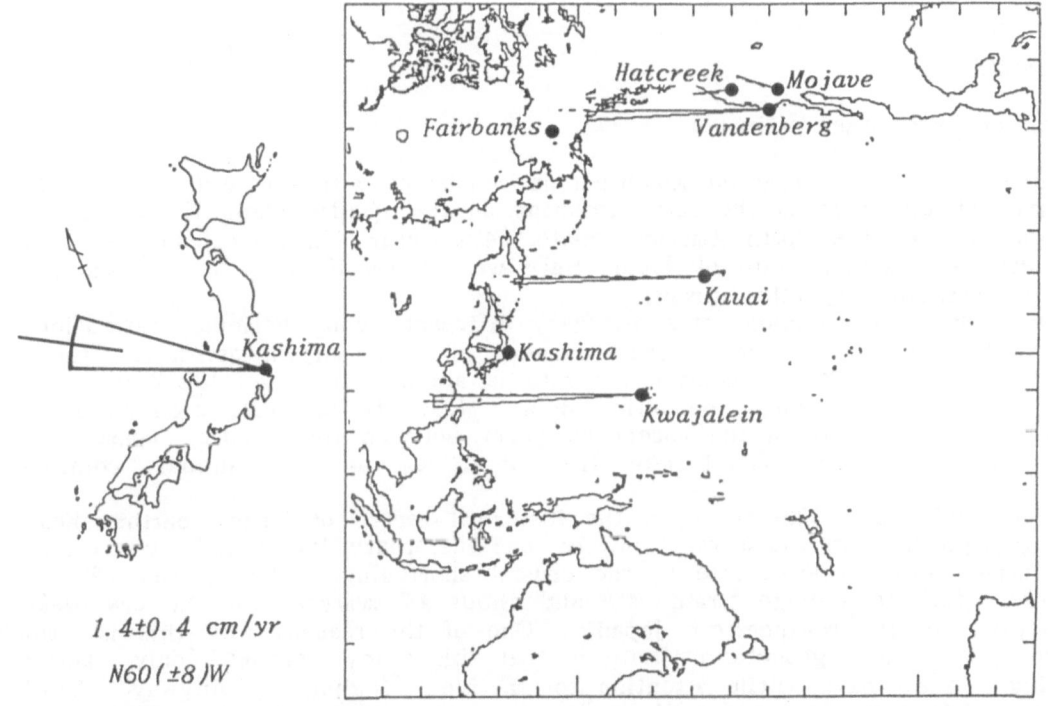

Fig.4. Estimated horizontal velocity vectors for VLBI sites viewd from the North American plate (Table 1). The projection pole is shifted to the Euler pole of the North American and Pacific plates and each vector is exaggerated by its secant of the latitude so that the expected vectors of three Pacific plate stations are parallel (exactly right to left) and of equal length. Broken lines are the expected vectors and fans are the observed vectors and their one-sigma formal errors. One division in the map corresponds to 10 degrees.

Haystack and Fairbanks stations are on the stable plate interior and are assumed to be fixed on the North American plate (velocity=0). Two Californian stations (Mojave and Hatcreek) are assumed to have certain velocity vectors (Table 1) with respect to the plate interior which can be interpreted either by the crustal extension of Basin and Ranges (Minster and Jordan, 1987) or an elastic-plastic megashear along North America and Pacific plate boundary (Ward, 1988). By assuming such a-priori velocity vectors for North American stations, those of other stations are estimated as the velocity viewed from the North American plate.

RESULTS

The estimated vectors and their formal errors are summarized in Table 1 and are shown in Fig.4. In this figure, the projection pole is shifted to the Euler pole of the Pacific and North American plates (pole position after RM-2 [Minster and Jordan, 1978]) so that the expected movement of Pacific plate is exactly right to left. We can see that the estimated movements of three Pacific plate stations (Vandenberg, Kauai and Kwajalein) are very coherent suggesting that Pacific plate has moved almost as a rigid body in this period.

The Japanese part is magnified to the left, where we can notice that Kashima is moving in the same direction as the Pacific plate. If Kashima were on the North American (Eurasian) plate and the plate were really rigid, Kashima should not move (should move eastward) with respect to the North American plate. Accordingly, the WNW-ward velocity of 1.4 cm/year indicates the existence of the on-going deformation of Japan.

DISCUSSION

Comparison with other estimates

The direction of the observed movement is consistent with our expectation that Kashima should move in the same direction as the Pacific plate. If we assume Kashima is on the North American plate, 1.4 cm/year will correspond to the east-west contraction rate of Japan. Here we discuss if this is consistent with the estimates by other means.

The estimated contraction rates are very different among various estimation methods. Conventional ground geodetic surveys give the largest estimates that the typical recent strain accumulation rate is about 1 to 3 x 10E-7/year in Japan. Based on such geodetic data, Mogi (1970) calculated that the culmulative shortening in the recent 60 years between the Pacific coast and the Japan Sea coast is ∼1 meter (1 to 2 cm/year as an annual contraction rate).

Wesnousky et al. (1982) surveyed the historical record of large earthquakes and geologically determined rates of slip on Quaternary faults and converted the seismic moment release rate to the crustal shortening. They present 2.4 x 10E-8/year for the average strain rate and about 4.7 mm/year for the east-west contraction of the Northeastern Japan. One of the reasons why this is much smaller than the geodetic estimate is that this study argues only active "faulting" and pays little attention to active "folding". Anyway, VLBI results of 1.4cm/year is not far discrepant from other estimates and possibly is the most reliable estimate of the current east-west shortening rate of Japan.

Plate Boundaries In Japan

We want to go back to the discussion "on which plate does Kashima reside?". This study assumes the model that the North America-Eurasia plate boundary runs along the back arc down to the central Japan and Kashima is on the North American plate (Fig. 5A). In this model, the estimated movement of Kashima directly corresponds to the contraction of Japan. Also there should be the plate convergence of about 1.1 cm/year along the back arc (after RM-2).

Another possibility is that this boundary ends in central Hokkaido (Fig.5B) and Kashima is on the Eurasian plate. In this model, RM-2 predicts that Kashima moves eastward by about 1.5cm/year. Then, intraplate deformation (difference between the predicted and observed vectors) is about 2.6cm/year in a almost E-W direction. As explained before, island arc lithosphere is the weakest along the back arc and deformation should concentrate along this zone to a certain extent. Fig.5A considers it "plate convergence" and Fig.5B considers it a part of "intraplate deformation". The two models are not so different essentially.

Anyway, such a discussion is meaningful only if the plate boundary can be defined. Now it is evident that, in an island arc, deformation occurs not along some distinct plate boundary but over a wide zone because of the weakness of the plate. In this context, a discussion about "plate boundary" might be useless in an island arc.

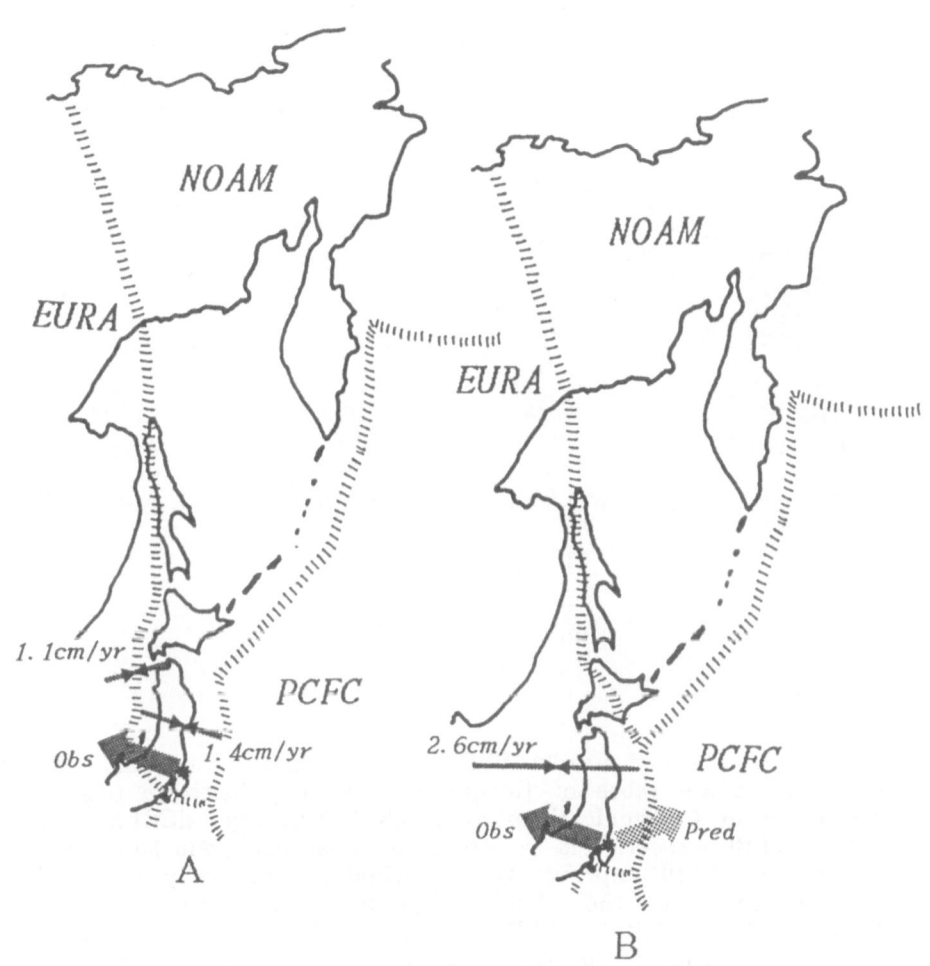

Fig.5. Two models of plate boundaries around Japan. A and B suppose Kashima (denoted by an asterisk) belongs to the North American and Eurasian plates respectively. Observed and predicted movements of Kashima with respect to the North American plate are shown by arrows (in model A, predicted motion is zero). For details, see text.

CURRENT VLBI ACTIVITIES IN JAPAN

The older 26m antenna in Kashima is now being replaced by a new 34m antenna. We also started geodetic VLBI experiments between Kashima and many mobile and fixed VLBI sites in Japan (Fig.6). Among these stations, Minamitorishima (Marcus Island) on the Pacific plate and Minamidaito Island on the Phillippine Sea plate are planned to be operated regularly in order to monitor the movement of these plate.

Also the stations in Fig.6 cover other island arcs such as Ryukyu Arc (Okinawa station) and Izu Bonin Arc (Chichijima station). We will be able to clarify deformation of these island arcs as well as the Northeast Honshu Arc.

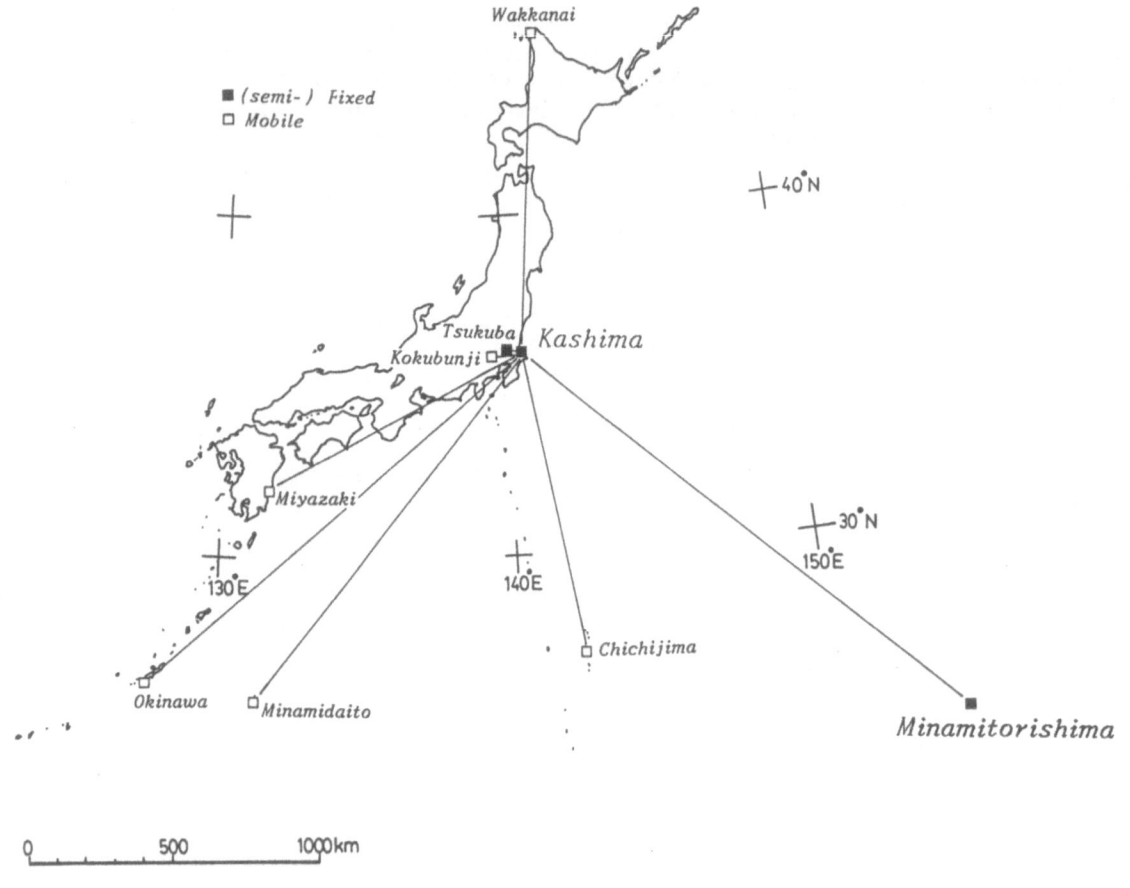

Fig.6. Fixed VLBI stations and mobile VLBI sites in Japan. Tsukuba, Miyazaki and Chichijima stations are those of Geographical Survey Institute (GSI) and the others are those of Communications Research Laboratory (CRL). Japanese "Western Pacific VLBI Network" is composed of Kashima, Minamitorishima and Minamidaito stations and VLBI experiments are scheduled regularly in order to monitor the movements of the Pacific plate (Minamitorishima) and the Phillipine Sea plate (Minamidaito). VLBI stations dedicated to astronomical researches are not included in the figure. Solid lines indicate that VLBI experiments have been already started for these baselines.

Acknowledgement. We wish to thank the VLBI group members of Kashima Space Research Center and the CRL headquater and all the stuffs of VLBI stations of the world. We also thank Foundation for C&C Promotion for its financial support to the travel fee to IAG Edinburgh 1989.

REFERENCES

Chapman, M. E. and Solomon, S. C. (1976). North American-Eurasian plate boundary in Northeast Asia, J. Geophys. Res., 81, 921-930.

Kobayashi, Y. (1983). On the initiation of subduction of plates, The Earth Monghly, 5, 510-514.

Ma, C., Ryan, J. W. and Caprette, D. (1988). Crustal Dynamics Project data analysis – 1988 VLBI geodetic results 1979-87, NASA Technical Memorandum, 100723, NASA, Goddard Space Flight Center, Greenbelt Maryland.

Minster, J. B. and Jordan, T. H. (1978). Present-day plate motions, J. Geophys. Res., 83, 5331-5354.

Minster, J. B. and Jordan, T. H. (1987). Vector constraints on western U.S. deformation from space geodesy, neotectonics and plate motions, J. Geophys. Res., 92, 4798-4804.

Mogi, K. (1970). Recent horizontal deformation of the earth's crust and tectonic activity in Japan (1), Bull. Earthq. Res. Inst., 48, 413-430.

Nakamura, K. (1983). Possible nascent trench along the eastern Japan Sea as the convergent boundary between Eurasian and North American plates, Bull. Earthq. Res. Inst., 58, 711-722.

Nakamura, K. and Uyeda, S. (1980). Stress gradient in arc-back arc regions and plate subduction, J. Geophys. Res. 85, 6419-6428.

Ward, S. N. (1988). North America-Pacific plate boundary, an elastic-plastic megashear: evidence from very long baseline interferometry, J. Geophys. Res., 93, 7716-7728.

Wesnowsky, S. G., Scholtz, C. H. and Shimazaki, K. (1982). Deformation of an island arc: rates of moment release and crustal shortening in intraplate Japan determined from seismicity and Quaternary fault data, J. Geophys. Res., 87, 6829-6852.

DEFORMATION IN THE PACIFIC BASIN FROM LAGEOS

Peter J. Dunn, John W. Robbins
ST Systems Corp., Lanham, MD 20706 USA

David E. Smith
NASA, Goddard Space Flight Center, Laboratory for Terrestrial Physics,
Code 920, Greenbelt, MD 20771 USA

ABSTRACT

Measurements to LAGEOS have provided the means to determine the relative positions as a function of time for six laser tracking sites in the Pacific Basin. These relative positions, and the determined relative motions, have been used to generate a motion model that is compared to geological predictions. The motion of the central station in this sub-network (from the global SLR network) on Maui, Hawaii agrees well with that suggested by Minster and Jordan's AM0-2 (1978) model when considered relative to the stations on the North American continent. The station at Yarragadee, Australia provides the longest record of continuous tracking (9 years). Its motion suggests a rate which is 11 mm/year slower than the 74 mm/year given by the AM0-2 prediction for the Austro-Indian plate at that location. The observed motion of Simosato, Japan does not conform to that predicted by the AM0-2 model for the Pacific, North American, or the Eurasian plate. The station at Huahine, in the Society Islands, appears to be moving south-west of the direction expected from the AM0-2 model by approximately 9° at a rate 11 mm/yr faster than AM0-2. The motion of Easter Island (on the Nazca plate) is expected to be better resolved when the 1988-1989 occupation is completed. The observed westward movement of Monument Peak in Southern California is an outcome of the so called "San Andreas anomaly" and provides an important link to sites located on the stable North American continent.

INTRODUCTION

Satellite Laser Ranging (SLR) sites around the Pacific Basin lie on at least six tectonic plates: Pacific, Indo- Australian, North American, South American, and Nazca, with Simosato, Japan possibly on an undetermined microplate. The international network of SLR sites in this region include eight stations with observation histories adequate for long-term monitoring of plate tectonic motion. There were brief occupations at American Samoa and Kwajalein, but, unfortunately, neither of these islands has been revisited by laser systems since November, 1980.

Global laser tracking data has been analyzed by the Geodynamics Branch, of NASA's Goddard Space Flight Center, to yield three-month averaged (quarterly) three-dimensional coordinates of the tracking sites relative to the Earth's center of mass. Geodesic distances are then determined for successive time intervals across the entire SLR network. Where the data quality and quantity were sufficient, a reconstruction of each station's motion and change in relative network position has been determined. A more detailed discussion of this analysis can be found in (Smith et al., 1989a, Smith et al., 1989b, and Christodoulidis et al., 1985).

DISCUSSION OF THE RESULTS

Figure 1 shows the time evolution of geodesic distances to Maui, Hawaii from three SLR stations with the longest observing histories. Error bars shown are at the 99% confidence level. The SL7.1 site velocities shown in Table 1 of Smith et al., (this issue) were used to compute the network implied geodesic rates along with their respective uncertainties. The slopes for these lines are in fair agreement with the geologically predicted values.

Geodesic rates for the regions surrounding each of these stations are shown in Figures 2,3, and 4. The projection used for these maps is a Lambert azimuthal equal area projection which is centered on each site. Geodesic rates (in mm/yr) for the SL7.1 model are shown (top values), along with AM0-2 predicted rates (bottom left values) and NUVEL-1 (DeMets et al., 1989) predicted rates (bottom right values).

Figure 5 expands upon this idea, giving the SL7.1 and AM0-2 geodesic rates between eight sites around the Pacific Basin. Figure 6 shows the vector motions across the Pacific Basin as determined from the observed intersite geodesic rates. The solution utilizes a least squares estimation procedure. In Figure 6, the black arrows with error ellipses are based upon the LAGEOS SL7.1 analysis. Vectors predicted by AM0-2 are shown as thinner arrows.

These figures suggest several things about deformation in the Pacific Basin. Huahine has a larger westward component of motion than predicted by the geologic models. Recently, it has been suggested that plate spreading along at least a portion of the East Pacific Rise is asymmetrical (Hey et. al, 1985). Our rates further suggest that the effects of asymmetric spreading may extend for hundreds of kilometers across the Pacific. The SL7.1 geodesic rate between Huahine and Easter Island is very near to that predicted by AM0-2 and is 13 mm/yr faster than NUVEL-1.

Other stations which appear to be affected by intra-basin deformation include Simosato and Monument Peak. Simosato appears to be highly influenced by the tectonics of the Eurasian-Pacific-Philippine convergence, and does not exhibit motion characteristic of any of the major plates in the region. Monument Peak (located on the Pacific plate, but near the San Andreas Fault in Southern California) exhibits motion which is significantly different from that predicted by the geologic models. Our results suggest that motion across the Pacific-North American plate boundary is about 20 mm/yr slower than predicted by rigid plate motion models and that extension on the order of a feww mm/yr may be occuring offshore of California as evidenced by the 4±2 rate for the Monument Peak to Maui line.

REFERENCES

Christodoulidis, D. C., D. E. Smith, R. Kolenkiewicz, S. M. Klosko, M. H. Torrence and P. J. Dunn, Observing Tectonic Motions and Deformations from Satellite Laser Ranging, *J. Geophys. Res.*, V.90, pp.9249-9263, 1985.

DeMets, C., R. G. Gordon, D. F. Argus and S. Stein, Current Plate Motions, *J. Geophys Int.*, in press, 1989.

Hey, R. N., D. F. Naar, M. C. Kleinrock, W. J. Phipps Morgan, E. Morales and J.-G. Schilling, Microplate Tectonics Along a Superfast Seafloor Spreading System Near Easter Island, *Nature*, V.317, pp.320-331, 1985.

Minster, J. B. and T.H. Jordan, Present-Day Plate Motions, *J. Geophys. Res.*, V.83, pp.5331-5254, 1978.

Smith, D. E., R. Kolenkiewicz, P. J. Dunn, M. H. Torrence, J. W. Robbins, S. M. Klosko, R. G. Williamson, E. C. Pavlis, N. B. Douglas and S. K. Fricke, Tectonic Motion and Deformation from Satellite Laser Ranging to LAGEOS, *J. Geophys. Res.*, in press, 1989a.

Smith, D. E., R. Kolenkiewicz, B. H. Putney, P. J. Dunn, S. M. Klosko, E. C. Pavlis, J. W. Robbins, M. H. Torrence, R. G. Williamson and S. K. Fricke, A Geodetic Earth Motion Model Derived from LAGEOS Observations: GSFC-SL7, *NASA Tech. Memo.*, in preparation, 1989b.

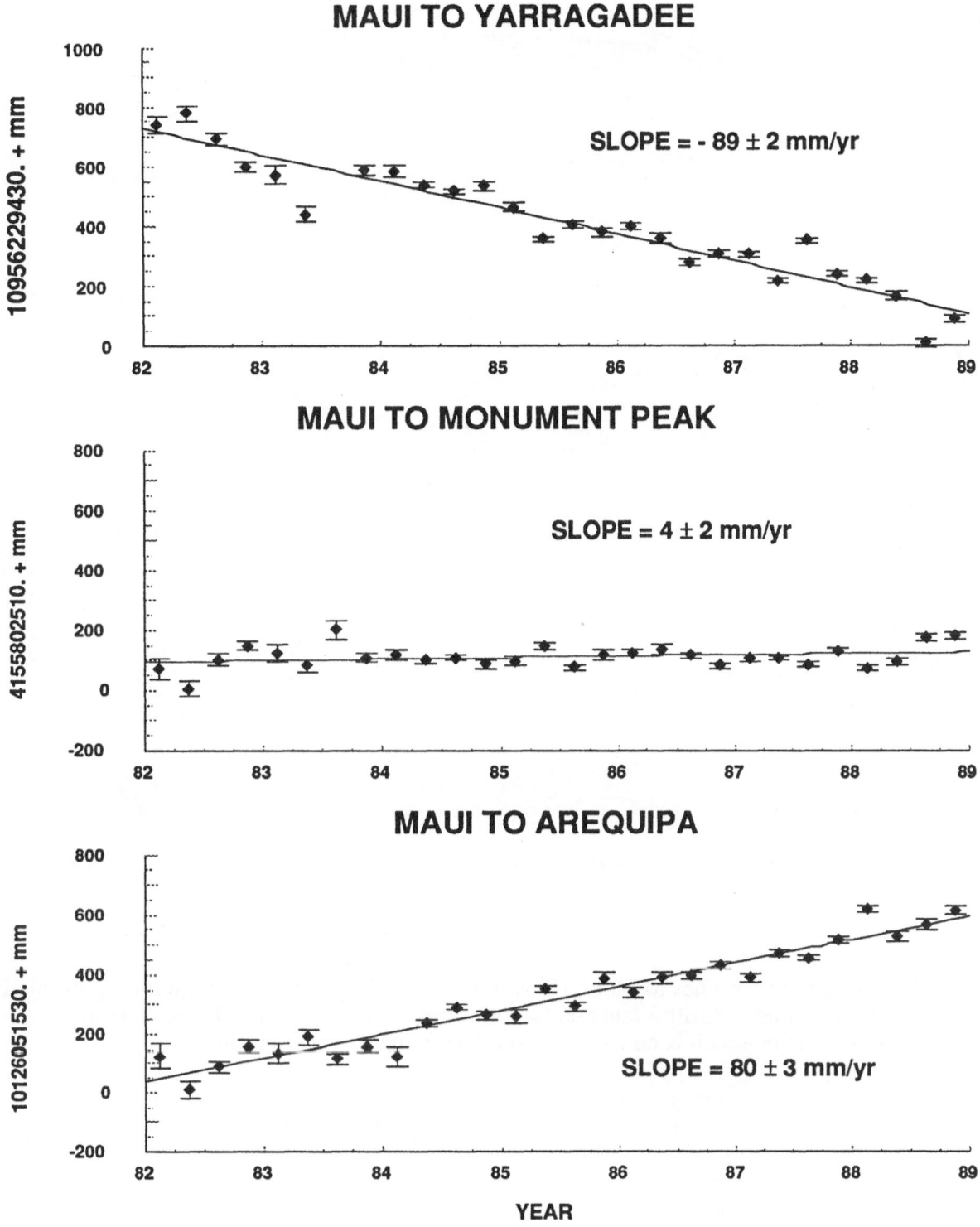

Fig. 1. Time evolution of geodesic distances to Maui.

Fig. 2. Geodesic rates for lines to Maui, Hawaii in mm/yr. Top value is for solution presented here, bottom left value is AM0-2 rate and bottom right value is NUVEL-1 rate. The line between Maui and Greenbelt is constrained to AM0-2 rate in our solution.

Fig.3. Geodesic rates for lines to Arequipa, Peru in mm/yr. Top value is for solution presented here, bottom left value is AM0-2 rate and bottom right value is NUVEL-1 rate.

Fig.4. Geodesic rates for lines to Monument Peak, Calfornia in mm/yr. Top value is for solution presented here, bottom left value is AM0-2 rate and bottom right value is NUVEL-1 rate.

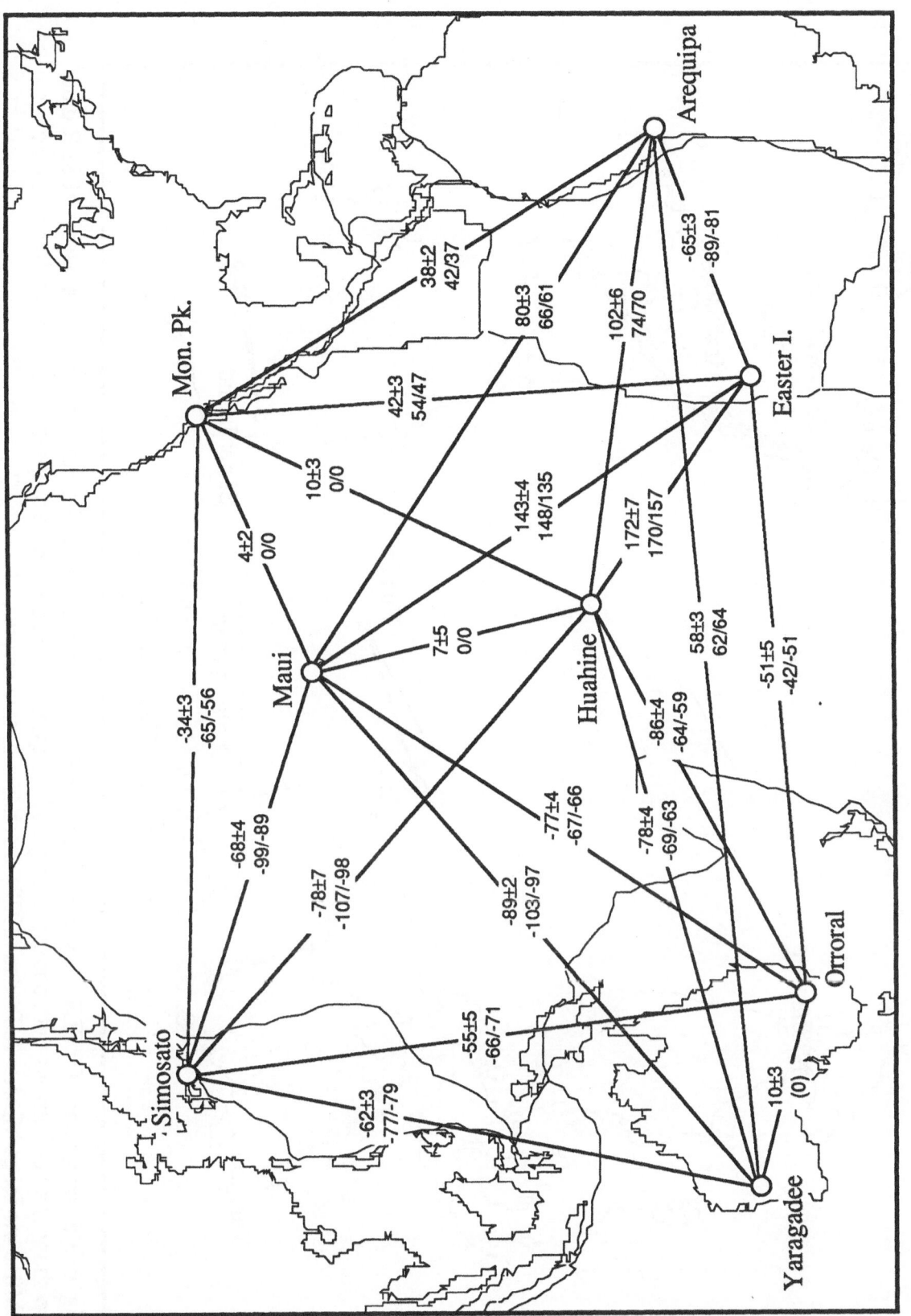

Fig. 5. Geodesic rates across the Pacific Basin in mm/yr. Top value is for solution presented here, bottom left value is AM0-2 rate, and bottom right value is NUVEL-1 rate.

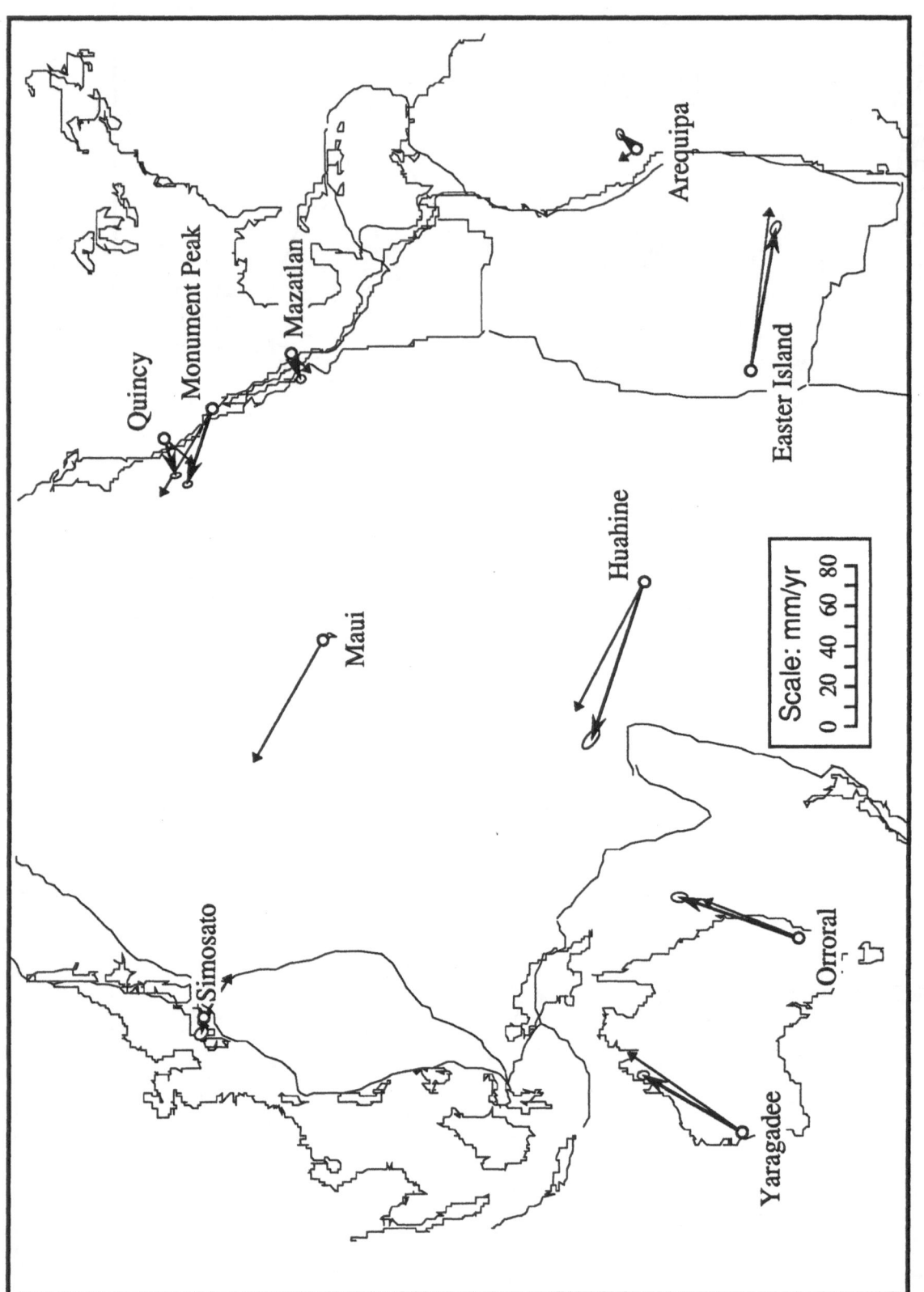

Fig. 6. Vector motions for SLR sites in and near the Pacific Basin. Motions determined from the analysis of SLR data are shown as bold arrows with corresponding error ellipses (67% confidence ellipses). Motions implied from the AM0-2 model of Minster & Jordan (1978) are shown as thin arrows.

FIRST RESULTS OF WEGENER/MEDLAS DATA ANALYSIS

B.A.C. Ambrosius, R. Noomen, K.F. Wakker
Faculty of Aerospace Engineering
Delft University of Technology
Delft, The Netherlands

INTRODUCTION

The WEGENER/MEDLAS project is aimed at the determination of crustal deformations in the Mediterranean area, the region where the Eurasian, the African and the Arabic tectonic plates meet (Reinhart, 1985). The deformations are to be derived from the very precise distance measurements to the geodetic satellite LAGEOS (Cohen, 1985), that are obtained by mobile satellite laser ranging (SLR) systems. Participating mobile laser systems are MTLRS-1 from Germany, MTLRS-2 from the Netherlands and the US TLRS-1 system. From 1985, they are being deployed at about 15 carefully selected locations in this region, occupying each site at intervals of about one or two years. From the SLR tracking data acquired at these sites, successive and very precise solutions for the positions of the mobile laser systems are computed. The small differences between these solutions are assumed to be related to the tectonic deformations in the region. This paper will mainly discuss the analysis results that have been computed from full-rate laser range observations. These results are based on the measurements taken during the first two observation campaigns, that were organized in 1986 and 1987. In addition, a preliminary result from the analysis of quick-look data of the 1989 campaign will also be presented.

ANALYSIS TECHNIQUE AND DATA OVERVIEW

The observations of LAGEOS were processed in the form of 2-minute normal points, which were computed at and provided by DGFI in Munich (Reigber, 1987- 1988). The total data yield during the two years consisted of almost 8500 passes, obtained by mobile and stationary laser systems at 39 different sites around the world. The data analysis was performed with the NASA GEODYN I software package for orbit determination and parameter estimation (Martin, 1978). The analysis relied on the so-called multi-arc technique, in which the total data span is subdivided into many successive one-week data arcs. Combinations of several of these one-week arcs are processed simultaneously, yielding very precise solutions for the laser station

Table 1. General information on the selected data intervals. Listed are the start and stop time of each data period, the total number of weeks covered by this period and the number of contributing stations outside respectively inside the Mediterranean area.

Solution	Data period		Weeks	Stations (global)	(Medit.)
1	January 5, 1986	March 15, 1986	10	16	3
2	March 16, 1986	May 24, 1986	10	17	3
3	May 25, 1986	August 30, 1986	14	15	5
4	August 31, 1986	October 25, 1986	8	16	5
5	October 26, 1986	December 27, 1986	9	16	2
6	January 4, 1987	March 14, 1987	10	15	2
7	March 15, 1987	May 23, 1987	10	17	5
8	May 24, 1987	September 12, 1987	16	16	8
9	September 13, 1987	December 26, 1987	15	17	6

coordinates in which all the information in the tracking data is accumulated, while at the same time separate solutions are obtained for the satellite-dependent arc parameters such as the state-vector at epoch. It is emphasized that in each separate solution, the full set of global laser station coordinates is adjusted.

The computation model that was used generally matches the MERIT standard (Melbourne, 1983). Notable excursions are the gravity field and the model for ocean tides: for both elements, a derivative of the NASA GEM-T1 solution was used (Marsh, 1987). The model for the gravity field was truncated at degree and order 20. As for the ocean tides model, only the constituents in the GEM- T1 solution that are also available in the older Schwiderski model were used in the computations.

The observations were processed in batches, each of them spanning a period from 8 to 16 weeks, which approximately corresponds to the periods in which the networks of mobile systems in the Mediterranean area remained fixed. Within each data period, the tracking observatories are considered to be stationary relative to each other. The monitoring of successive solutions for individual station positions or interstation distances will yield evidence for crustal deformations. An overview of the selected data periods is given in Table 1. It shows, for each period, the start and stop time, the number of weeks that are covered, the number of laser stations outside the Mediterranean area that contributed data in this period, and, finally, the total number of fixed and mobile lasers in this region. The Table clearly shows that the number of sites in the Mediterranean area is much smaller than the number of global stations. This suggests that the positions of the former sites may be determined relative to a framework which is defined by the global laser stations. This property will be used later on in this study, when the various individual global coordinates solutions are further processed.

Table 2: Overview of the laser sites in the Mediterranean area and their contribution of data passes in the different data periods.

Station		Solution								
		1	2	3 (1986)	4	5	6	7 (1987)	8	9
7510	Askites, Greece			52					131	
7512	Kattavia, Greece				56		60			
7515	Dionysos, Greece			49					83	64
7517	Roumelli, Greece			68	2				55	55
7520	Karitsa, Greece		46							
7525	Chrisokellaria, Greece				23					56
7530	Bar Giyyora, Israel			3	11	22	15	15	15	
7540	Matera, Italy	33								
7541	Matera, Italy	62								
7544	Lampedusa, Italy									83
7550	Basovizza, Italy		29							
7575	Diyarbakir, Turkey							29		
7580	Melengiclik, Turkey							41		
7585	Yozgat, Turkey								26	
7587	Yigilca, Turkey								39	
7831	Helwan, Egypt								76	7
7939	Matera, Italy	68	54	40	24	36	28	90	151	87

The data yield of the mobile and stationary laser systems in the Mediterranean area is summarized in Table 2, which lists the amount of passes that were acquired at the various sites for each of the selected data periods. Bar Giyyora, Helwan and Matera are stationary laser observatories. The remainder of the sites listed here was occupied by either MTLRS-1, MTLRS- 2 or TLRS-1. The Table clearly shows that the 1987 campaign has been much more successful than the one in 1986. This is a consequence of the fact that in 1987 three mobile laser systems participated, versus only two in 1986, and the more favorable orbital constellation of LAGEOS relative to the Sun, resulting in bcttcr (nighttime) tracking conditions. The Table also clearly shows that most of the Greek sites were occupied both in 1986 and in 1987.

INTERSTATION DISTANCE VARIATIONS

The analysis resulted in a series of 9 individual and independent global network solutions. Crustal deformations may be observed directly from the results, for instance by comparing the solutions of linear interstation baselines. Figure 1 shows an example, in which the solutions of the baseline between the observatories in Graz in Austria and Herstmonceux in the United Kingdom are plotted as a function of time. From a linear fit of the data, the distance between the two stations seems to increase at a rate of +1.1

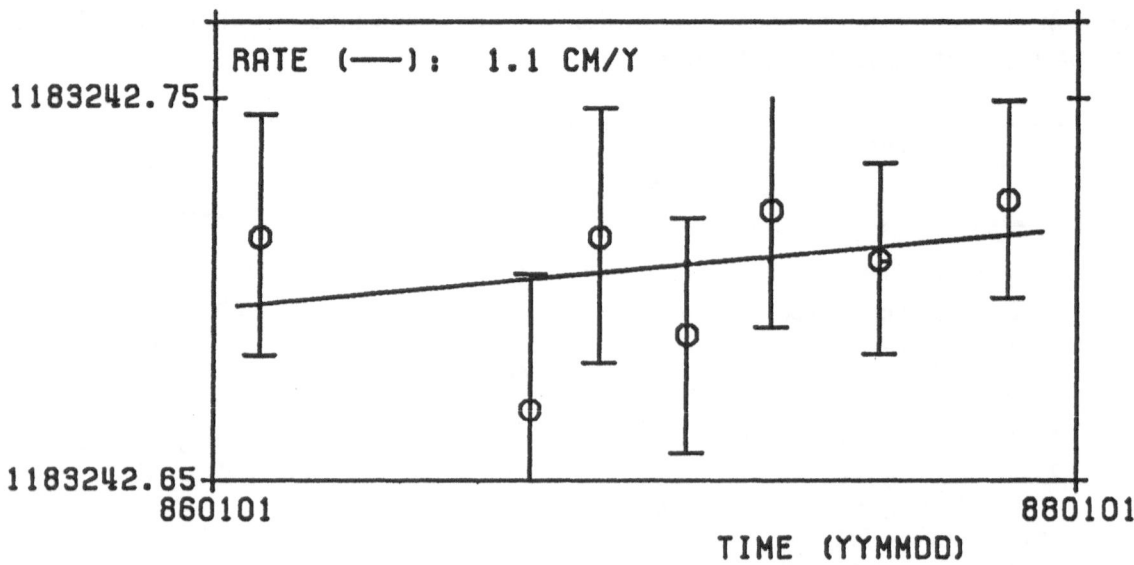

Fig. 1. Observed variation in the baselines between Graz in Austria and Herstmonceux in the United Kingdom. The solutions were obtained from independent batches of full-rate observations of LAGEOS taken during 1986 and 1987. The straight line represents a weighted least squares fit through the individual baselines.

cm per year. It should be noted, however, that this number is much smaller than the standard deviation of 2.6 cm per year, which means that it is not significantly different from zero, as would be expected for this baseline. Nevertheless, the Figure does show the very good repeatability of the baseline solutions, which demonstrates the accuracy of the analysis technique.

INDIVIDUAL STATION MOTIONS

Instead of looking at interstation distances, it is generally more elucidative to study the apparent motions of individual laser stations. To this aim, a special computer program was developed to derive these motions from the separate successive global coordinates solutions.

There are a number of particular problems related to this procedure. A poor quality of SLR-derived global networks is their absolute accuracy, in particular for those that are based on data from a relatively small period of time (like the series of solutions presented here). Generally, assumptions are made for certain components of this network to stabilize the estimation process. The relative positioning accuracy is quite good, however, as was demonstrated in the previous Section. Experience has learned that using the analysis technique discussed here, the linear interstation baselines have an accuracy of approximately 3 cm. This value holds for network solutions based on approximately 10 weeks of tracking data. Thus, each of the 9 network solutions may be regarded as a collection of coordinates of well-defined reference points on the

Earth, with a fairly well-known position relative to each other but with a rather poor absolute accuracy. Each of the position solutions is, of course, accompanied by its own three-dimensional error ellipsoid.

It has already been pointed out that the majority of the laser stations in the network solutions are located outside the Mediterranean area. In addition, these "outer" stations are generally situated in regions where the plate tectonic deformations are well known from geological studies (with a few exceptions). The combination of the aforementioned factors makes it possible to fit the "outer" part of the successive network solutions to the "real" world, which can be described with high accuracy by a model for global tectonic deformations. In this way, the tectonic model provides the absolute accuracy that individual SLR network solutions lack. Once the absolute positions of the "outer" stations are known as a function of time (the corresponding solutions are fitted to the modeled trend), the positions of the sites in the Mediterranean region are also accurately available in this absolute reference system, since their relative positions with respect to the "outer" stations are known with high accuracy within each solution.

In practice a model of the motion of the global stations is fitted through the successive solutions of the individual laser station coordinates in a weighted least squares sense. The motion of the "outer" stations is taken after an accurate model for global tectonic deformations (Minster, 1978). The coordinates of these stations are fitted as well as possible to this model through the estimation of solution-specific parameters that affect the scale, origin and orientation of the individual network solutions: so-called Helmert parameters. The unknown tectonic motions of the sites in the Mediterranean area are approximated by linear functions. The velocities of these sites are estimated simultaneously with the other parameters. An identical approach is adopted for the two stations for which the motions according to the geological models differ significantly from the observed deformations: Monument Peak in California and Simosato in Japan. In addition, the initial positions of all stations at the epoch January 1, 1986, are also estimated.

The results are presented in Table 3. This Table gives the estimated latitudinal, longitudinal and vertical velocity components of the sites in the Mediterranean area from which at least two independent coordinates solutions are available. The velocities are relative to a fixed Eurasian plate. It is stressed here once again that these estimates are based on two successive years of LAGEOS tracking data only, and that consequently these velocity estimates have a very preliminary character. Note the standard deviations, which approximate in many cases the (absolute) value of the corresponding velocity components. Nevertheless, the results appear to be quite reasonable, judging from the fact that most velocities are smaller than 5 cm per year, which is about the highest velocity to be expected in this area. An exception is the solution for Kattavia on Rhodes, but it is probably ill-determined since the two occupations of this site were no more than 6 months apart.

To ease the interpretation, the results are also presented in a graphical form. Figure 2 is a plot of the area that is covered by the current set of solutions. It shows the horizontal vector components of the velocity estimates and the associated error ellipses. A few interesting phenomena can be observed. First, the station in Southern

Table 3. First estimates of the yearly motions of the sites in the Mediterranean area. The velocity components in latitude, longitude and height are given together with the corresponding standard deviations. The velocities are relative to a non-moving Eurasian tectonic plate.

Station	Location	Latitudinal velocity (cm/y)	Longitudinal velocity (cm/y)	Vertical velocity (cm/y)
7510	Askites, Greece	-1.0 ± 4.0	0.6 ± 3.9	2.1 ± 0.5
7512	Kattavia, Greece	-12.3 ± 9.2	-11.5 ± 9.0	-2.7 ± 1.0
7515	Dionysos, Greece	-3.4 ± 3.0	0.6 ± 2.9	0.4 ± 0.4
7517	Roumelli, Greece	-3.2 ± 3.0	5.5 ± 2.9	-0.8 ± 0.4
7525	Chrisokellaria, Greece	-4.5 ± 4.3	-0.2 ± 4.1	-5.7 ± 0.6
7939	Matera, Italy	2.7 ± 2.0	3.0 ± 1.8	2.6 ± 0.2

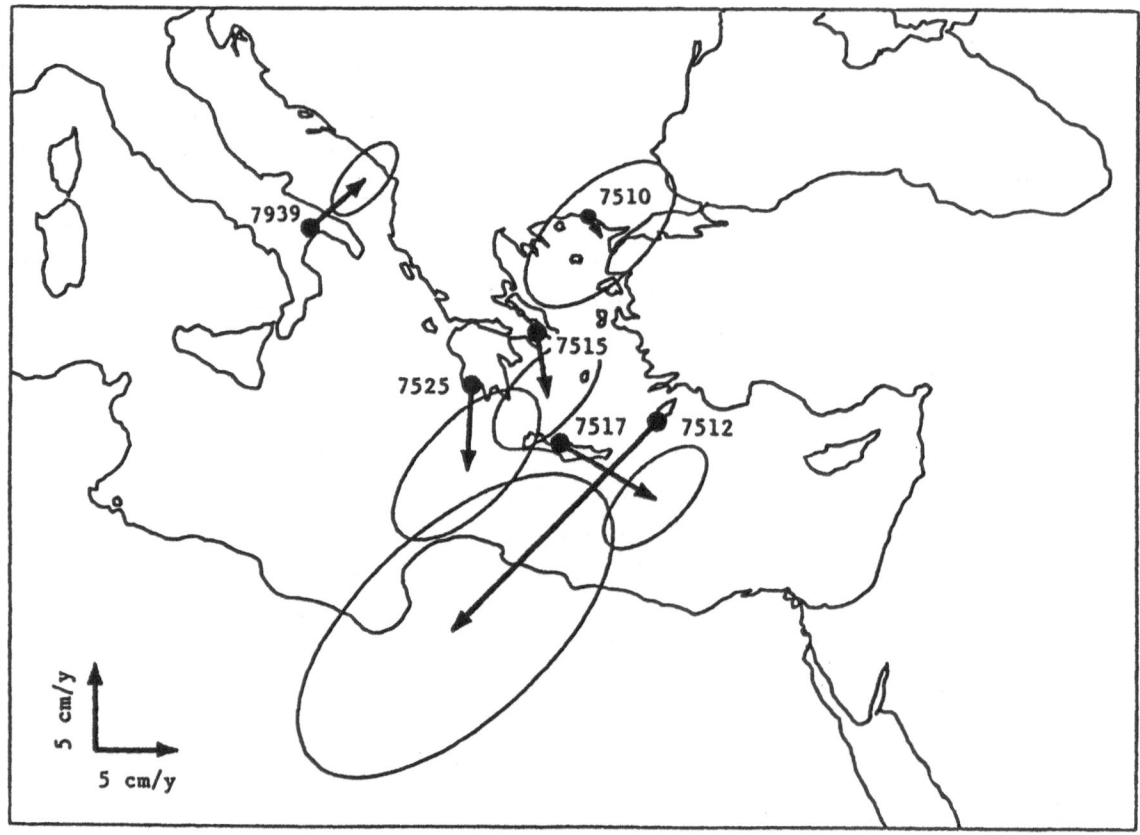

Fig. 2. First estimates of the horizontal motions of sites in the Mediterranean area. The results are computed at DUT/SOM from LAGEOS laser ranging data taken during 1986 and 1987, and are relative to the stable Eurasian plate.

Table 4. Relative velocities between sites in the Mediterranean area. The rates in the second column were based upon the solutions for the three-dimensional yearly motion, whereas the values in the third column were computed directly from successive solutions for the geodesic distances.

Baseline			Velocity from 3D-motion (cm/y)		Velocity from geodesics (cm/y)	
Askites	-	Kattavia	6.9	± 9.2		
Askites	-	Dionysos	2.2	± 5.8	3.5	± 5.9
Askites	-	Roumelli	1.7	± 5.4	1.6	± 5.4
Askites	-	Chrisokellaria	3.5	± 7.2		
Askites	-	Matera	-2.6	± 3.7	-6.6	± 4.2
Kattavia	-	Dionysos	-4.5	± 5.6		
Kattavia	-	Roumelli	-18.3	± 9.0		
Kattavia	-	Chrisokellaria	-9.2	± 6.8		
Kattavia	-	Matera	-5.0	± 5.4	-6.0	± 6.8
Dionysos	-	Roumelli	0.9	± 3.8	1.0	± 3.3
Dionysos	-	Chrisokellaria	1.4	± 6.2		
Dionysos	-	Matera	0.3	± 2.0	-2.0	± 2.2
Roumelli	-	Chrisokellaria	4.2	± 2.9		
Roumelli	-	Matera	5.4	± 2.0	3.5	± 2.0
Chrisokellaria	-	Matera	2.6	± 2.8	3.3	± 3.1

Italy, Matera, appears to move in a Northeasterly direction, obviously as a consequence of the collision between Eurasia and Africa, which moves due North with respect to the former plate. The motion of Askites, station 7510, is very small. This suggests that this part of Northern Greece is on the rigid Eurasian plate. The dimension of the error ellipsoid, however, is much larger than the recovered velocity, which means that this conclusion is still rather uncertain. The three stations directly to the West and South of the Aegean Sea, Chrisokellaria (7525), Dionysos (7515) and Roumelli (7517), exhibit velocities with a significant Southerly trend. These results seem to confirm the spreading theory of the Aegean Basin. The origin of the significant longitudinal velocity component that was recovered for Roumelli still remains unclear. Although it might be caused by problems in the data, this is contradicted by the most recent results that will be presented in the next Section. Another explanation could be local deformations on the island of Crete. The very large vector for the motion of Kattavia on Rhodes (station 7512) is related to the fact the the occupations of this site were only 6 months apart in time. This is reflected by the very large error estimate for this solution.

The solutions for the individual station velocities that are listed in Table 3, are easily converted into relative distance variations between pairs of stations. The results are presented in Table 4. For comparison, the Table also includes the estimates for the yearly geodesic change that may be computed directly from the successive network solutions (and the corresponding observed geodetic distances). The agreement between the two solutions is generally quite good. The comparison of the velocity estimates does show an additional advantage of the technique that was used: individual station velocities can be converted into interstation distance variations for each arbitrary pair of stations, whereas the direct technique is dependent on the availability of simultaneous occupations of the pairs of sites.

THE 1989 OBSERVATION CAMPAIGN

The results described thus far are all based on just two years of laser ranging observations of LAGEOS, i.e. 1986 and 1987. This year, the third WEGENER/MEDLAS observation campaign is being conducted. All sites that were occupied during the previous campaigns are scheduled to be revisited this year.

In addition to performing analysis of the full-rate data, Delft University also acts as the Quick-Look Data Analysis Center (QLDAC) for WEGENER/MEDLAS. The main task of QLDAC is to monitor the performance and the quality of the measurements of the mobile laser systems in semi real-time. This is accomplished by weekly analyses of samples of the observations, that are taken by all global laser systems. These so-called quick-look data are generally received in Delft within one week after their acquisition.

The avalaibility of these very recent observations facilitates a first, preliminary look into the Mediterranean situation for 1989. To this aim, a batch of quick-look observations was processed in the same way as the full- rate data of the previous years. The data period that was investigated stretches from May 7 to June 18, and covers a total time-span of 6 weeks. During this period, MTLRS-1, MTLRS-2 and TLRS-1 were deployed at Lampedusa (Italy), Basovizza (Italy) and Roumelli (Crete, Greece), respectively. Here, only some results pertaining to Roumelli will be presented. Figure 3 shows the observed variation of the baseline between Roumelli and Matera in Southern Italy. The first three solutions are based on the full-rate observations taken during 1986 and 1987. The fourth solution, however, has been obtained from the quick-look observations collected during the campaign of 1989. The results look very interesting: the data of the 1986 and 1987 campaigns revealed an increase of this distance at a rate of 3.6 cm per year. However, this result was based on just two years of data. The latest solution, now, seems to confirm this extension since it fits the earlier trend. The slope of the trend, indicated by the (best-fit) straight line, amounts to 4.7 cm per year, with an estimated uncertainty of 0.9 cm per year. Although it is not possible from this result to differentiate between the individual contributions of Matera and Roumelli, it is quite encouraging to see that the trend that was found from the first two campaigns is continued here. It suggests that the individual station motions that are depicted in Figure 2 may well agree with the actual deformations that are taking place in the Mediterranean area.

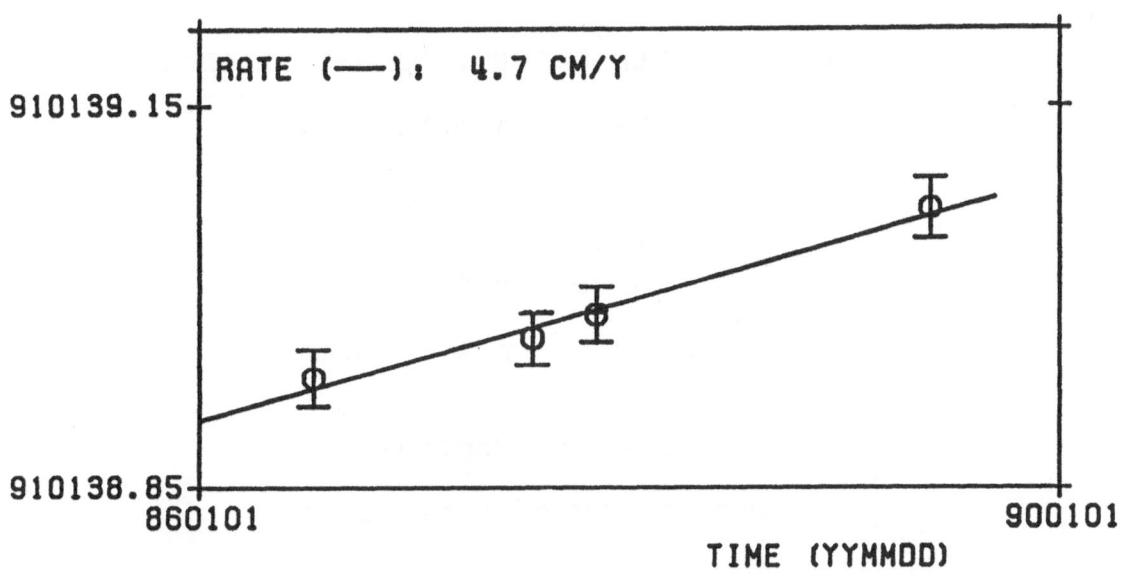

Fig. 3. Observed variation in the baseline between Roumelli on Crete and Matera in Italy. The solutions were obtained from independent batches of full-rate observations of LAGEOS taken during 1986 and 1987 and a 6-week period of quick-look observations of the same satellite taken during May/June 1989. The straight line represents a weighted least squares fit through the individual baselines.

REFERENCES

Cohen, S.C. and Smith, D.E. (1985). LAGEOS Scientific Results: Introduction. *J. Geophys. Res.*, **90**, 9217-9220

Marsh, J.G. et. al. (1987). An Improved Model of the Earth's Gravitational Field: GEM-T1. NASA Technical Memorandum 4019, Goddard Space Flight Center, Greenbelt.

Martin, T.V. (1978). GEODYN descriptive summary, report contract no. NAS 5-22849, Washington Analytical Services Center, Riverdale.

Melbourne, W. et. al. (1983). Project MERIT Standards, United States Naval Observatory Circular No. 167, U.S. Naval Observatory, Washington.

Minster, J.B. and Jordan, T.H. (1978). Present-day Plate Motions. *J. Geophys. Res.*, **83**, 5331-5354.

Reigber, C. (1987-1988). Full-rate Data Report. Monthly circular of the WEGENER-MEDLAS Full-rate Data Analysis Center at DGFI, Munich.

Reinhart, E. et. al. (1985). The WEGENER Mediterranean Laser Tracking Project: WEGENER-MEDLAS. *CSTG Bull.*, **8**, 145-162.

EUROPEAN BASELINES DETERMINED WITH LAGEOS:

THE TPZ-88.1 SOLUTION

Alberto Cenci
Marco Fermi

Telespazio S.p.A. , Rome

Alessandro Caporali

University of Bari, Bari

1. INTRODUCTION

The possibility of monitoring a large scale geodetic network in southern Europe and East Mediterranean Area represents one of the most far reaching perspectives in modern experimental research in geodynamics. In the WEGENER/MEDLAS project the approach to such a monitoring is based on the LAGEOS satellite being tracked by a number of fixed and mobile stations which locally densify the worldwide laser tracking network. The analysis includes, but is not limited to, the repeated estimation of the baselines. Our main objective is to reliably determine the deformation of the WEGENER geodetic network (see fig.1) and in this case the time series of the baselines estimates are analysed in conjunction with other results, such as polar motion, LAGEOS kinematical and dynamical model, geological survey and local field measurements including the monitoring of the performance of LASER stations partecipating to the WEGENER project.
In the following we will focus on what we believe to be the most important results of the 1985/1987 WEGENER campaign, that is having inserted a number of sites occupied by mobile stations in the network of fixed stations, thereby initiating to collect baseline estimates which, if repeated with extensive LAGEOS tracking and combined with additional geodetic and geological information, will permit the network deformations to be reliably determined and interpreted. To specify the geodetic framework in which the mobile stations are beginning to operate, we will first present the time series of the european baselines between fixed stations determined with LAGEOS in the period September 1983 - December 1987. The time span is yet too short for a reliable

detection of signal of clear geodynamic nature. However, for some lines, the data are very good, the baselines estimates have very little spread, and we are motivated to expect that crustal deformation, if present, should become to be visible in a reasonable short lapse of time. The baselines between mobile and fixed european stations obtained from the data collected during the first WEGENER/MEDLAS campaign from 1985 to 1987 will be also presented.

2. METHOD OF ANALYSIS

To meet the very demanding goals set by the WEGENER project an adequate orbital model is needed. This is not just simply an analytical expression for the force field. The definition of the origin and orientation of the Conventional Terrestrial Reference System, the spherical harmonics expansion of the gravity field of the Earth and the tidal model in the past have been treated almost independently from each other. Now they are considered as parts of a unique kinematical and dynamical model to represent high precision satellite data.
In comparison with earlier models such as those recommended for the MERIT Campaign (Melbourne et al., 1983),the NASA-GSFC gravity model GEM-T1(Marsh et al., 1988) represents an improvement in several aspects. GEM-T1 is a gravity field complete to degree and order 36 obtained from the analysis of data from 17 satellites with inclination distributed from 15 to 90 degrees. Simultaneously with the constant part of the field, a subset of 66 ocean tide coefficients has been estimated and 550 other ocean tide coefficients have been modelled, the body tides being constrained by Wahr's theory.
As to the definition of a terrestrial reference system, the Z-axis was assumed to coincide with the LAGEOS derived mean pole in the period 1979-1984. In terms of r.m.s. (root-mean-square) of the post-fit residuals, for 30-day LAGEOS arcs GEM-T1 is credited to yield, on average, an improvement of several cm relative, for example, to GEM-L2 (Lerch et al., 1985), which is a gravity field "tailored" on LAGEOS.
In the past, our group has been using a variety of models for the analysis of LAGEOS data so that we had solutions for station coordinates, baselines, polar motion and orbital dynamics of LAGEOS obtained under different assumptions. At this stage, to assemble multi-year time series of these quantities in support to
the WEGENER project, a model as GEM-T1 was felt to offer a substantial increase in quality and consistency for our LAGEOS solutions. We therefore decided to re-analyze our data set, at this time extending from September 1983 to December 1987, using systematically a unique approach. Table 1 summarizes the basic setup of GEODYN used for this

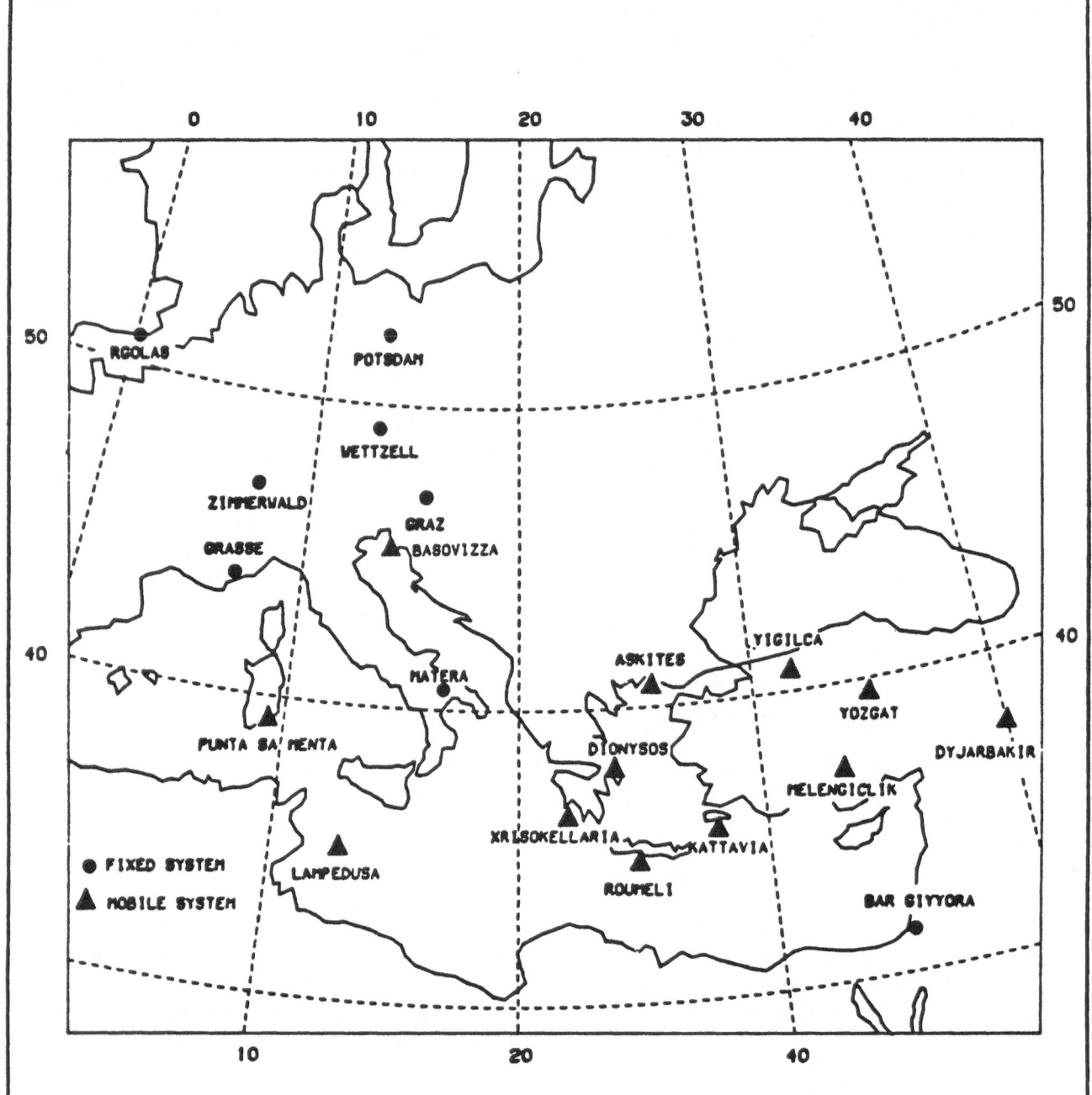

FIG. 1 FIXED EUROPEAN STATIONS AND WEGENER/MEDLAS SITES OCCUPIED
BY MOBILE SYSTEMS DURING THE FIRST MEASUREMENTS CAMPAIGN

analysis which will hereafter be called TPZ88.1 solution. We have chosen to work with relatively short arcs (15 days) combined in multi-arc solutions covering 6 months each. This approach was selected because 1) it gives a frequency resolution of 1 cycle/30 day (Nyquist frequency) of particular interest for analysis of orbit perturbations and typical r.m.s.'s below 10 cm for each arc, and 2) because the common parameters, particularly station coordinates and baselines, are estimated with a sizeable number of passes, while keeping an acceptable time resolution of 6 months. The tidal model consists of the MERIT subset of the earth and ocean tides and is not the full GEM-T1 model.

3. COORDINATES AND BASELINES FOR EUROPEAN FIXED AND MOBILE STATIONS.

3.1 Time series analysis

Figures Fig.2/a and Fig.2/b represent the present of our time series of some european baselines.
Our plotting package supports the capability of fitting a straight line to the time series. The resulting estimated slope, it must be stated very clearly, does not necessarily represent the actual rate of change of the baseline length. It simply gives the most likely trend of a straight line which "best" fits the data. So, for instance, in the line WETTZEL-RGO (see Fig. 2/b) subtracting a straight line of small slope (-0.6 cm/yr), neglegibly decreases the RMS (from 3.2 to 3.0 cm). Things are different e.g. on the line WETTZEL-MATERA, GRAZ-MATERA (Fig. 2/a) or GRAZ-RGOLAS (Fig. 2/b), where there might be some indication of contraction at a rate of 1-2 cm/yr.
The conclusion we can draw at this stage is that the repeatibility of the network of fixed station gives a reasonable guarantee of stability which is an essential prerequisite at the stage when we incorporate the Mediterranean mobile stations into the network.

3.2 Results from the 1985-1987 WEGENER/MEDLAS campaign.

Table 2 summarizes baselines between mobile sites occupied twice in 1986 and 1987. Baselines to mobile stations are referred to the main eccentricity marker of the pad. Columns STD DEV show the formal error (one-sigma) associated to the

baselines, scaled with the RMS of the orbital post-fit residuals. The last column shows the differences between the two yearly determinations. The average of these differences

gives an RMS of about 3 cm (excluding lines to GRASSE and BAR GIYYORA) that is, considering the limited amount of data generally available from mobile station occupations, quite satisfactory. Worse results for baselines to GRASSE are probably due to the bad observation statistics for this station in 1986.
Only a limited amount of observations is yet available for BAR GIYYORA.

4. CONCLUSIONS

The time series we have presented cover a time span still too short to draw conclusive statements on centimetric baselines changes, but we have enough successive estimates of baselines to assess the repeatibility and thus the reliability, up to a constant bias not affecting the rates, of our time series. On the average the repeatibility is in the order of 2-3 cm RMS, and is probably controlled by systematic effects (e.g. model errors, data distribution or pass geometry) rather than random noise.
In order to achieve subcentimetric accuracy in baselines, improvements in data analysis and observation statistics and geometry are necessary. Indeed, development of new sophisticated force-models and calibration procedures will help us in achieving this goal, but perhaps more important the extension and densification of mobile site occupations, improvement of day-light capability and increase of satellite constellation are really also needed to constrain more tightly the baseline solutions, in particular for the mobile site occupations for which a limited amount of data is in general available.

ACKNOWLEDGMENT: This work is supported by the Agenzia Spaziale Italiana.

REFERENCES

Cenci A. and M. Fermi:
"Matera Colocation Activity and data Analysis Results";
Telespazio Internal Report, February 1987

Lerch, F.J.,S.M. Klosko, G.B. Patel and C.A Wagner:
"A Gravity Model for Crustal Dynamics (GEM-L2)";
Journ. of Geophys. Res. B11, 9301-9311 (1985).

Marsh, J.G.,F.J. Lerch,B.H. Putney,D.C. Christodoulidis,T.L.
Felsentreger, B.V. Sanchez,D.E. Smith,S.M. Klosko,T.V.
Martin, E.C. Paulis,J.W. Robbins, R.G. Williamson, O.L.
Colombo,N.L. Chandler,K.E. Rachlin,G.B. Patel,S. Bhati and
D.S. Chinn:
"An Improved Model of the Earth's Gravitational Field: GEM-
T1";
NASA-TM4019 (1988).

Melbourne, W.,R. Anderle,M. Feissel,R. King,D. Mc Carthy,D.
Smith,B. Tapley,R. Vicente:
"Project MERIT Standards";
US Naval Observatory Circular n. 167 (1983)

NASA/GSFC publication:
"Report on MATLAS, MTLRS1 and MTLRS2 Collocation at Matera
January through March 1986";
NASA/GSFC, (1988)

Fig.2a

GRAZ   -   MATERA   BASELINE (719400.+U)

U (M)   slope -1.5 cm/Yr   WRMS 2.2 (3.2) cm TPZ-88.1   JAN 1989

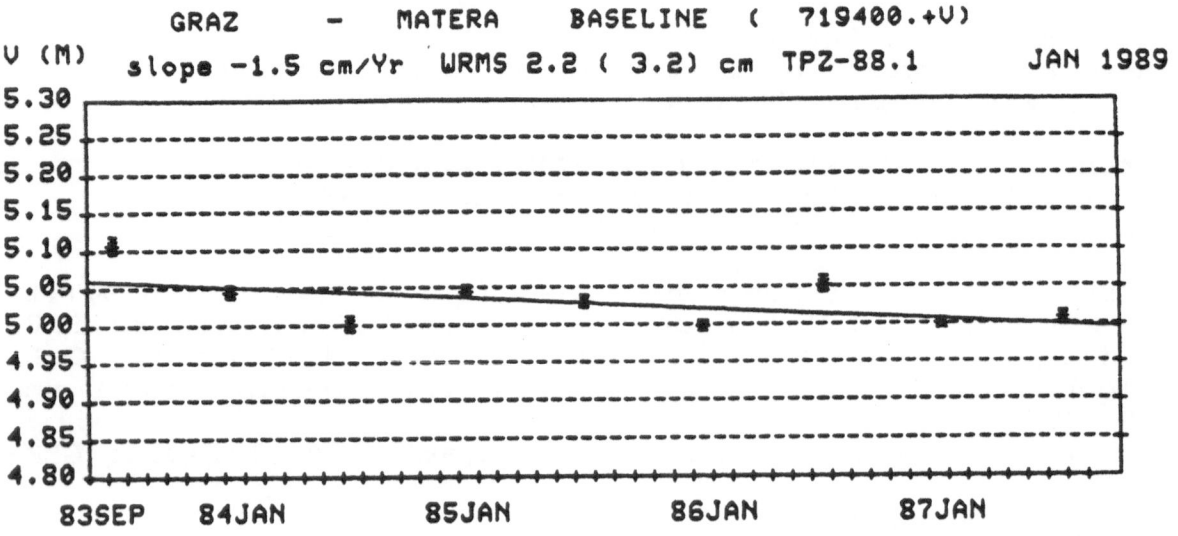

WETZEL   -   MATERA   BASELINE (990110.+U)

U (M)   slope -1.6 cm/Yr   WRMS 1.3 (3.0) cm TPZ-88.1   JAN 1989

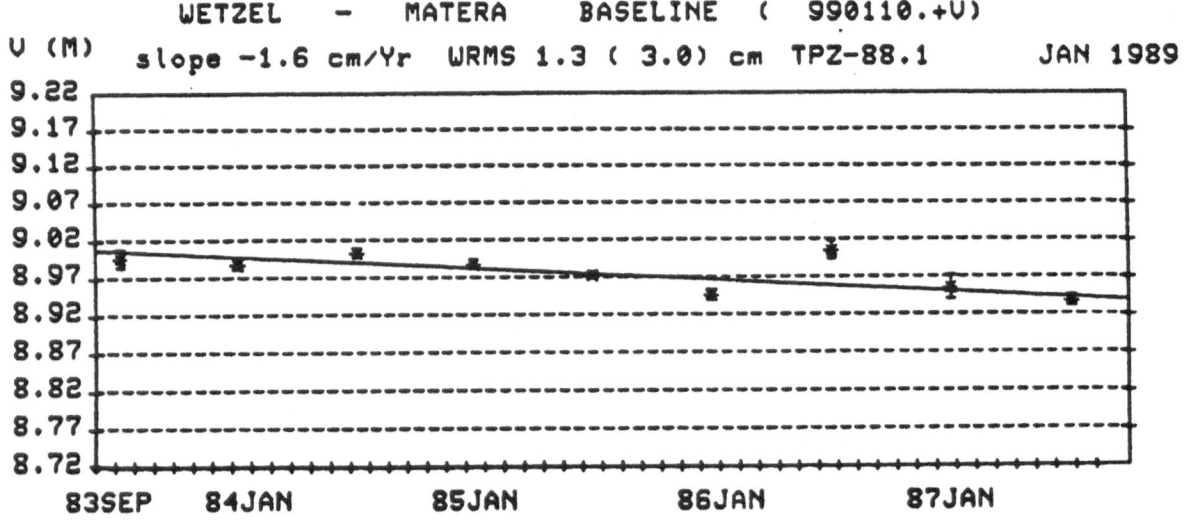

WETZEL   -   GRAZ   BASELINE (302130.+U)

U (M)   slope -0.6 cm/Yr   WRMS 2.4 (3.8) cm TPZ-88.1   JAN 1989

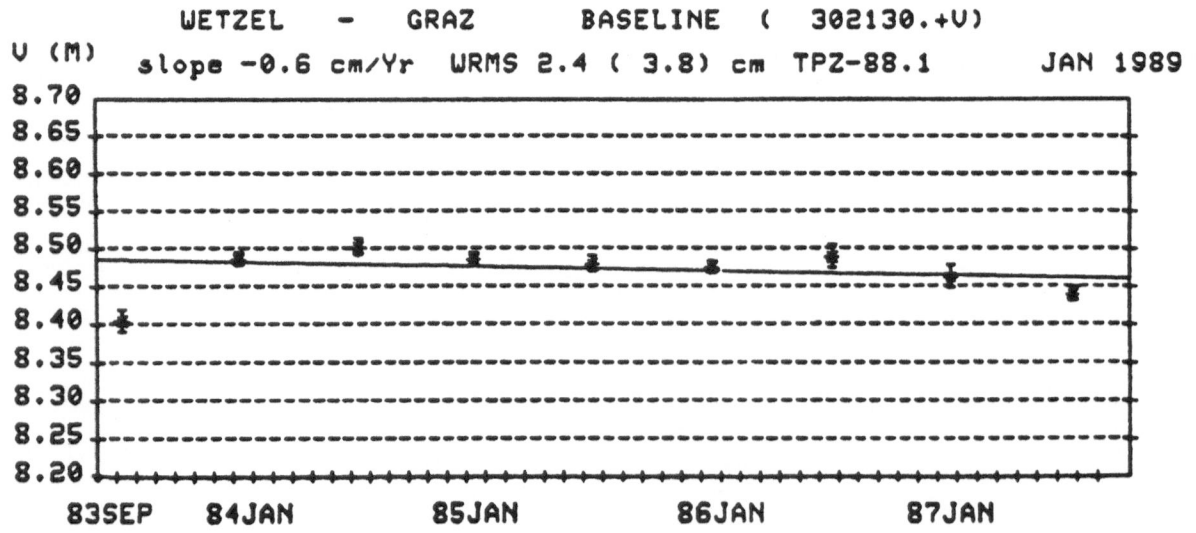

Fig. 2b

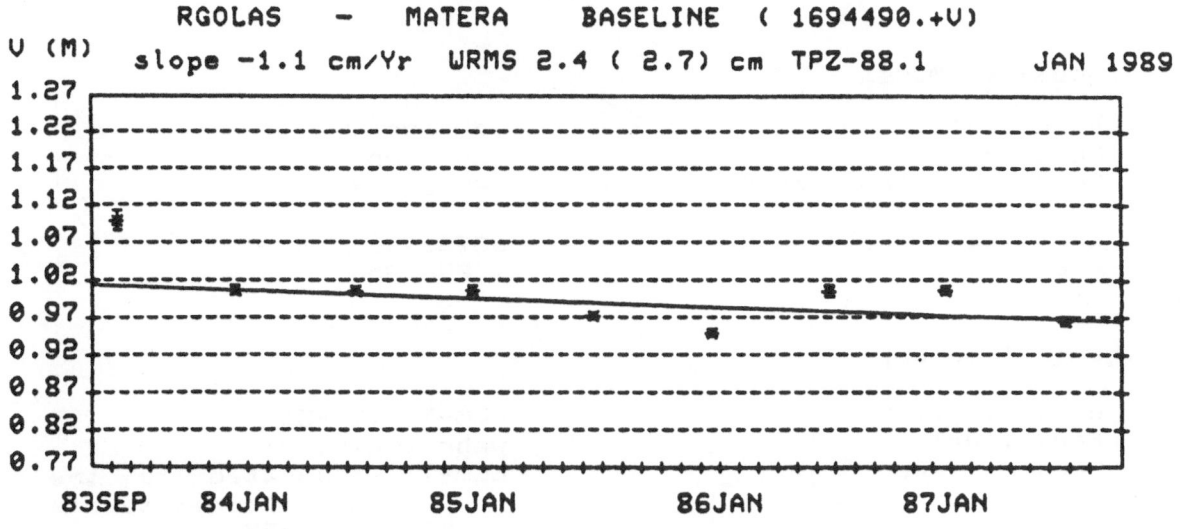

RGOLAS — MATERA BASELINE (1694490.+V)
V (M)   slope −1.1 cm/Yr   WRMS 2.4 (2.7) cm   TPZ−88.1   JAN 1989

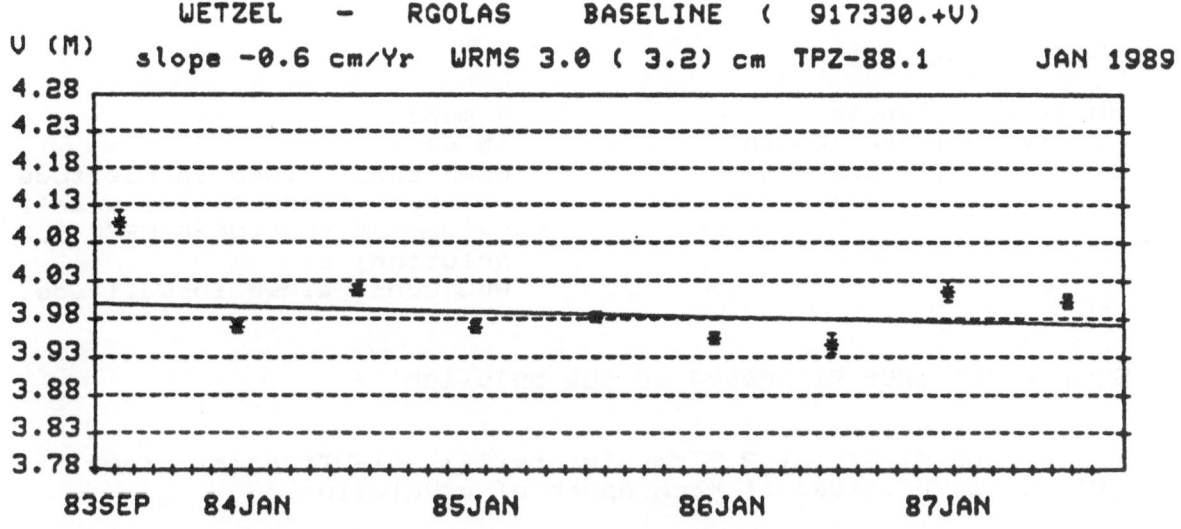

WETZEL — RGOLAS BASELINE (917330.+V)
V (M)   slope −0.6 cm/Yr   WRMS 3.0 (3.2) cm   TPZ−88.1   JAN 1989

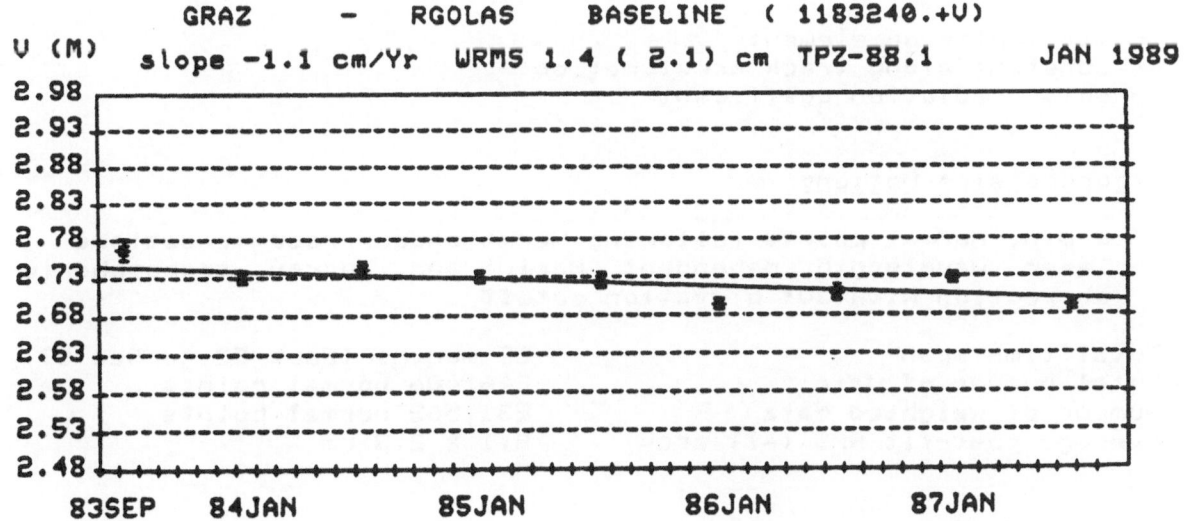

GRAZ — RGOLAS BASELINE (1183240.+V)
V (M)   slope −1.1 cm/Yr   WRMS 1.4 (2.1) cm   TPZ−88.1   JAN 1989

Tab. 1

SETUP FOR THE TPZ88.1 SOLUTION

Kinematical model

Precession	IAU 1976
Nutation	IAU 1980
Lunar and Planetary Ephemeris	JPL DE118
Reference System	1950.0
Earth semi major axis	6378144.11 m
Flattening	1/298.255

Dynamical model

Gravity field	GEM-T1 (36x36)
Body tides	Wahr ($K_2 = 0.30$)
Ocean Tides	GEM-T1 (truncated to the MERIT constituents)
h_2, l_2	0.6090 , 0.0852
Gm	$3.98600440 * 10^{14}$ m^3/s^2

Method of Analysis

Multi-arc length	6 months
Individual arc length	15 days
"fixed" stations	GORF constrained in Latitude and Longitude to the yearly values of the NASA/GSFC SL7 solution; RGO constrained in Latitude.

Common parameters estimated in the solution

- station coordinates
- pole and A1-UT1 at 2.5 day intervals; A1-UT1 constrained at BIH circ. D values at each epoch of osculation.

Arc parameters estimated in the solution

- six Keplerian elements
- constant along track acceleration
- solar radiation coefficent

Preprocessing Options

- 2 min. normal points following Herstmonceux recommendations
- laser wavelength dependent Marini and Murray tropospheric correction with 20° elevation cutoff

Total time span	52 months (Sept 83 to Dec 87)
Total number of data	246,406 normal points
Number of weighted data	231,563 normal points
Average post-fit RMS (all arcs)	8.1 ± 2.8 cm

Tab. 2

BASELINES BETWEEN MOBILE AND FIXED EUROPEAN STATIONS
OPERATIONAL IN 1986 AND 1987
SOLUTION TPZ88.1

		** 1986 **		** 1987 **		
BASELINES		LENGTH (m)	STD DEV (m)	LENGTH (m)	STD DEV (m)	DIFF. (cm)
POTSDA 1181	ASKITE 7510	1583729.796	0.051	1583729.821	0.038	2.5
POTSDA 1181	KATTAV 7512	2152405.755	0.044	2152405.777	0.036	2.2
POTSDA 1181	DYON1 7515	1793598.320	0.046	1793598.316	0.032	-.4
POTSDA 1181	ROUMEL 7517	2089408.352	0.050	2089408.381	0.035	2.9
POTSDA 1181	KRISOK 7525	1858425.993	0.051	1858425.967	0.034	-2.6
ASKITE 7510	KATTAV 7512	585020.182	0.035	585020.201	0.031	1.9
ASKITE 7510	DYON1 7515	346119.058	0.031	346119.096	0.022	3.8
ASKITE 7510	ROUMEL 7517	617642.775	0.035	617642.799	0.025	2.4
ASKITE 7510	KRISOK 7525	559522.125	0.039	559522.101	0.027	-2.4
ASKITE 7510	BARGIO 7530	1328033.273	0.043	1328033.144	0.035	-12.9 *
ASKITE 7510	ZIMMER 7810	1588473.799	0.036	1588473.834	0.027	3.5
ASKITE 7510	WETZEL 7834	1348310.745	0.034	1348310.734	0.030	-1.1
ASKITE 7510	GRASSE 7835	1560942.263	0.031	1560942.135	0.024	-12.8 *
ASKITE 7510	GRAZ 7839	1054732.954	0.032	1054732.997	0.027	4.3
ASKITE 7510	RGOLAS 7840	2220675.434	0.027	2220675.493	0.022	5.9
ASKITE 7510	MATERA 7939	747881.868	0.028	747881.884	0.023	1.6
KATTAV 7512	DYON1 7515	415834.032	0.027	415834.033	0.023	0.0
KATTAV 7512	ROUMEL 7517	285917.319	0.032	285917.235	0.027	-8.4
KATTAV 7512	KRISOK 7525	537653.067	0.031	537652.986	0.025	-8.1
KATTAV 7512	BARGIO 7530	822224.125	0.038	822224.024	0.035	-10.1 *
KATTAV 7512	ZIMMER 7810	2068211.096	0.034	2068211.093	0.029	-.3
KATTAV 7512	WETZEL 7834	1895393.758	0.028	1895393.732	0.030	-2.6
KATTAV 7512	GRASSE 7835	1969045.998	0.026	1969045.819	0.022	-17.9 *
KATTAV 7512	GRAZ 7839	1596967.558	0.027	1596967.579	0.026	2.1
KATTAV 7512	RGOLAS 7840	2724687.145	0.021	2724687.164	0.021	1.9
KATTAV 7512	MATERA 7939	1097753.241	0.022	1097753.215	0.022	-2.6
DYON1 7515	ROUMEL 7517	304459.581	0.034	304459.601	0.023	2.0
DYON1 7515	KRISOK 7525	231251.642	0.037	231251.564	0.024	-7.8
DYON1 7515	BARGIO 7530	1236412.146	0.039	1236412.045	0.030	-10.1 *
DYON1 7515	ZIMMER 7810	1659892.617	0.036	1659892.607	0.025	-1.0
DYON1 7515	WETZEL 7834	1512373.078	0.030	1512373.042	0.026	-3.6
DYON1 7515	GRASSE 7835	1558063.640	0.027	1558063.458	0.019	-18.2 *
DYON1 7515	GRAZ 7839	1212010.773	0.029	1212010.779	0.023	0.6
DYON1 7515	RGOLAS 7840	2324305.084	0.023	2324305.093	0.017	0.9
DYON1 7515	MATERA 7939	684581.979	0.023	684581.951	0.018	-2.8
ROUMEL 7517	KRISOK 7525	296654.842	0.036	296654.854	0.024	1.2
ROUMEL 7517	BARGIO 7530	1046174.601	0.043	1046174.448	0.034	-15.3 *
ROUMEL 7517	ZIMMER 7810	1912764.725	0.040	1912764.765	0.029	4.0
ROUMEL 7517	WETZEL 7834	1800247.231	0.034	1800247.234	0.029	0.3
ROUMEL 7517	GRASSE 7835	1775146.889	0.032	1775146.767	0.022	-12.2 *
ROUMEL 7517	GRAZ 7839	1501676.224	0.033	1501676.265	0.026	4.1
ROUMEL 7517	RGOLAS 7840	2582776.441	0.027	2582776.493	0.020	5.2
ROUMEL 7517	MATERA 7939	910138.966	0.027	910139.000	0.022	3.4
KRISOK 7525	BARGIO 7530	1336476.832	0.044	1336476.679	0.033	-15.3 *
KRISOK 7525	ZIMMER 7810	1630373.527	0.043	1630373.534	0.029	0.7
KRISOK 7525	WETZEL 7834	1550127.733	0.036	1550127.696	0.029	-3.7
KRISOK 7525	GRASSE 7835	1482086.919	0.034	1482086.773	0.022	-14.6 *
KRISOK 7525	GRAZ 7839	1255181.208	0.035	1255181.202	0.025	-.6
KRISOK 7525	RGOLAS 7840	2305570.475	0.028	2305570.488	0.019	1.3
KRISOK 7525	MATERA 7939	620666.368	0.030	620666.377	0.022	0.9

* The bad repeatibility is related to the poor statistics of
GRASSE and BAR GIYYORA

123

MODELLING IN NON-CARTESIAN REFERENCE FRAMES

Haim B. Papo
Department of Civil Engineering
Technion – Israel Institute of Technology

DEFORMATION ANALYSIS, CONCEPTS AND OBJECTIVES

Deformation analysis has to cope with a trivial problem which is common to most other geodetic activities, namely discretization of a continuous field. The time-like behaviour of the earth crust or of man-made structures is, in general, continuous in space and in time while, by contrast, geodetic measurements are usually discrete. As a result, in every specific case a singular deformation monitoring network has to be designed and measured. However, even an optimal program of measurements can at best only sample the continuous deformation field at discrete epochs and locations.

This paper deals with a problem which is typical to the analysis of continuous deformation fields. Ideally we would like to know $\underline{z}$, – the continuously varying velocities of points in the monitoring network, with respect to some absolute (preferably inertial) reference frame. Integration of those velocities over any time interval could lead to assessment of the respective deformations. In most cases, however, this is impossible. Absolute frames are inaccessible as there is no practical way to test and guarantee their absolute stability. A conventional substitute which is relatively easy to form, can be established on the basis of a selected subset of the network points. The primary characteristics of those points is to remain congruent during a given time interval. The conventional reference frame defined by the subset of reference points is related to the hypothetical absolute frame at a given epoch by a vector of translational and rotational velocity components, $\underline{m}$. In order to qualify for membership in the subset a reference point has to be motionless (to have a zero velocity) with respect to the conventional reference frame. Thus the absolute velocity of a point is partitioned into a rigid (translational and rotational) component and into a conventional velocity as shown in the following equation:

$$\underline{z} = E \cdot \underline{m} + \underline{w} \tag{1}$$

The above partitioning is easily achieved by applying the following linear constraints to the conventional velocities:

$$E^T \cdot \underline{w} = \underline{0} \tag{2}$$

where E in the above two equations is the basis of null space of the velocities, known also as Helmert's transformation matrix.

Least squares processing of geodetic measurements under the above constraints would normally produce estimates of the conventional velocities $\underline{w}$, while the vector $\underline{m}$ which contains datum parameters remains nonestimable. If the reference points have been properly selected, their conventional velocities would come out insignificantly different from zero. The velocities of the remaining points – the object points – are defined, thus, with respect to the rigid whole of the reference points. Assuming that the points in the reference subset retain their status over a given time interval, deformations between any two epochs within that interval can be evaluated by integrating the velocities of the respective object points. The object point velocities are usually modelled as constants (kinematic model) or as variable in time quantities (dynamical model) depending on the geodetic and auxiliary data which are available.

In this paper we focus on a problem associated with the establishment of a conventional reference frame in the presence of significant system noise. There are many cases where it is practically impossible to select a sufficient number of reference points with insignificant velocities as needed for the establishment of a conventional reference frame.

REFERENCE POINTS, SELECTION AND TESTING

The basis needed for the establishment of a conventional reference frame is formed by geodetic measurements which have been made between the reference points. Good spatial and time distribution of those measurements is vital for defining a useful and well balanced reference frame. With the introduction of each successive batch of measurements the velocities of all the points in the network are updated. Following each update the reference points' subset is tested for possible significant motion of its members. This is accomplished by subjecting the reference points' conventional velocities $\underline{w}_r$ to a statistical test. The null hypothesis $H0: \underline{w}_r = \underline{0}$ is tested against the alternative $H1: \underline{w}_r \neq \underline{0}$ by applying an F-test of hypothesis. The vector $\underline{w}_r$ contains the velocity components of the reference points which are obtained, in turn, by taking all non-zero elements of the product $P_r \cdot \underline{w}$. P_r is normally a diagonal positive semidefinite matrix whose non-zero elements are associated with the velocity components of the reference points. The test quantity k is compared by the following inequality to an appropriate value taken from F-distribution tables at the α significance level and for r_1 and f degrees of freedom:

$$k = [(\underline{w}_{r1} \cdot \hat{Q}_{r1}^{-1} \cdot \underline{w}_{r1}) / r_1] < F(r_1, f, \alpha); \quad \underline{w}_r = \begin{bmatrix} \underline{w}_{r0} \\ \underline{w}_{r1} \end{bmatrix} \quad (3)$$

where vector $\underline{w}_{r1}$ stands for the r_1 estimable velocity components of the reference points (Papo and Perelmuter, 1984);

$\hat{Q}_{r1}$ is the respective variance-covariance matrix which has been scaled by the estimated variance of unit weight; and finally

f if the number of degrees of freedom of the measurements which have participated in the adjustment.

The null hypothesis $H0$ is rejected whenever the above inequality is not satisfied (Hamilton, 1964). In case of rejection of the null hypothesis a reference point suspected in having non-zero velocity is removed from the subset (the respective diagonal elements in P_r are set to zero), the velocities are S-transformed (Papo and Perelmuter, 1984) and the test is repeated. The testing procedure is continued until the null hypothesis is finally accepted or until there are no more points left in the subset.

Let us assume that the velocity field of the reference points is a Gaussian stationary random process with the following first two moments (Gelb, 1984):

$$E(\underline{w}_r) = \underline{g}(\underline{q}, \underline{y}_r) \tag{4}$$

$$E\{[\underline{w}_r - E(\underline{w}_r)] \cdot [\underline{w}_r - E(\underline{w}_r)]^T\} = C_r \tag{5}$$

where C_r is the covariance matrix of the process. In the special case of $\underline{g}=\underline{0}$ the matrix C_r may play an important role in softening the F-test (Koch, 1984);

$\underline{g}$ is a model of the velocity field based on a-priori information. In the absence of such information the model can be developed ad-hoc by inspection and by subsequent experimentation;

$\underline{y}_r$ is a vector of the reference points' approximate coordinates;

and $\underline{q}$ is a vector of global deformation parameters.

If $\underline{q}$ is significantly non-zero, not a single pair of reference points would remain congruent. As a result, the null hypothesis $H0: \underline{w}_r = \underline{0}$ will be rejected repeatedly and it would be impossible to establish a conventional reference frame. If the establishment of a conventional (Cartesian) frame of reference is impossible we can look for a non-Cartesian substitute. An attempt can be made to identify reference points which move consistently with respect to each other according to a certain model. A vector of non-Cartesian velocity components can be defined by splitting out the global deformation component as follows:

$$\underline{x} = \underline{w} - \underline{g}, (\underline{q}, \underline{y}) \tag{6}$$

The null and the alternative hypotheses are defined now in terms of $\underline{x}_r$ so that $H0: \underline{x}_r = \underline{0}$ is tested against the alternative $H1: \underline{x}_r \neq \underline{0}$.

MODELLING THE VELOCITY FIELD IN NON-CARTESIAN REFERENCE FRAMES

The solution proposed in this paper is based on modelling the velocity field of the reference points by the $\underline{g}$ function. Linear or nonlinear $\underline{g}$ models can be designed, applied and tested as an integral part of a general adjustment procedure. Ideas for such models can come from other disciplines. In the absence of such a-priori information, ad-hoc models can be applied, provided they stand the test of time i.e. that they persist for a reasonable time interval and result in an $\underline{x}_r$ vector which satisfies the F-test. In what follows we will limit the discussion to linear models (in $\underline{q}$) of the velocity field. In a linear model the $\underline{g}$ function is presented as a product of a matrix $G(\underline{y})$ and the $\underline{q}$ vector as follows:

$$\underline{g}(\underline{q}, \underline{y}) = G(\underline{y}) \cdot \underline{q} \tag{7}$$

Before proceeding with details of modelling and estimation we should pay attention to some conceptual oddities of the proposed solution. Going back to the absolute velocities $\underline{z}$ we see that they are split now into three components:

$$\underline{z} = E \cdot \underline{m} + G \cdot \underline{q} + \underline{x} \tag{1'}$$

where
$\quad E \cdot \underline{m}$ — is the nonestimable datum component
$\quad\quad G \cdot \underline{q}$ — is the global deformation component
and $\quad\quad \underline{x}$ — is the residual component

In order to define a substitute conventional reference frame we minimize the $\underline{x}_r^T \cdot \underline{x}_r$ quadratic form instead of $\underline{w}_r^T \cdot \underline{w}_r$. The $\underline{x}_r^T \cdot \underline{x}_r$ minimum is obtained by constraining linearly the vector $\underline{x}_r$ in accordance with the extended free net approach (Papo, 1985):

$$\begin{bmatrix} E^T \\ -G^T \end{bmatrix} \cdot P_r \cdot \underline{x}_r = \bar{H}^T \cdot \underline{x}_r = \underline{0}; \quad \bar{H} = P_r \cdot H; \quad H = (E, -G) \tag{8}$$

The observation equations of a batch measured at epoch t_i ($\Delta t_i = t_i - t_0$) are:

$$\underline{v}_i + \underline{l}_i = [A_i, \Delta t_i \cdot A_i, \Delta t_i \cdot A_i \cdot G] \cdot \begin{bmatrix} \underline{y} \\ \underline{x} \\ \underline{q} \end{bmatrix} \tag{9}$$

The vector of corrections to point coordinates at the zero epoch $(\underline{y})$ is of no interest in deformation analysis. Its normal matrices are folded into those of $\underline{x}$ and $\underline{q}$. The "fold-in" operation can be done only after minimum linear constraints are applied to $\underline{y}$ as needed for datum definition at the t_0 epoch.

The essence of the extended free net adjustment techniques is that while retaining the properties of a minimally constrained solution ($\underline{v}$ — the corrections to the measurements remain invariant) we are at almost complete liberty to define the velocity model. The adjustment procedure produces estimates and respective variance-covariance matrices of the model parameters $\underline{q}$ and of the residual (non-Cartesian) velocities $\underline{x}$.

In addition to the rationale which leads to equation (8) we note that $\underline{x}$, H and the parameter vectors $\underline{m}$ and $\underline{q}$ can be interpreted as follows:

$\underline{x}$ — is a vector of velocity components in a non-Cartesian reference frame;

H — is the basis of null space of the $\underline{x}$ velocities field (Koch and Papo, 1985);

$\underline{m}$ and $\underline{q}$ — form together a vector of datum parameters of the above non-Cartesian reference frame which define its motion and behaviour with respect to the hypothetical absolute frame. Note that only $\underline{q}$ is estimable.

It is well known that depending on the information contained in the measurements only a part of the datum parameters in an ordinary Cartesian frame can be estimated. Scale and orientation angles of a conventional Cartesian frame are good examples of such estimable parameters.

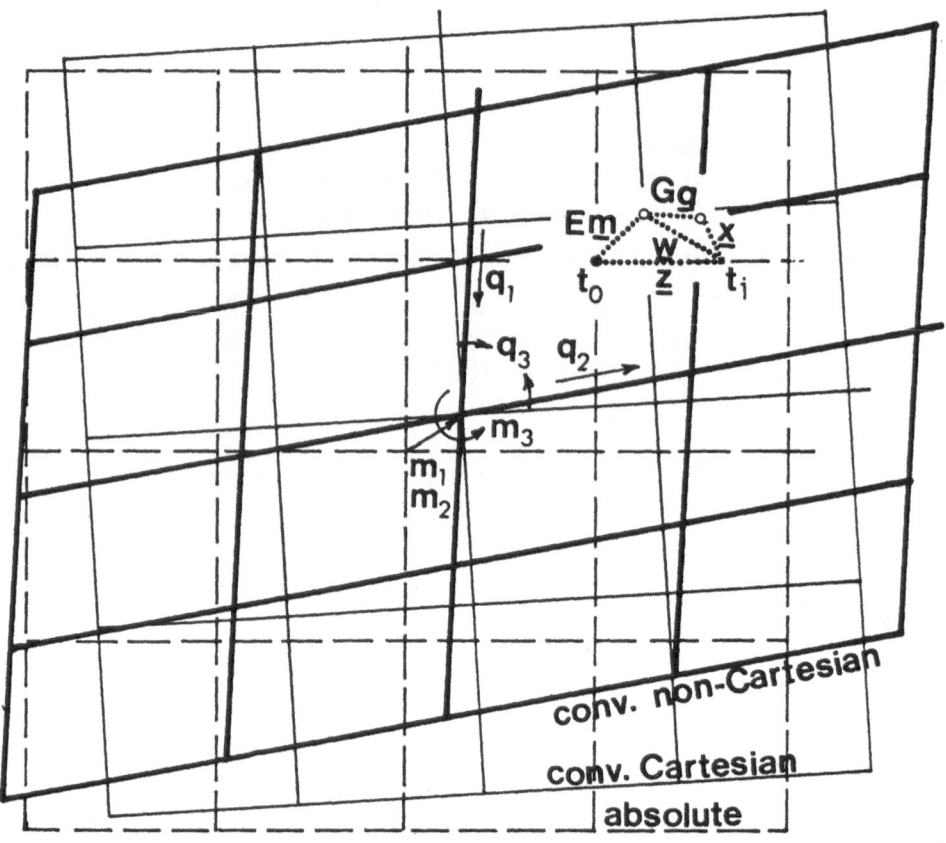

Fig. 1.   Absolute and conventional reference frames.

Figure 1 shows a simple conventional non-Cartesian frame in 2-D whose parametric lines are marked by heavy solid lines. The corresponding absolute and conventional Cartesian frames are shown in the background. The rigid motion (translation and rotation) of the conventional frames with respect to the absolute frame is given by the components of the $\underline{m}$ vector. The rates of deformation of the non-Cartesian frame with respect to the conventional Cartesian frame are given by the components of the $\underline{q}$ vector. Figure 1 illustrates also how decomposition of the absolute velocity $\underline{z}$ of a point depends on the $\underline{m}$ and $\underline{q}$ vectors and on $\underline{y}$ - the approximate coordinates of that point.

More sophisticated and useful non-Cartesian reference frames can be defined along the following lines:
* scale need not be constant along the axes;
* the axes need not be straight lines;
* the angle between intersecting parametric lines need not be constant.

The estimated velocities $\underline{x}_r$ are subjected to an F-test in order to check the reference points for significant departure from the model motion. Once the reference frame has been established we can proceed with the analysis of deformations i.e. with estimation of object point velocities.   We should note that the residual velocities of the object points represent the departure of the object points from the global model motion of the reference points.   Local disturbances in the deformation pattern may be important by themselves as indicators of specific local conditions and phenomena. At the end the conventional velocities $\underline{w}$ of all the points in the network can be evaluated by summing the global $(G \cdot \underline{q})$ and the residual $(\underline{x})$ velocity components. Thus the conventional velocities of the object points are defined in a conventional Cartesian frame of reference which is based on the non-congruent but consistent behaviour of the reference points.

SIMULATION AND ANALYSIS OF A VELOCITY FIELD

A simple test field in 2-D was simulated where two batches of measurements, each consisting of 17 horizontal distances, have been measured between 7 reference points using an EDM instrument (see Figure 2). The two epochs of measurement were exactly one year apart so that point velocities in millimeters per year would come out numerically equivalent to the respective displacements between the two measurement epochs. The absolute point coordinates at the second epoch were simulated without system noise $(C_r = 0)$ but with a well pronounced bias in the form of a simple linear velocity field $G \cdot \underline{q}$ as follows:

$$\underline{g} = \begin{bmatrix} q_1 & q_3 \\ q_3 & q_2 \end{bmatrix} \cdot \begin{bmatrix} y_1 \\ y_2 \end{bmatrix} = \begin{bmatrix} y_1 & 0 & y_2 \\ 0 & y_2 & y_1 \end{bmatrix} \cdot \begin{bmatrix} q_1 \\ q_2 \\ q_3 \end{bmatrix} = G \cdot \underline{q} \qquad (9)$$

where point coordinates $(\underline{y})$ are given in kilometers and the elements of the $\underline{q}$ vector have the following meaning:

$q_1 = -4$; axis 1 is getting shorter at the rate of 4 ppm per year;

$q_2 = 5$; axis 2 is getting longer at the rate of 5 ppm per year;

$q_3 = -3$; the angle between the two axes is getting smaller at the rate of 1.2 secarc per year.

The velocity of point #7 was disturbed by introducing into it additional inconsistent components.   It was destined to become a "misfit" point which would be removed eventually from the subset of reference points. Initially P_r was defined as an identity matrix. The measurements were simulated by adding noise taken from a normally distributed random sequence with zero mean and with a standard deviation of 1 ppm of the distance.

The six velocity components of points #1, #2 and #3 were declared as dependent (nonestimable) parameters according to the extended free net approach (Papo, 1985). The estimated variance of unit weight came out equal to 1.23 with $f = 12$ degrees of freedom. The F-test was applied initially to the velocity components of points #4 through #7 $(r_1 = 8)$. Due to the disturbance introduced into point #7 the null hypothesis $(HO: \underline{x}_r = \underline{0})$ was rejected. After removing point #7 from the reference subset and transforming the solution, the F-test was applied to the velocity components of points #4, #5 and #6. The test quantity came out smaller than the value taken from the tables:

$k = 0.743 \ll F (6, 12, 0.05) = 2.9961$

Table 3 Results of the adjustment of two measurement batches

Point #	Approx. position [km]		simulated		$G \cdot \hat{q}$		$\hat{x}$		$\hat{w} = G \cdot \hat{q} + \hat{x}$	
						velocities in mm per year				
1	0.9	-0.9	-0.9	-7.2	-1.2	-7.2	-0.5	-1.0	-1.7	-8.2
2	0.5	-0.2	-1.4	-2.5	-1.5	-2.4	-0.3	0.7	-1.8	-1.8
3	-1.3	-0.7	7.3	0.4	7.2	-0.1	0.0	-0.9	7.2	-1.0
4	0.4	0.8	-4.0	2.8	-3.8	3.1	0.6	-0.3	-3.2	2.8
5	-0.3	-0.7	3.3	-2.6	3.1	-2.9	1.0	1.6	4.2	-1.3
6	-0.8	0.3	2.3	3.9	2.4	3.8	-0.9	-0.1	1.6	3.7
7	1.1	0.7	-2.5	6.2	-6.4	0.7	4.8	3.0	-1.6	3.7

$$\hat{q}^T = [-4.08, 5.25, -2.74]$$

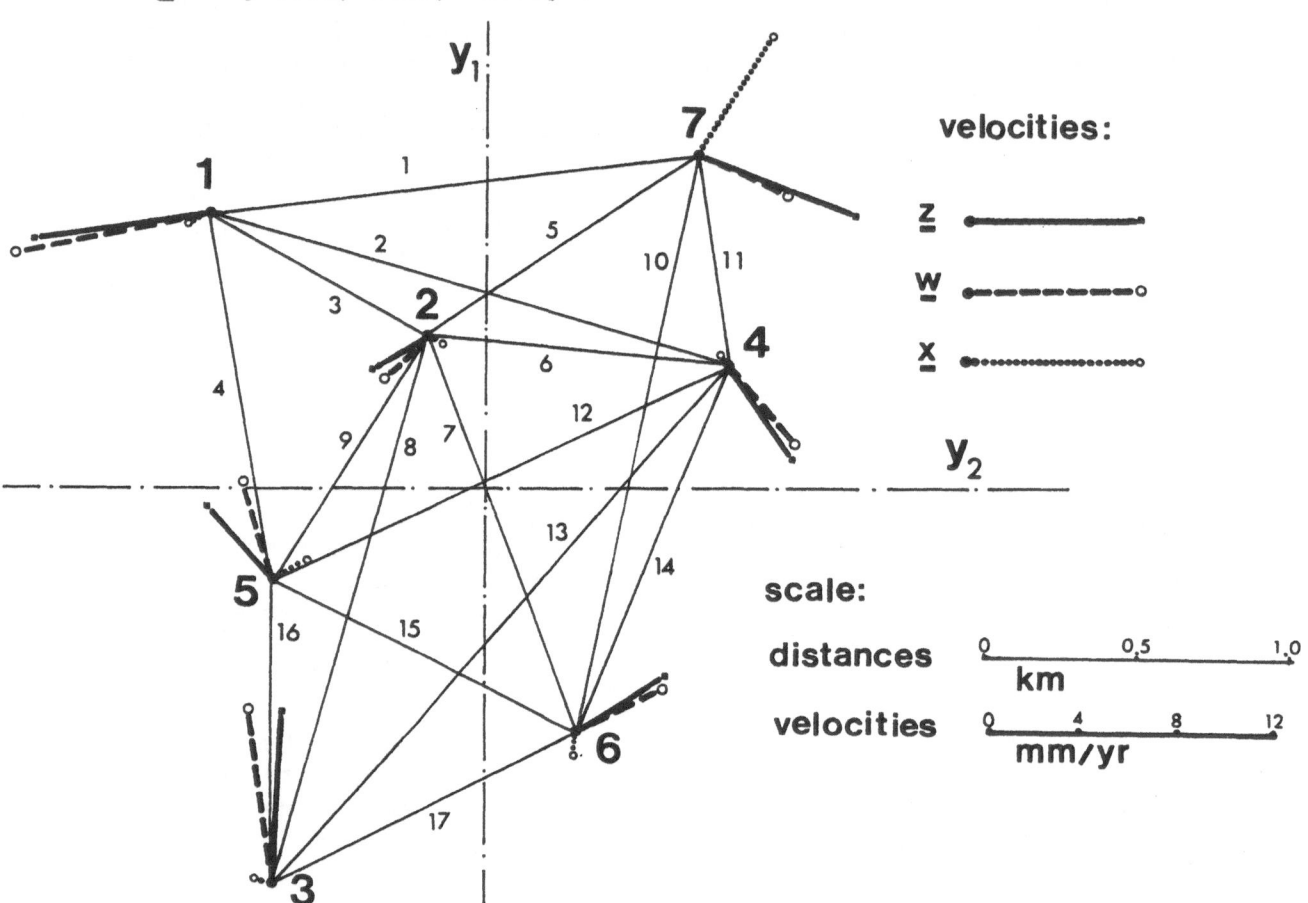

Fig. 2. Simulation of a 2-D deformation monitoring network

and so points #1 through #6 were approved and accepted as reference points while point #7 was declared to be a moving (object) point.

Table 3 summarizes the results of the adjustment. Using the estimated values of the parameters $\hat{q}$ and $\hat{x}$ the Cartesian velocities $\hat{w}$ of all 7 points were evaluated. Points #1 and #3 seem to have quite large velocities as compared to the velocity of point #7 and yet according to the principles of modelling in non-Cartesian reference frames points #1 and #3 are regarded as reference points while point #7 is an object points. Figure 2 shows all three types of velocities, namely:

$\underline{z}$ — the simulated (absolute) velocities;
$\underline{w}$ — the estimated conventional Cartesian velocities;
$\underline{x}$ — the estimated non-Cartesian (residual) velocities.

SUMMARY AND CONCLUSIONS

There are cases where deformation monitoring networks do not contain a sufficient number of reference points which remain congruent in time. In those cases we are compelled to resort to unorthodox techniques in trying to establish a conventional reference frame.

The essence of the solution proposed in this paper is to choose the most appropriate velocity model which fits the behaviour of the reference points' subset. If a-priori information from other disciplines is not available there is a large variety of plausible ad-hoc models which can fit almost any situation. Properties of lower order terms of such models are under study with the objective of creating a library of elementary geometric tools for non-Cartesian modelling.

Non-Cartesian modelling in deformation analysis provides the research worker with analytical tools which are otherwise unavailable. A reference frame can be established and subsequent analysis can be performed despite not having sufficient basis of congruent points in the network. If the motion of the selected subset of reference points is consistent and can be modelled by a finite number of parameters a sound reference frame can be established in which the deformations of all network points can be predicted with a high degree of confidence.

Acknowledgement. The study reported in this paper was supported by the Technion V.P.R. fund and by L. Edelstein research fund.

REFERENCES

Gelb, A. (1984). *Applied optimal estimation*, The MIT Press, MIT, Cambridge, Mass.

Hamilton, W.C. (1964). *Statistics in physical sciences*, The Ronald Press Co., New York.

Koch, K.R. (1984). Statistical tests for detecting crustal movements using Bayesian inference, *NOAA Tech. Rep.* NOS NGS 29. Nat. Info. Center, NOAA, Rockville, Md.

Koch, K.R. and Papo, H.B. (1985). Erweiterte freie Netzausgleichung, *ZfV*, 110 Jahrg., Heft 10, 451-457.

Papo, H.B. (1985). Extended free net adjustment constraints, *Bulletin Geodesique*, 59, (4), 378-390.

Papo, H.B. and Perelmuter, A. (1984). Densification of geodetic networks in four dimensions, *AVN*, 91 Jahrg., Heft 11-12, 450-458.

PRECIPITATION, GROUNDWATER AND GROUND DEFORMATION

T. Tanaka, E. Shimojima, K. Mitamura, Y. Hoso and Y. Ishihara
Disaster Prevention Research Institute, Kyoto Univ.
Gokasho, Uji, Kyoto-fu, 611 Japan

INTRODUCTION

Strainmeters and tiltmeters installed on or at a shallow part under the ground surface record ground deformations caused by rainfall. Such deformations are detected to a depth of several ten meters with these instruments, and accordingly they are main noises to observations of crustal movements and/or earth tides with these instruments.

From this fact some structures such as bench marks or piers are also suspected to be inclined or displaced by the ground deformations, especially in such a case that they are built on unconsolidated ground or alluvial plain under which aquifers exist. Recent geodetic and astrometric measurements with very high precision require more and more stable bases for observations and this kind of deformations should be taken into consideration, when a tilt of order of second of arc or a vertical displacement being smaller than a few mm is required as the object of measurement.

Ground tilts and strains by rainfall at Yura were investigated by Tanaka and Hoso(1986). They found that there are seasonal and secular

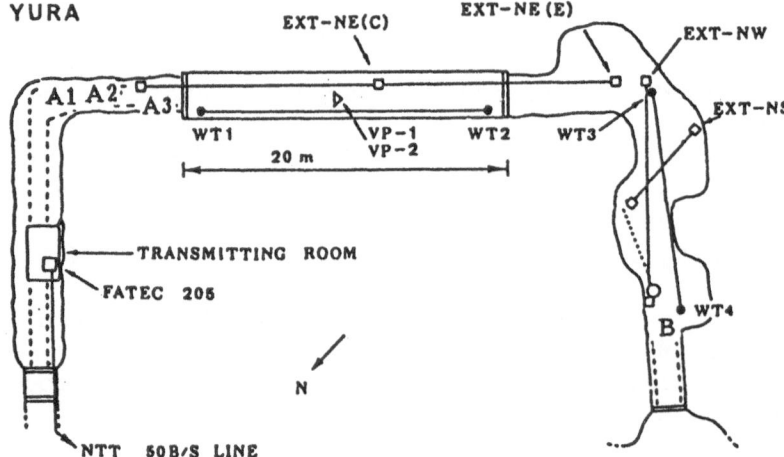

Fig.1. Observation vault of Yura Station. Entrances are located at the both sides of the bottom. A1,A2,A3 and B;Funnels, WT2-1 and WT4-3;Water-tube tiltmeters, EXT-NE, -NW and -NS;Strainmeters in respective directions, VP-1 and VP-2;Vertical pendulum tiltmeters.

changes in the tilt response to rainfall. Shichi and Iida(1973) suggested that formation of confining pore air pressure by downward moving water front is the main cause of the deformation by rainfall. Ishii et al.(1973) tried to explain the deformation by the theory of a porous medium. Tanaka(1979) and Yamauchi(1981) applied "tank model" to represent the time change of the deformation. Kümpel(1986) explains the ground strain and tilt observed after rainfall as deformations due to differential water-loading by an gradient of groundwater table, and carried out model calculations.

In order to investigate the recharge process of rainwater into aquifer and elucidate the runoff process of groundwater in the hill,we have started an intensified observation of groundwater discharge at different depths in the tunnel of Yura Station, southwestern Japan. Accordingly it became possible to do comparative research of the ground strains and tilts with precipitation and discharge at different depths in the same tunnel. From the present provisional analysis we have found a seasonal change in the discharge of groundwater and obtained some clue to make clear the generation mechanism of deformation of the ground by comparing the onset of the discharge with strains and tilts at different depths.

YURA STATION AND OBSERVATION OF GROUNDWATER DISCHARGE

Yura Station is in a tunnel under the flank of a small hill(33°57′N, 135°07′E), as shown in Fig.1, in Wakayama Prefecture, southwestern Japan. The tunnel was dugged into strata of weathered sandstone and shale having fissures. The surface of the hill is covered with thin soil, and grass and trees. Three strainmeters, two water-tube tiltmeters and one set of vertical pendulum tiltmeters were installed in the tunnel in 1983, and the data from these instruments, together with some meteorological data, are transmitted to the Institute by an exclusive telephone line. Since the depth of the tunnel is about 10m even at the

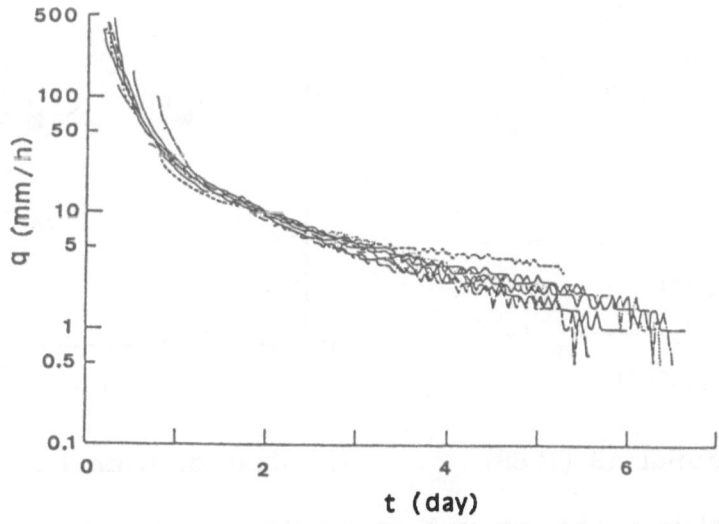

Fig.2. Decrease of discharge on funnel B after rainfall.

deepest point where strainmeter EXT-NE, water-tube tiltmeter WT2-1(NE component) and horizontal pendulum tiltmeter VP-1 and -2 are installed as shown in Fig.1, deformation of the ground is observed when a precipitation exceeds about 10mm(Tanaka,1979).

In order to measure the amount of the flow of seeping groundwater from the ceiling of the tunnel, four funnels have been installed at the different locations with different areas as shown in Fig.1. Areas of the funnels A1, A2, A3 and B are 1.00, 1.00, 3.73 and 3.43m^2, respectively and they are connected to rain gauges of 0.5mm step. The data of discharge are also telemetered to the Institute and therefore precise comparisons of the temporal change in discharges of groundwater with those of strains and tilts are possible. Onset time of the increase and the maximum in discharge of groundwater are closely related to the depth from the ground surface. Since the start of observation of discharges in February 1988 several changes in discharges and deformations were observed and analysed.

A SEASONAL CHANGE IN THE RESPONSE OF GROUNDWATER DISCHARGE

Groundwater discharge increases rapidly after rainfall, reaching their peak values and decreases exponentially afterwards. In this case peak values are dependent on the pattern of each hyetograph, but the exponential decreases show a common tendency as shown in Fig.2. This is interesting to understand the relation between flow and hydraulic properties of the ground. Each funnel is estimated to receive groundwater corresponding to rainwater fell on the ground surface, for example, about a half of the aperture in the case of funnel B.

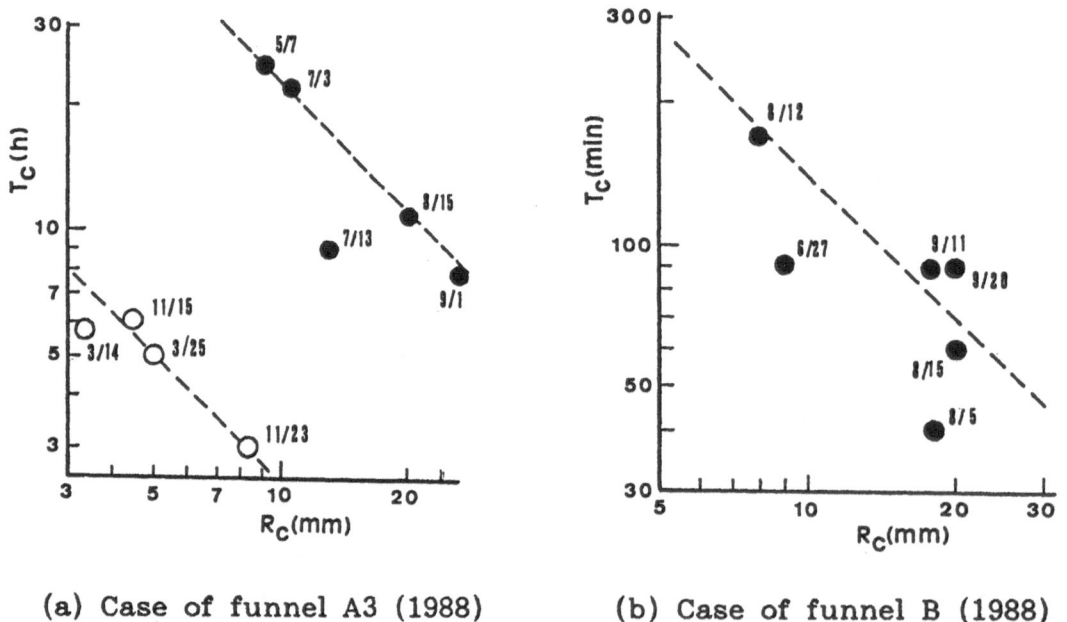

(a) Case of funnel A3 (1988)   (b) Case of funnel B (1988)

Fig.3. Relation between the amount of rainfall R_c and time duration T_c from the end of the specified duration of each rainfall to the increase of discharge. The month and date are given for each rainfall.

Fig.3 shows the relation between amounts of rainfall for a specified duration and the lag time to the increase of discharge. It is apparent that the larger the amount of rainfall is, the faster the response appears on the water discharge, and moreover the response of groundwater is faster in cold seasons than that in warmer seasons. Tanaka and Hoso(1986) found a seasonal change in the tilt response to precipitation at Yura: In cold seasons larger tilts are generated by the same amount of rainfall than in warmer seasons. They considered that this seasonal change in the response was caused by thermal effect on gaps between sand grains or fissures in rock at rather a shallow part under the ground, namely by a seasonal change in permeability. A similar change was reported by Yamauchi(1981) on the observation at Mikawa Crustal Movement Observatory, and this phenomenon is not unique at Yura Station but concluded as general one.

COMPARISON OF GROUNDWATER DISCHARGES WITH STRAINS AND TILTS

Fig.4 shows hourly amounts of rainfall on June 12, 1988 and the discharges of groundwater in the tunnel. The increase in discharge on the funnel B at the shallower point in the tunnel appeared 5 hours after the commencement of rainfall, reached its peak and began to decrease as soon as the rain stopped. Discharges recorded with the other 3 funnels A1, A2 and A3 installed at deeper part of the tunnel, show a different behavior. Although A3 shows a more complex change because of the large amount of discharge, its fundamental behavior is similar to the other two funnels A1 and A2; the discharges began to increase about 7 hours after the peak of rainfall and reached their maxima around 12 hours after the onset. Since the depths of funnel B and the group A are estimated to be about 5 and 10 meters, respectively, the different behaviors in their discharge are partly due to the difference of the depth from the ground surface.

The changes in strains and tilts caused by this rainfall are summarized in Fig.5. In this figure the amplitudes are normalized to give equal deflections. Comparing these two figures it is apparent that almost all of the strain and tilt components began to deform around the peak of the precipitation and reached their peak 10 or 15 hours after the commencement. In other words, the deformations observed with strainmeters and tiltmeters neither correspond to the increase nor peak of the groundwater discharge at the place where those instruments are installed, but probably are related to the flow of groundwater at a shallower level under the surface. It is not clear in Fig.4 and 5,but detailed investigations show that strain and tilt changes sometimes appear earlier than the increase in discharge on the funnel B, the shallower one. Therefore it is deduced from this fact that the ground deformation is generated when rainwater infiltrating into the ground reached at some depth, probably 3 or 4 m, namely at an equal or shallower level than that of the ceiling of the tunnel at B.

On March 5, 1989 an impulsive heavy rain was observed. The peak intensity reached 38mm/hour and as apparent in Fig.6 the increase of discharge occurred within one hour after the peak of the rainfall. Again the onset of strain and tilt disturbances appeared as soon as the rainfall reached its maximum intensity, and the discharge on A1, A2 and

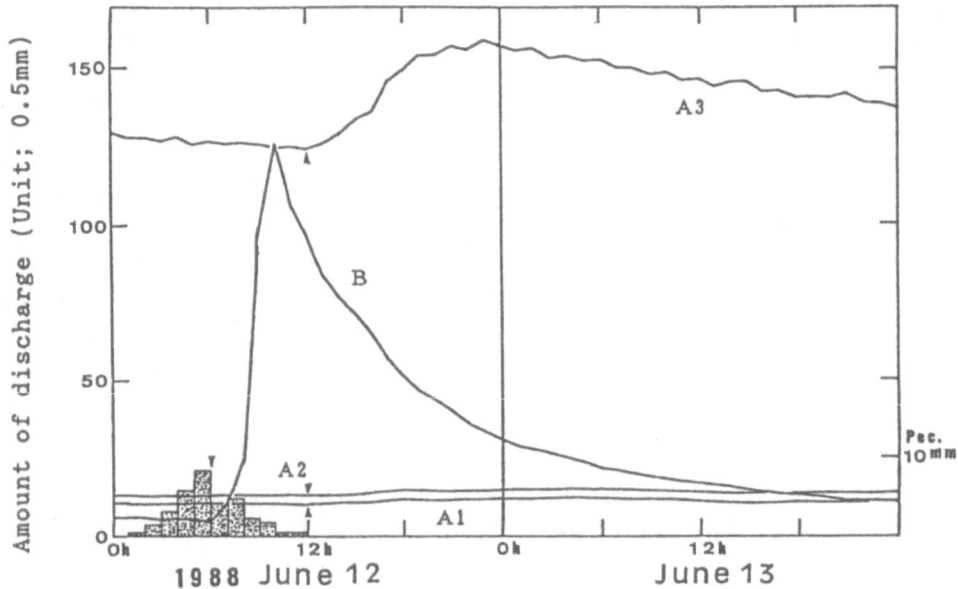

Fig.4. Hourly amount of rainfall(dotted columns) and groundwater discharge on June 12 and 13, 1988. Triangles show the onset time of increase.

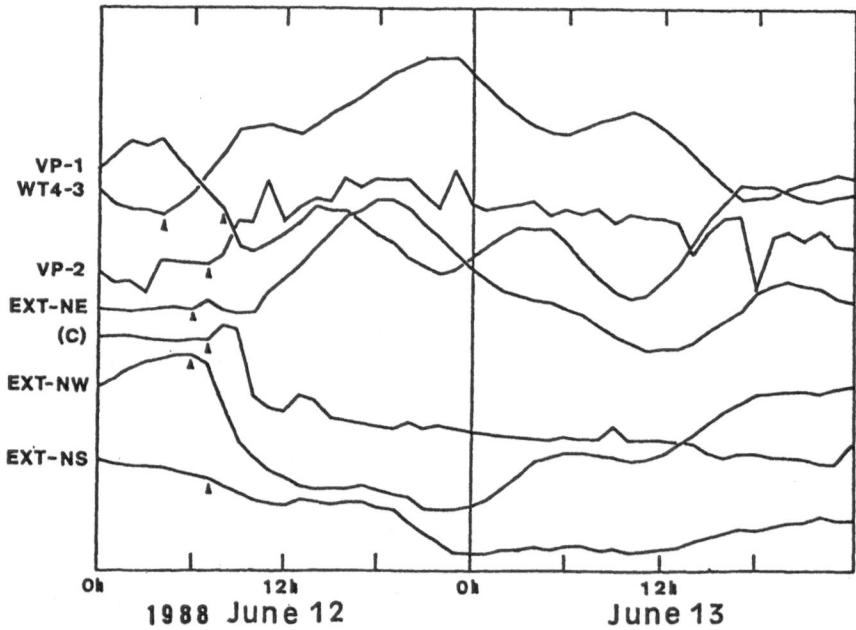

Fig.5. Strain and tilt changes due to the rainfall on June 12 and 13, 1988. Triangles show the onset time of the deformation. (C) is the strain observed by a sensor attached at the center of the rod of strainmeter EXT-NE.

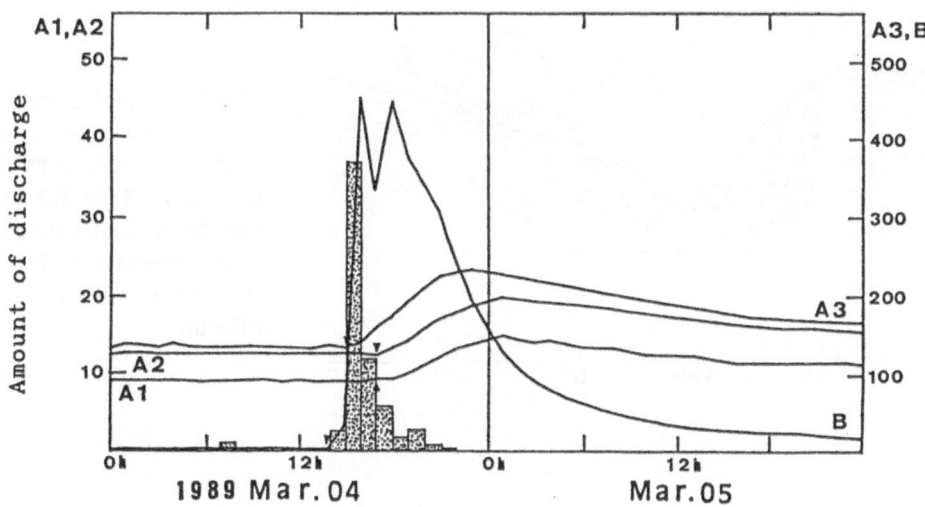

Fig.6. Hourly amount of rainfall(dotted columns) and groundwater discharge on March 5, 1989. The amount of rainfall is read by the left scale in the unit of mm. Triangles show the onset time of increase.

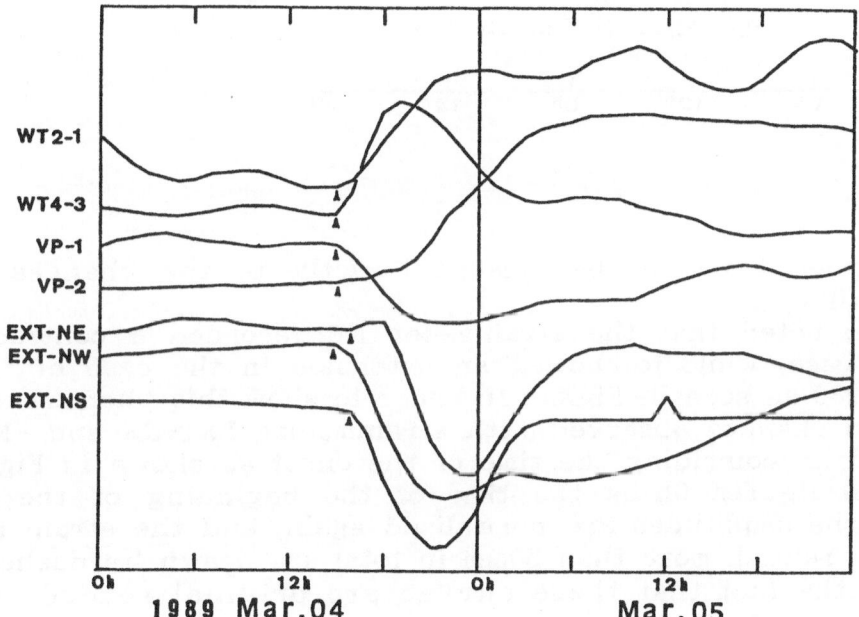

Fig.7. Strain and tilt changes due to the rainfall on March 5,1989. Triangles show the onset time of the deformation.

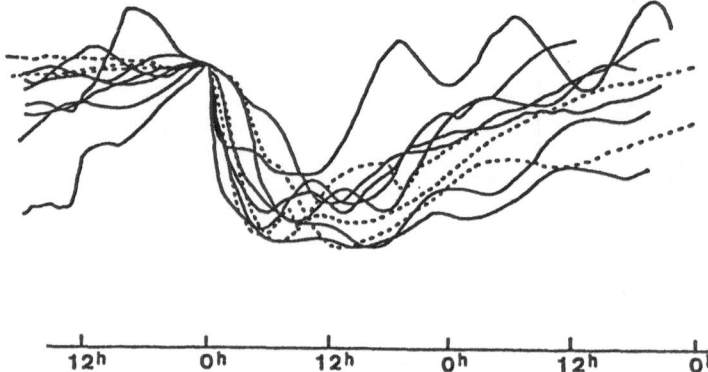

Fig.8. Comparison of the shape of the strain changes observed with strainmeter EXT-NS. Dashed curves are the the changes due to rainfall of more than 50mm.

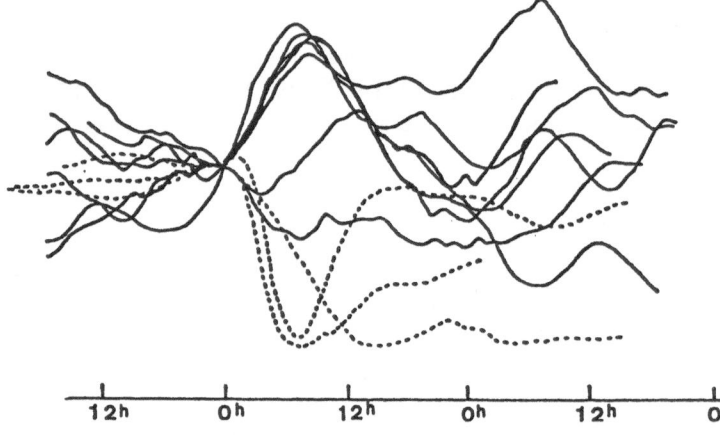

Fig.9. Comparison of the shape of the strain changes observed with strainmeter EXT-NE. Dashed curves are the the changes due to rainfall of more than 50mm.

A3 are considered not to be related directly to the changes in the strains and tilts.

It is to be noted that the strainmeter NE recorded a contraction in the present case, while it showed an extension in the case of the rain on June 6, 1988 as seen in Fig.5. In order to show this phenomenon more clearly strain changes observed with strainmeters EXT-NS and -NE were superimposed by coinciding the time of the onset as shown in Figs.8 and 9. On these figures 0h is the time of the beginning of the rainfall effect. Here the amplitudes are normalized again, and the strain changes due to heavy rainfall more than 50mm in total are drawn by dashed lines. Considering the fact that these curves are original records and no reductions for drift or linear changes have been given to draw these figures, we may say that the response to different rains is similar to each other to some degree. On the component NE the ground contracts for precipitation larger than 50mm. This should be interpreted as follows: When a rain starts and exceeds 10mm the ground begins to extend and it

138

continues until the amount of precipitation reaches to some critical value. If the rain stops before the amount exceeds the critical value, then the deformation stops, and decreases exponentially to recover the original level. Alternatively if the rain continues and exceeds the critical amount, the ground will then show rapid contraction, like the dashed curves in Fig.9. The contraction is probably caused by a different source or mechanism from the one that generates the preceding extension. It is likely that such sources are distributed under the ground and they act according to their respective characteristics. If we assume the mechanism as formation of confining pore air pressure by wetting front, a process such that different amounts of rain generate this pressure of different magnitude at different locations is a plausible mechanism. Recorded ground deformation should be a convolution of this kind of deformations.

CONCLUSIONS

We have investigated the relation among precipitations, groundwater discharge and ground deformations at the level of about 10m under the surface.

On the seepage of groundwater a seasonal change has been observed, namely the response of the groundwater to precipitation is faster in the cold seasons and larger for heavy rainfall.

The ground deformation due to rainfall is concluded to be generated by sources or some mechanism at shallow parts in the ground, since the deformation is observed when the groundwater reaches a few meters under the surface. Moreover the deformation is considered to be a convolution of deformations caused by several sources which are located at different places and act in different property.

The present work is the results from a provisional analysis. We are continuing to accumulate more data, and the mechanism of the ground deformation due to rainfall will be investigated in more detail in the near future.

REFERENCES

Ishii,H.,Sato,T. and Tachibana,K.(1973).Observation of crustal movement at the Akita Geophysical Observatory(2), - On crustal strain change caused by a rainfall -,*J.Geod.Soc.Japan*,19,135-144.(in Japanese)

Kümpel,H.-J.(1986).Model calculations for rainfall induced tilt and strain anomalies,*Proc.Tenth Intern.Symp.Earth Tides*,889-903.

Shichi,R. and Iida,K.(1973).Observation of crustal deformation at Inuyama(III),*J.Geod.Soc.Japan*,19,8-21.(in Japanese)

Tanaka,T.(1979).Effects of rainfalls on tiltmetric and extensometric records and their simulation,*J.Geod.Soc.Japan*,25,91-100.(in Japanese)

Tanaka,T. and Hoso,Y.(1986).Effect of rainfall on a continuous observation of ground tilts,*Bulletin of the Royal Society of New Zealand*, 24,19-28.

Yamauchi,T.(1981).Simulation of strain responses to rainfall,*J.Geod. Soc.Japan*,27,40-49.(in Japanese)

LITHOSPHERIC DEFORMATION
AND ASTHENOSPHERIC PRESSURE

Martine Amalvict and Hilaire Legros
Institut de Physique du Globe
Strasbourg, France

INTRODUCTION

We present here a model of lithospheric deformation caused by a radial pressure acting at the base of the lithosphere. Such a pressure may arise, for instance, as a result of convective motions occurring in the Earth's mantle. As a first approximation, we consider a rather simple Earth model consisting of a fluid interior, the mantle or asthenosphere, overlain by an elastic lithosphere. The model as a whole is assumed to be incompressible and have a constant mass density throughout. These simplifying assumptions allow us to find analytical solutions to the equations of gravito-elasticity which we use here in a peculiar way. We believe that these solutions provide a sufficiently meaningful description of the global yield of the actual terrestrial lithosphere, and thus may be used to model the evolving shape of the lithosphere.

In the original theory of plate tectonics, the lithospheric plates are assumed to drift over a viscous asthenosphere without undergoing deformation. However, the concept of perfectly rigid plates is valid only in a very rough approximation. Detailed investigations show that moving plates behave in fact like elastic, or visco-elastic, bodies and, as such, may be strained quite significantly. This straining is particularly important near plate boundaries.

In this paper we do not consider tectonic straining at the plate boundaries, which results from shear stresses and leads essentially to tangential motions. We consider rather the elastic deformation of the lithosphere as a whole which we model as a continuous and unique shell limited below by the asthenosphere. Similar work has been performed by other authors in an attempt to study the global yielding of the Earth's crust or lithosphere either as a consequence of crustal wandering or drifting of plates over an ellipsoidally constrained surface (Vening Meinesz, 1947; Turcotte, 1974), or as a consequence of the secular deceleration of terrestrial or planetary rotation caused by tidal friction (Melosh, 1977; Denis, 1986). Notice that in both latter cases, the source of deformation is a volume force, whereas in the case considered by us the source of deformation is a surface force. Indeed, in this study we wish to estimate the order of magnitude of elastic strains which may possibly be set up by a pressure at the lithosphere-asthenosphere boundary (in the following abbreviated to "LAB"). Some definite questions we wish to address are the following: How does the shape of the Earth's surface get altered by such a pressure acting across the LAB? What is the resulting perturbation of the geoid? Can such a pressure cause significant vertical motions?

THEORETICAL FORMULATION

Consider then an elastic lithosphere resting upon a fluid asthenosphere which we assume to extend right down to the centre of the Earth (Fig. 1). Let the whole model consist of uniform incompressible material, the density being the average density of the Earth. This model is similar to that considered a long time ago by Kelvin and allows a simple mathematical formulation leading to straightforward solutions. Numerical calculations show that, owing to the small thickness of the lithosphere, the results would not be strongly modified if we would consider a more realistic Earth model (in which properties vary with depth both in the lithosphere and the asthenosphere).

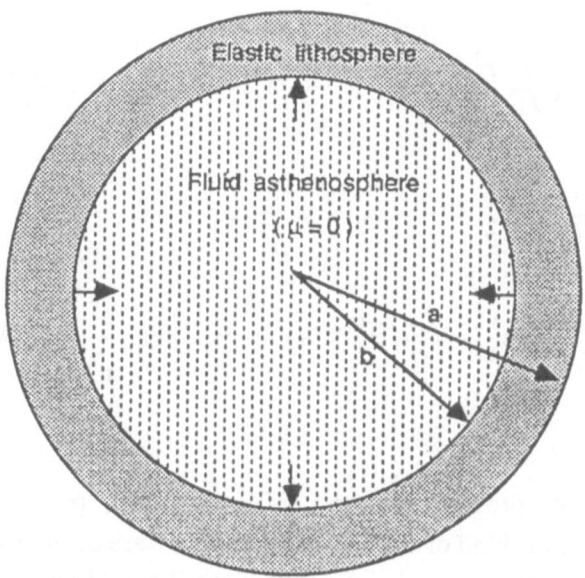

Fig. 1

Moreover, in this preliminary investigation, we retain only terms of the second degree in the expansion of the perturbing pressure field into spherical harmonics. Clearly, spherical harmonics of higher degrees are necessary to reach phenomena of smaller horizontal wavelengths but, apart from introducing more intricacy into the formulation, do not fundamentally change the theoretical approach taken in this paper. Work considering higher degree harmonics is in progress.

The general equations are presented in Amalvict and Legros (1986, 1989). The displacement vector is expanded as follows:

$$u = \frac{5}{42\mu} r^2 \nabla [p_2^+ \frac{r^2}{a^2} Y_2(\theta, \lambda)] - \frac{2}{21\mu} p_2^+ \frac{r^2}{a^2} Y_2(\theta, \lambda) \; r + \nabla [\varphi_2^+ \frac{r^2}{a^2} Y_2(\theta, \lambda)] +$$

$$+ \frac{1}{2\mu} p_2^- \frac{a^3}{r^3} Y_2(\theta, \lambda) \; r + \nabla [\varphi_2^- \frac{a^3}{r^3} Y_2(\theta, \lambda)]$$

where $Y_2(\theta, \lambda)$ is the general spherical surface harmonic of degree 2, i.e. a linear combination of the functions P_2, R_{21}, S_{21}, R_{22} and S_{22} given in the appendix. The parameter μ denotes the shear modulus, the variables θ and λ represent colatitude and longitude, respectively. From this we easily derive the expressions of the radial component, u_r, and of the tangential component, u_t, of the displacement vector $\mathbf{u}$, as well as the normal components of the elastic stress tensor, i.e. the normal elastic traction, σ_{rr}^{el}, and the normal elastic shear stress, σ_{rt}^{el}. They are

$$u_r = \frac{1}{2\mu} \left(\frac{2}{7} \frac{r^3}{a^2} p_2^+ + 4\mu \frac{r}{a^2} \varphi_2^+ + \frac{a^3}{r^2} p_2^- - 6\mu \frac{a^3}{r^4} \varphi_2^- \right) Y_2(\theta, \lambda)$$

$$u_t = \frac{1}{2\mu} \left(\frac{5}{21} \frac{r^3}{a^2} p_2^+ + 2\mu \frac{r}{a^2} \varphi_2^+ + 2\mu \frac{a^3}{r^4} \varphi_2^- \right) \nabla_t Y_2(\theta, \lambda)$$

$$\sigma_{rr}^{el} = \left(\frac{6}{7} \frac{r^2}{a^2} p_2^+ + \frac{4\mu}{a^2} \varphi_2^+ - 2 \frac{a^3}{r^3} p_2^- + 24\mu \frac{a^3}{r^5} \varphi_2^- \right) Y_2(\theta, \lambda)$$

$$\sigma_{rt}^{el} = \left(\frac{8}{21} \frac{r^2}{a^2} p_2^+ + \frac{2\mu}{a^2} \varphi_2^+ + \frac{1}{2} \frac{a^3}{r^2} p_2^- - 8\mu \frac{a^3}{r^5} \varphi_2^- \right) \nabla_t Y_2(\theta, \lambda)$$

[Eqns.1]

The unknowns of our problem are the coefficients p_2^+, p_2^-, φ_2^+, and φ_2^-. Their expressions depend on the kind of excitation function we are dealing with. This excitation function, or source of deformation, is introduced into [Eqns.1] by writing down explicitly the boundary conditions at the LAB and at the outer surface.

We notice that we may apply this method to quite different situations such as the straining by a rotation potential, the straining by an atmospheric pressure acting at the Earth's outer surface, or the straining caused by imposing a given displacement (i.e. a given shape) to the LAB, ... We have already considered cases 1 and 3 in previous papers. Here we investigate the effect of a perturbing pressure or traction originating in the asthenosphere and acting across the LAB.

Let P^I denote this perturbing pressure acting at the base of the lithosphere. We assume that the displacement is virtual inside the asthenosphere, such that the asthenosphere follows the shape of the interface and remains in contact with the lithosphere which is being deformed by the pressure P^I. The perturbation of the gravity potential caused by the deformation extends of course from the surface down to the centre. Moreover, let from now on u_r and u_t denote only the radial factors of the radial and tangential displacement components, respectively, and let T_r and T_t denote the radial factors of the tractional and shear stress components, respectively. Under these circumstances, it may be shown that the following boundary conditions are to be fulfilled:
- at the lithosphere-asthenosphere boundary, $r=b$:

$$T_r(b) = -P^I$$
$$T_t(b) = 0$$

142

- at the outer lithospheric boundary, r=a :

$$T_r(a) = -\rho\, g\, u_r(a)$$
$$T_t(a) = 0$$

ρ is density, and g is gravity at the Earth's surface.

Expressing these four conditions in terms of the coefficients p_2^+, p_2^-, φ_2^+ and φ_2^- by means of [Eqns.1], and solving for this set of unknowns, we obtain

$$p_2^+ = -\frac{3}{5}\frac{P^I}{\mu}\frac{20 - 88\,\varepsilon - 3\gamma\,(1-6\,\varepsilon)}{D}$$

$$p_2^- = \frac{8}{5}\frac{P^I}{\mu}\frac{5 - 33\,\varepsilon + 2\gamma\,(1-6\,\varepsilon)}{D}$$

$$\varphi_2^+ = \frac{2\,a^2}{5\,\mu}\frac{P^I}{\mu}\frac{6 - 33\,\varepsilon - 2\gamma\,(1-6\,\varepsilon)}{D}$$

$$\varphi_2^- = \frac{a^2}{5\,\mu}\frac{P^I}{\mu}\frac{37 - 286\,\varepsilon + 6\gamma\,(1-6\,\varepsilon)}{14\,D}$$

where

$$D = 11\,\gamma - 12\,\varepsilon\,(4\,\gamma - 1) \qquad \text{and} \qquad \gamma = \frac{\rho\, g\, a}{5\,\mu}.$$

The parameter ε represents the normalized lithospheric thickness, 1 - b/a. Here we have assumed that the lithosphere is thin, i.e. $\varepsilon \ll 1$, and limited all derivations to quantities involving only first order terms in ε.

Thus the components of the surface displacements are

$$u_r(a) = \frac{P^I}{\mu}\frac{11\,a\,(1-6\,\varepsilon)}{2\,D} \quad , \quad u_t(a) = \frac{P^I}{\mu}\frac{a\,[\,3 - 22\,\varepsilon - \gamma\,(1-6\,\varepsilon)\,]}{2\,D}.$$

[Eqns.2a]

The components of the displacement at the LAB are

$$u_r(b) = \frac{P^I}{\mu}\frac{a\,(11 - 62\,\varepsilon + 6\,\varepsilon\gamma)}{2\,D} \quad , \quad u_t(b) = \frac{P^I}{\mu}\frac{a\,[\,3 - 14\,\varepsilon - \gamma\,(1-7\,\varepsilon)\,]}{2\,D}.$$

[Eqns.2b]

Similarly, the stress components, which are interesting quantities for geophysical purposes, could easily be derived (cf. Amalvict and Legros, 1989).

In the case of a vanishing lithospheric thickness ($\varepsilon=0$), we have

$$u_r(a) = \frac{P^I}{\mu}\frac{a}{2\,\gamma} \quad , \quad u_t(a) = \frac{P^I}{\mu}\frac{a\,(3-\gamma)}{22\,\gamma}.$$

Typical numerical values, assuming $\rho = 5.5\times10^3$ kg.m^{-3}, a = 6.371×10^6 m, $\mu = 0.6\times10^{11}$ Pa, $\varepsilon = 2.3\times10^{-2}$ (= 150 / 6371), and thus g = 9.8 m.s^{-2}, $\gamma = 1.14$, D = 11.6, are (in metres)

143

$u_r(a) = 4.3 \times 10^{-5} \, P^I, \ u_t(a) = 6.9 \times 10^{-5} \, P^I, \ u_r(b) = 5.1 \times 10^{-5} \, P^I, \ u_t(b) = -3.2 \times 10^{-5} \, P^I.$

[Eqn.3]

If we consider a perturbing pressure at LAB of 10^5 Pa (i.e. 1 bar), which seems to be a plausible value, we obtain $u_r(a) = 4.3$ m, $u_t(a) = 6.9$ m, $u_r(b) = 5.1$ m, and $u_t(b) = -3.2$ m.

SOME GEOPHYSICAL APPLICATIONS

The formulae and results established in the previous section may be used to discuss a number of topics of some definite geophysical interest. Let us therefore first define the geophysically meaningful parameters that can be derived from the formulae above, and then discuss problems involving different types of harmonics of degree two: zonal, tesseral, and sectorial.

Geophysical parameters

Among the geophysically significant quantities which are addressed by our study, we may retain the radial surface displacement, $u_r(a)$, the variation of the gravity potential, χ, at the Earth's outer surface and in space above, and finally the variation of the lithospheric thickness, δ.

The radial displacement at the surface, $u_r(a)$, depicts the departure of the Earth's shape from the hydrostatic equilibrium shape, as a consequence of the supplementary internal pressure P^I acting at the LAB. We have seen above that it amounts to 4.3 P^I metres, if P^I is expressed in bars.

Directly linked to this geometric height difference is the perturbation of the gravity potential at the surface,

$$\chi(a) = \frac{3 \, g}{5} \, u_r(a) = N \, g,$$

where N is the height of the geoid. Again, if P^I is expressed in bars, the geoidal height amounts to $N = 2.6 \, P^I$ metres. It is clear, from Laplace's equation, that the continuation of this perturbation into outer space is provided by

$$\chi(r) = \frac{a^3}{r^3} \, \chi(a)$$

and will, more or less significantly, influence the trajectory of artificial satellites. For a perturbing pressure of 1 bar at LAB, the ensuing effect is about 22 $m^2.s^{-2}$ at a flight altitude of 300 kilometres.

Finally, we may define the variation of the thickness of the lithosphere as

144

$$\delta = u_r(a) - u_r(b) = -\frac{P^I}{\mu}\frac{a\varepsilon(2+3\gamma)}{D}.$$

This quantity is seldom reached analytically in such an immediate way. We notice that the pressure effect compresses the lithosphere, as it should. Introducing the values given above, the change of lithospheric thickness amounts to -0.8 m for $P^I = 1$ bar.

Application to a zonal problem

Let us first consider a zonal repartition of the pressure expressed in the classical Earth reference system, i.e. we simply substitute the Legendre polynomial $P_2(\cos\theta)$ to the general spherical surface harmonic $Y_2(\theta, \lambda)$. In this case, the additional geometrical surface flattening of the Earth is, to first order terms, directly proportional to the radial displacement at the surface:

$$\Delta f = \frac{3}{2}\frac{u_r(a)}{a}.$$

[Eqn.4]

We may use this result to explain the difference between the observed geometrical flattening, $f_{geom} = 1/298.257$ (Moritz, 1980), and the hydrostatic flattening, $f_{hyd} = 1/299.68$ (Denis, 1989), as a consequence of an additional pressure acting across the LAB. If this interpretation is correct, then [Eqn.4] leads to $u_r(a) = 68$ m, and [Eqn.3] yields $P^I = 1.6\times10^6$ Pa $(= 16$ bar$)$. It may be noticed that if we assume that such a pressure perturbation may build up within a time span of 10 to 100 million years, which seems to be indeed a plausible time span over which significant changes in mantle convection can occur, then the pressure effect could possibly account, besides tidal friction, for small but significant secular changes of the Earth's flattening.

Application to a tesseral problem

Assume now that the axis of symmetry of the LAB pressure repartition is not the same as the axis of rotation. This case, which corresponds to a tesseral problem, can easily be settled with the formulae given above. Let "state 1" correspond to the case where the pole of $P_2(\cos\theta)$ coincides with the Earth's pole of rotation, and let "state 2" correspond to the case where the pole P_0 of $P_2(\cos\theta)$ is located at the point (θ_0, λ_0). Let ψ be the colatitude of a point P at the Earth's surface with respect to P_0 (Fig.2). All the earlier results hold when substituting ψ for θ. This may be achieved by means of the classical relation

145

$$\cos \psi \ = \ \cos \theta \ \cos \theta_0 \ + \ \sin \theta \ \sin \theta_0 \ \cos (\lambda - \lambda_0) \, ,$$

which enables us to come back to the geographic coordinates (θ, λ) of point P.

Thus, for an infinitesimal shift $\delta\theta$ of the shell over the ellipsoidal Earth, we have

$$u_r(a) \ = \ \frac{P^I}{\mu} \ \frac{11 \, a \, \varepsilon}{2 \, D} \ \sin \theta \ (2 \cos \theta - 1) \ \delta\theta .$$

[Eqn.5]

If we assume that the perturbing pressure P^I is of the same order of magnitude as the equivalent pressure necessary to maintain the difference between the hydrostatic and actual flattenings (16 bars), we find (with $\delta\theta = 1°$) for $u_r(a)$ a value of about -1.5 m at the equator, 0.5 m at $\theta = 45°$, and 0 at the poles. The other parameters defined above (variation of lithospheric thickness, geoidal height, etc.) may be derived in a similar way as [Eqn.5].

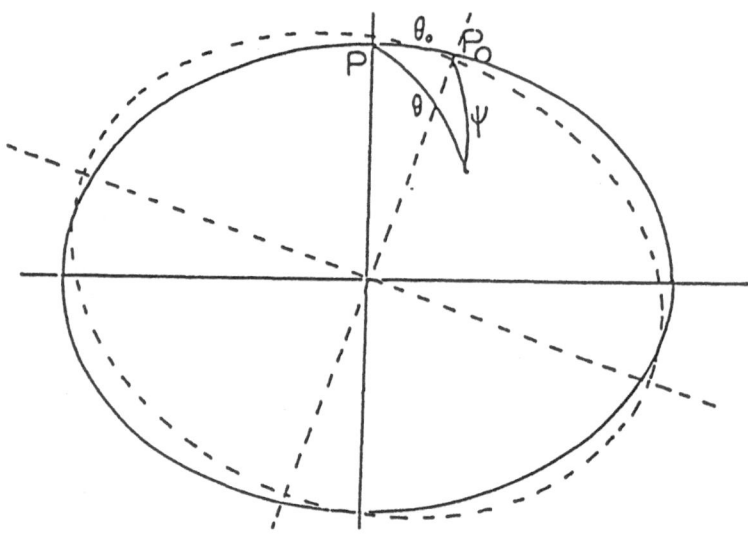

Fig. 2

Application to a sectorial problem

Using the same formalism as previously, we may locate the pole of reference at the equator and solve in some way an inverse problem by calculating the pressure at the LAB which could explain the excess ellipticity of the terrestrial equator. Thus, let us assume that the sectorial excess flattening of the Earth, i.e. the excess ellipticity of the equator, $f_e = 1/92800$, is due to an asthenospheric pressure $P^I P_2(\cos \psi)$. Replacing in $P_2(\cos \psi)$ the value of $\cos \psi$ by means of the trigonometric formula recalled above, where we take $\theta_0 = 90°$, we obtain $P^I = 1.6 \times 10^6$ Pa.

146

It is clear that all the other analytical expressions that we have derived previously can be adapted in the same way. For instance, we have

$$\delta = - \frac{P^I}{\mu} \frac{a \varepsilon}{D} (2 + 3\gamma) \frac{3 \sin^2\theta \, \sin^2\lambda - 1}{2}$$

and

$$N = \frac{P^I}{\mu} \frac{33 \, a \, (1 - 6\varepsilon)}{10 \, D} \, .$$

These few examples show that the action of an additional pressure acting across the interface between asthenosphere and lithosphere may constitute an important factor to consider when studying the shape of the Earth or the geoidal anomalies.

Let us remark, finally, that in this paper we have considered a perturbing internal pressure of the order of magnitude of 1 to 10 bar (10^5 to 10^6 Pa). Such values are consistent with the "topography" induced by convective motions in the mantle (Ricard et al., 1984).

CONCLUSION

We propose a modelling of vertical displacements associated with a non-hydrostatic pressure acting at the base of the lithosphere. Such a pressure may have its origin in mantle or asthenospheric dynamics, and may ultimately have a thermal origin. Using the concept of virtual displacements in the internal part of the Earth, assumed for simplicity to be fluid, it is very easy, in the framework of gravito-elasticity, to derive the elastic deformation of a relatively thin lithosphere. In this paper, we have considered only spherical surface harmonics of order 2, but in further work we intend to include into our analysis higher orders to be able to deal with more realistic tectonic problems.

At last, let us mention that the modelling we have presented here holds for quasi-static processes, involving time scales that are long with respect to the resonant modes of the Earth. It seems clear, indeed, that for a quasi-static response of the lithosphere, slow temporal changes of the internal pressure will cause a correspondly slow evolution of the shape and gravity field of the Earth. Such an evolution may possibly be traced in the future by means of precise orbit tracking of artificial satellites. This aim does not look very much out of reach if we consider that minute changes in the Earth's polar moment of inertia, and concomitant time variations of the gravitational coefficient J_2, which are probably related to post-glacial uplift, are presently being observed by LAGEOS. Thus, we believe that the existence and evolutive behaviour of an excess pressure deep inside the Earth could be detected, via changes in χ which affect satellite trajectories, by means of terrestrial orbitography extended over about a decade.

REFERENCES

Amalvict, M. and Legros, H. (1986). Lithospheric stresses: gravito-elastic model and geometric deformation, *Manuscripta Geodaetica*, **11**, 197-206.

Amalvict, M. and Legros, H. (1989). Deformation of the lithosphere and induced stresses, in: *Physics and Evolution of the Earth's Interior*, (ed. R. Teisseyre), vol. **5**, PWN, Warszawa and Elsevier, Amsterdam (in preparation).

Denis, C. (1986). On the breakup of the original lithosphere, *Geophys. Rep. Publ. Inst. Astroph. Liège*, **6**, 139-143.

Denis, C. (1989). The hydrostatic figure of the Earth, in: *Physics and Evolution of the Earth's Interior*, (ed. R. Teisseyre), vol. **4**, Chap. 3, pp. 111-186, PWN, Warszawa and Elsevier, Amsterdam.

Melosh, H.J. (1977). Global tectonics of a despun planet, *Icarus*, **31**, 221-243.

Moritz, H. (1980). Geodetic Reference System 1980, *Bull. Géodés.*, **54**, 395-405.

Ricard, Y., Fleitout, L. and Froidevaux C. (1984). Geoid heights and lithospheric stresses for dynamic Earth, *Annales Geophysicae*, **2**, 267-286.

Turcotte, D.L. (1974). Membrane tectonics, *Geophys. J. R. astr. Soc.*, **36**, 33-42.

Vening Meinesz, F.A. (1947). Shear patterns of the Earth's crust, *Trans. Am. Geophys. Un.*, **28**, 1-61.

APPENDIX

Spherical surface harmonics of order 2

Zonal harmonic :

$$P_2(\cos \theta) = (3 \cos^2\theta - 1) / 2$$
$$\partial_\theta P_2 = - 3 \sin \theta \cos \theta$$

Tesseral harmonics :

$$R_{21}(\theta,\lambda) = 3 \sin \theta \cos \theta \cos \lambda$$
$$S_{21}(\theta,\lambda) = 3 \sin \theta \cos \theta \sin \lambda$$
$$\partial_\theta R_{21} = 3 (\cos^2\theta - \sin^2\theta) \cos \lambda$$
$$\partial_\theta S_{21} = 3 (\cos^2\theta - \sin^2\theta) \sin \lambda$$
$$\partial_\lambda R_{21} = - 3 \sin \theta \cos \theta \sin \lambda$$
$$\partial_\lambda S_{21} = 3 \sin \theta \cos \theta \cos \lambda$$

Sectorial harmonics :

$$R_{22}(\theta,\lambda) = 3 \sin^2\theta \cos 2\lambda$$
$$S_{22}(\theta,\lambda) = 3 \sin^2\theta \sin 2\lambda$$
$$\partial_\theta R_{22} = 6 \sin \theta \cos \theta \cos 2\lambda$$
$$\partial_\theta S_{22} = 6 \sin \theta \cos \theta \sin 2\lambda$$
$$\partial_\lambda R_{22} = - 6 \sin^2\theta \sin 2\lambda$$
$$\partial_\lambda S_{22} = 6 \sin^2\theta \cos 2\lambda$$

A GPS SURVEY IN THE

YUNNAN EARTHQUAKE EXPERIMENTAL FIELD

OBJECTIVES AND FIRST RESULTS

Günter   Seeber

Institut für Erdmessung (IFE)

D-3000 Hannover 1, Nienburger Str. 5

Federal Republic of Germany

Lai   Xian

Institute of Seismology (IOS)

State Seismological Bureau   (SSB)

Wuhan, PR China

INTRODUCTION

GPS has been developed in recent years to be a powerful means for the determination of precise coordinate differences. It can thus be used for monitoring local and regional crustal motions through repeated observations.

In 1988 the Institute of Seismology (IOS), State Seismological Bureau of the People's Republic of China and the Institut für Erdmessung (IFE), University of Hannover, Federal Republic of Germany, agreed upon the realization of a joint project on the use of GPS for investigations on earthquake predictions. A first epoch control survey could be established in October 1988 in the province of Yunnan in South West China.

Yunnan is an area of very high seismic activity. This is why a testing ground, the "Western Yunnan Earthquake Prediction Study Area (WYEPSA)" was installed by the State Seismological Bureau (SSB) in the west of Yunnan province between 24.5 - 28.0 North and 98.5 - 101.5 East.

The purpose of the joint project between our institutes is to study the relation between the surface deformation field, the gravity field and earthquakes, and to understand the interaction between various fields and the relation to earthquake occurance. Observations of the relative and absolute gravity are planned for 1990, partly on identical stations.

As a first step, a first epoch measurement of the three-dimensional coordinates of the geometrical deformation field was performed with GPS in October 1988. A network of 20 stations was installed using 4 dual frequency receivers (Fig. 3).

The purpose of this contribution is to give some informations on the geological and seismological background of the network area, and to report on first results of the GPS survey.

GEOLOGICAL AND SEISMOLOGICAL BACKGROUND OF THE TEST AREA

The test area with the GPS network, being located in the eastern edge of the tectonic zone of the Himalaya-Burma arc, is located in the convergent zone of three first-order tectonic elements. The test field is also located in the conjunctive area of the western edge of the seismic belt from north to south in China and the Himalaya-Mediterranean Seismic Belt. The seismicity in the area has rather obvious correlation with that in the Around-Pacific Seismic Belt and the Himalaya-Mediterranean Seismic Belt (Fig. 1). Crust there is subjected to strong compression under the interaction of the Indian plate and the Eurasian plate, with well developed tectonic faults and frequent earthquakes (Lai and Shao, 1985).

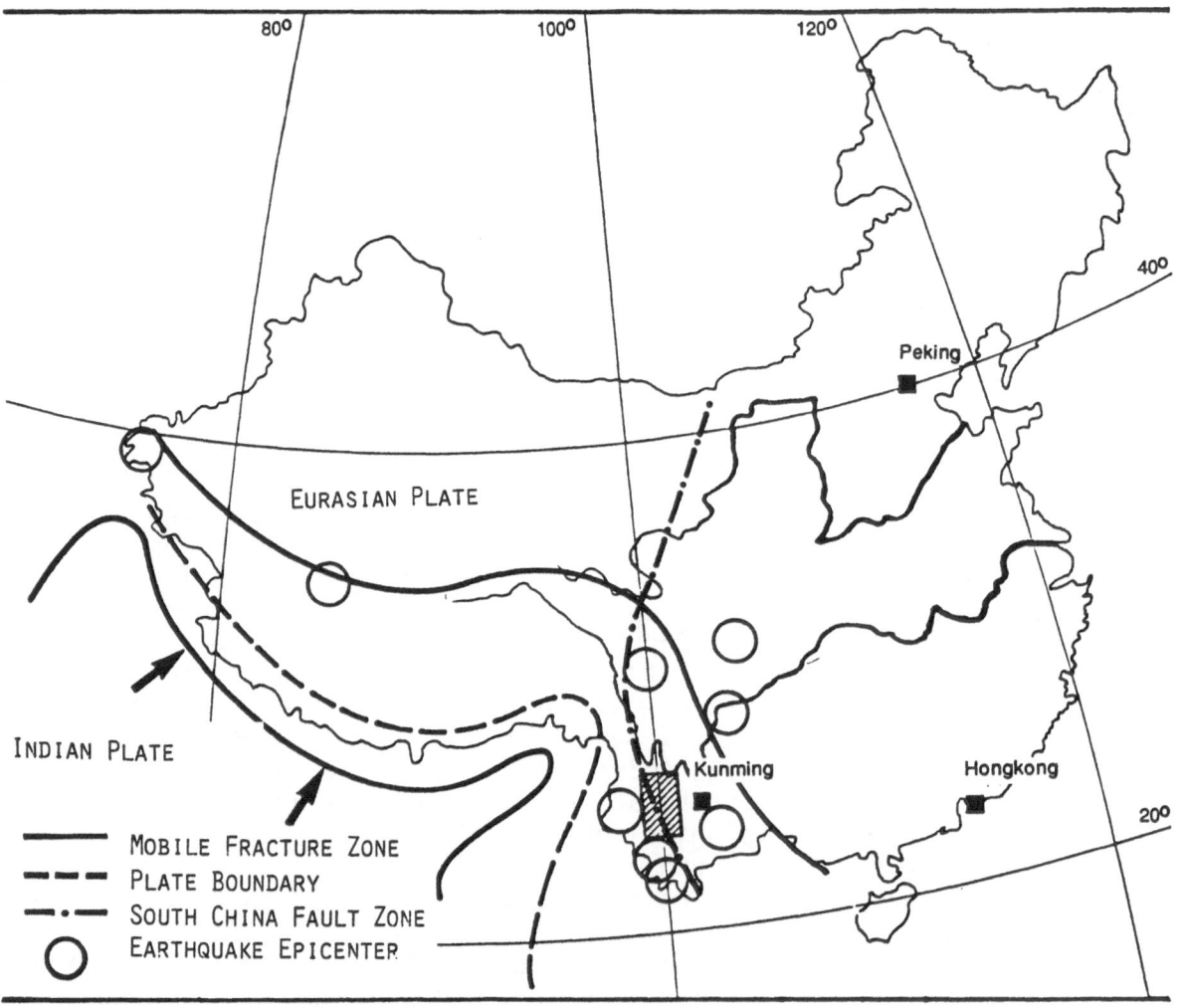

Fig. 1. General location of the Yunnan Test Field

Deep fractures are distributed vertically and horizontally. These fractures are strongly active and control the seismic activities of the area. In particular this holds for the Honghe fracture zone, the biggest fracture zone of the area, which crosses the GPS network obliquely from the north to the south (Fig. 2, Fig. 3).

Since the beginning of recorded history, the southern section of the Honghe fracture zone is relatively quiet, while the northern part is strongly active. 15 large earthquakes with M≥6 have occured since 1481, once in 31 years on the average. The Jianchuan-Dali segment at the northern section of Honghe fracture zone (Fig. 2) has been indicated by the Yunnan Seismological Bureau to be one of the dangerous regions where earthquakes with M≥7 may occur in the near future. This part is well covered by the GPS control network. Figure 2 indicates some of the main faults and the location of GPS stations in the central part of the network.

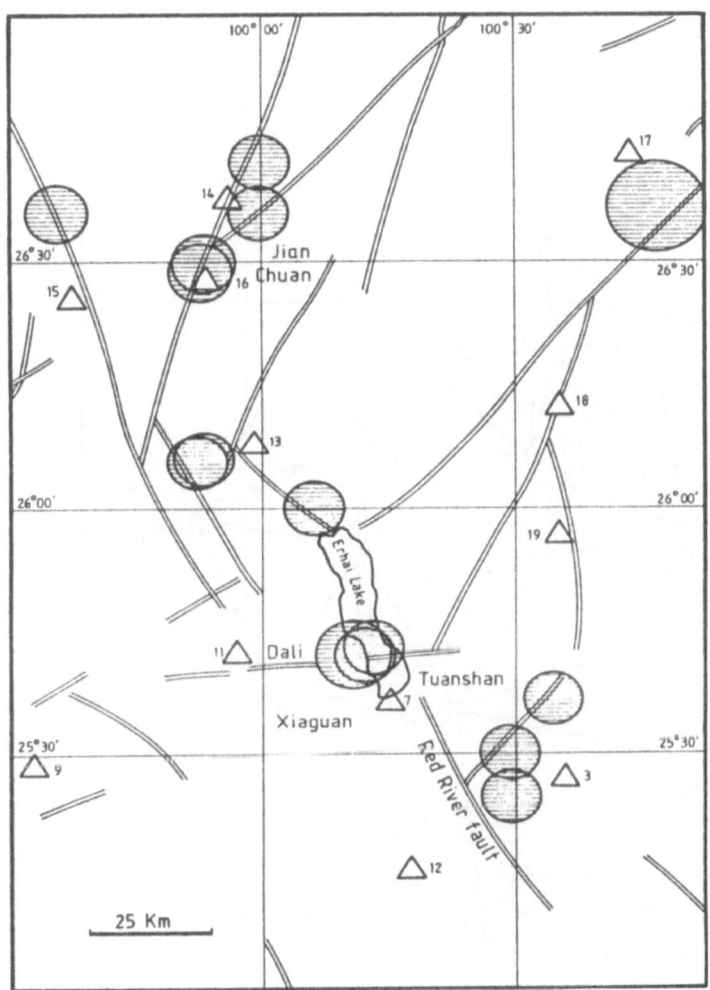

Fig. 2. GPS stations, main faults and epicenters of large earthquakes in the central part of the test field

151

With respect to increasing seismic activities in recent years, the test area is well suited for studying the relation between earthquakes and its precursory phenomena. The analysis of geometrical deformation fields, provided by repeated GPS observations, will form one of the important input parameters for probing earthquake prediction methods. Surface motion rates are predicted to be around 10 mm per year.

THE GPS FIELD WORK

GPS observations in the test area have been carried out on 17 days with 4 dual frequency TI 4100 receivers between October 7 and 23, 1988. The whole network covers an area of about 250 x 300 km (Fig. 3) and contains 20 stations. Station 7 (Tuanshan) near the city of Xiaguan at Lake Erhai was a permanent station. At Tuanshan the SSB is operating a seismological research center with convenient facilities which serves as the base of WYEPSA.

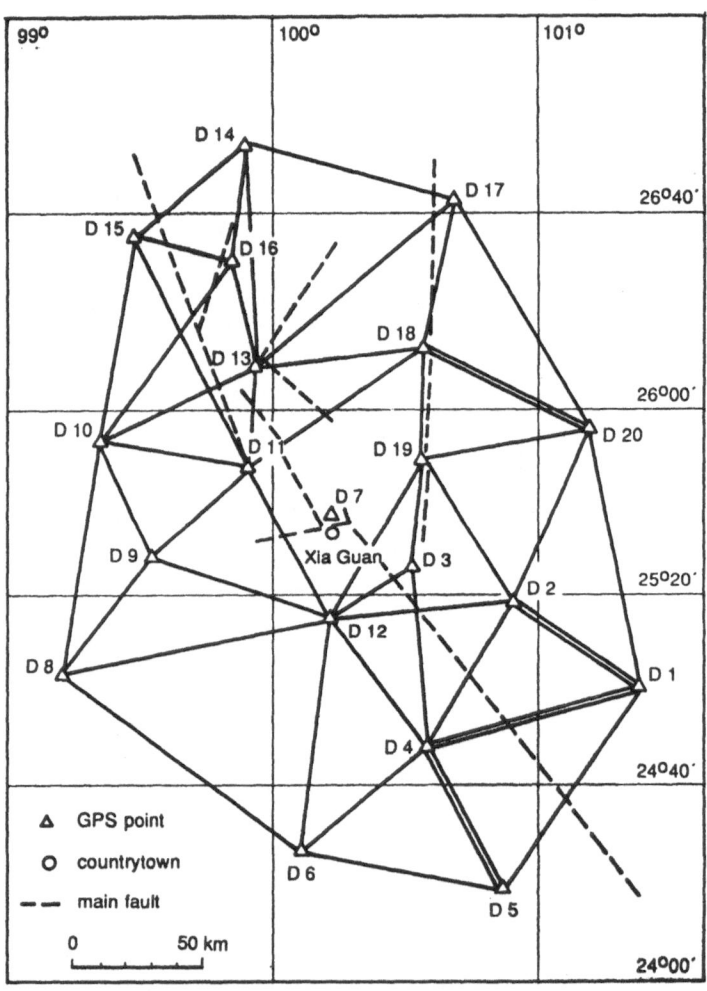

Fig. 3. GPS Network in the Western Yunnan Earthquake Prediction Study Area. Point locations are only approximate.

DAY

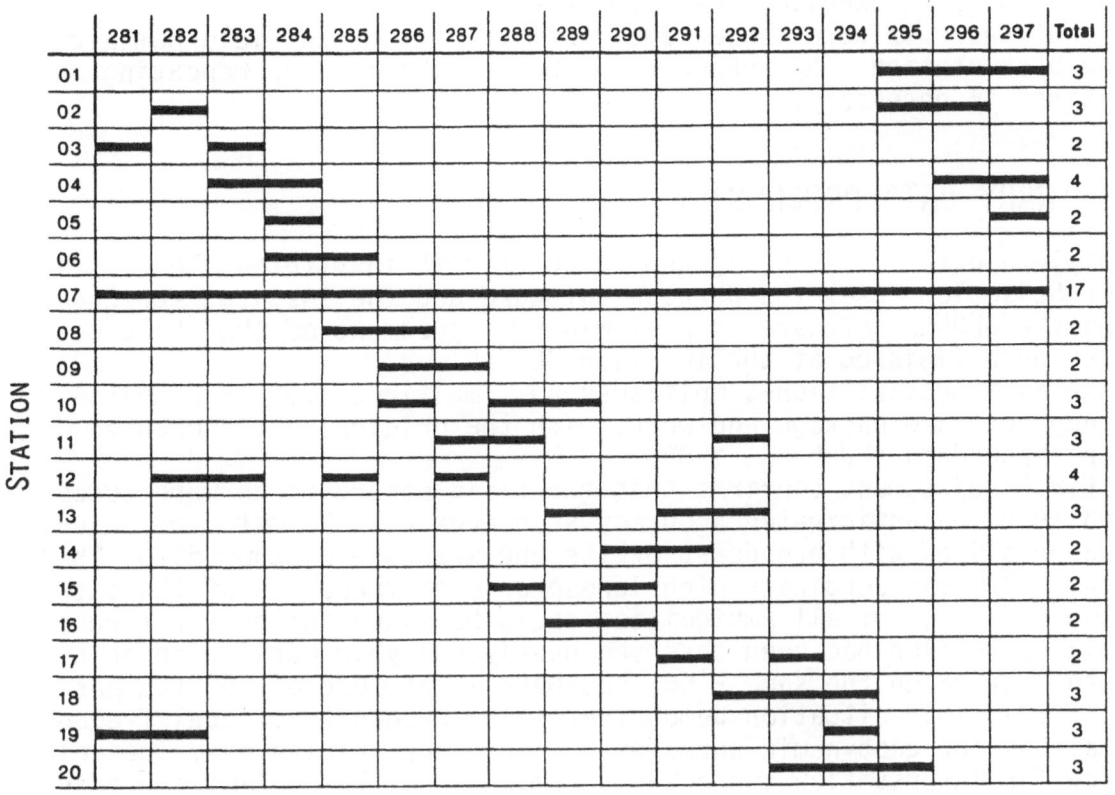

Fig. 4. Observation schedule

Three roving receivers were operated in the field. They usually formed triangles which were related to the permanent station. All simultaneously observed stations,without the central station, are connected with lines in Figure 3. Some lines have been observed two times. The distances between adjacent stations is from 20 to 100 km. The average distance is 50 to 60 km. Via the central station, it was possible to create a high degree of redundancy in the network and to increase the reliability. Most stations were occupied two or three times. The observation plan is shown in Fig. 4.

The network design and station selection was carried out by the Institute of Seismology. 11 from 20 stations were already pre-existing stations as triangulation-, levelling- or gravimetric points. 9 new points were established; the demarcation followed first order triangulation standards.

As a total it was tried to establish the network with equal spacing, good accessibility and free horizontal sight. This was not always possible because of the topography and the traffic conditions. Since Honghe fracture zone is a major zone of earthquake occurance, it was tried to establish control points on both sides of the fracture zone in order to monitor the relative motion of the blocks.

The test area turned out to be rather difficult from the logistical point of view. Since adjacent stations were not always connected by direct roads, rather long transportation ways were necessary and evoked some loss of observation time. Caused by these difficulties, only one session could be observed per day. In order to compensate reduced observation periods on individual stations, the general daily observation span was 4 hours, thus providing an excellent data coverage.

Most of the data were copied and verified during the field campaign in the base station. This check gave the possibility for repeating observations if necessary.

PRELIMINARY DATA PROCESSING

The complete data set is being processed at both institutes. The Institute of Seismology uses the Bernese software and the Institut für Erdmessung the GEONAP software. A preliminary adjustment of the whole network has been finished at the IFE with GEONAP.

GEONAP is a multistation-, multisession-, multireceiver- and multifrequency software package and works with the original undifferenced carrier phase data (Wübbena, 1989).

The first adjustment confirms that the field observations were successful and that the anticipated accuracy standards will be met. The adjustment was realized with broadcast orbits and standard weather data. The residuals show extraordinary high ionospheric disturbances of the data. As a consequence, not all ambiguities could be fixed in the first runs. A similar situation had been observed nearly one year earlier in an area of South America on the same - but southern - latitude of 25° (Campos etal., 1989). This situation demonstrates the necessity of dual frequency observations for geodynamic purposes in near equatorial areas. The relative accuracy for interstation connections coming from the preliminary adjustment is 0.3...1.0 ppm. After recovery of all ambiguities, the 1 cm accuracy level for all connections between control points is expected.

FURTHER PLANS

Deformation fields can only be recovered after re-measurements of the whole network. A second epoch GPS survey is planned for the end of 1990 or beginning of 1991. With an expected signal of 1 cm/year average motion, the resolution of the GPS could provide first indications. However, since a major earthquake occured near the observation area only a few weeks after the campaign, and with seismic activity going on, the expectation for considerable deformation signals is high and encourages a re-measurement in the near future. For spring 1990 first epoch measurements of absolute and relative gravity are planned, thus giving a basis for a further detailed study of the interrelation of regional deformation - gravity fields, fault motion and earthquakes.

REFERENCES

Campos, M., Seeber. G. and Wübbena, G. (1989). Positioning with GPS in Brazil. Proc. 5th Int.Geod.Symp. Satellite Positioning, 526-535.
Wübbena, G. (1989). The GPS Adjustment Software Package - GEONAP - Concepts and Models. Proc. 5th Int.Geod.Symp. Satellite Positioning, 452-461.
Lai, X. and Shao, Z.(1985). Testing Ground for Earthquake Prediction and its Testing Plan for Research into Crustal Deformation in Dianxi. Institute of Seismology, State Seismological Bureau, Wuhan.

TANGO:TRANSATLANTIC GPS NET FOR GEODYNAMICS AND OCEANOGRAPHY

H. Landau and G. W. Hein
Institute of Astronomical and Physical Geodesy
University FAF Munich, 8014 Neubiberg, F. R. Germany

M. L. Bastos and J. P. Osório
Observatório Astronómico, Centro de Astronomia
Universidade do Porto, Monte da Virgem, 4400 Vila Nova de Gaia, Portugal

ABSTRACT

The TANGO GPS campaign, carried out from November 25 to December 5, 1988, was the first step towards the realization of a project whose main goals are: (i) Definition of a unified 3D-datum of the Portuguese and the European network and connection to the American continent via the Azores and Bermuda islands by using GPS and satellite altimetry; (ii) Establishment of a high precision network to monitor the Kinematics of the Azores volcanic region, located in the area of convergence of the North American, African and Eurasian plates (Azores triple junction), and to assess GPS potentialities to measure continental drift.

For the measurement of epoch zero 10 TI 4100 GPS receivers were used at selected stations and tide gauges in Portugal, Spain, French Caribbean and on the Bermuda island. Through international cooperation, connection to the CIGNET network was established by pre-optimized observation schedule and selected satellite tracking for the purpose of precise orbit determination using phase data.

The paper presents the detailed description of the measurements and discusses the methodology adopted in the processing of the data. The specific problems due to intercontinental size of the network (non-intervisibility) and the bad satellite geometry at the Azores islands are outlined. Preliminary results are reported.

INTRODUCTION

The TANGO network, established with stations in four main plates, Eurasian, American, Caribbean and African, aims to assess GPS capabilities in Geodesy and Geodynamics both in a regional and global scale. The project has two main goals: Contribution to the definition of a unified 3D - datum between Europe and America via the Azores and Bermuda islands using GPS and satellite altimetry; establishment of a high precision network to monitor tectonic movements associated with plate boundaries and test the potential of GPS to measure continental drift.

Because of its geodynamic features the region of the Azores triple junction is worth special reference. The archipelago is located in the area of convergence of the American, the Eurasian

155

and the African plates (Fig. 1). The origin of its nine islands seems to be related with the volcanic activity associated with plate boundaries. The islands are spread in a NW-SE direction from latitude 37° to 40° N and longitude 228° to 235° E, with the Mid-Atlantic Ridge (MAR) located between the western and central group of islands.

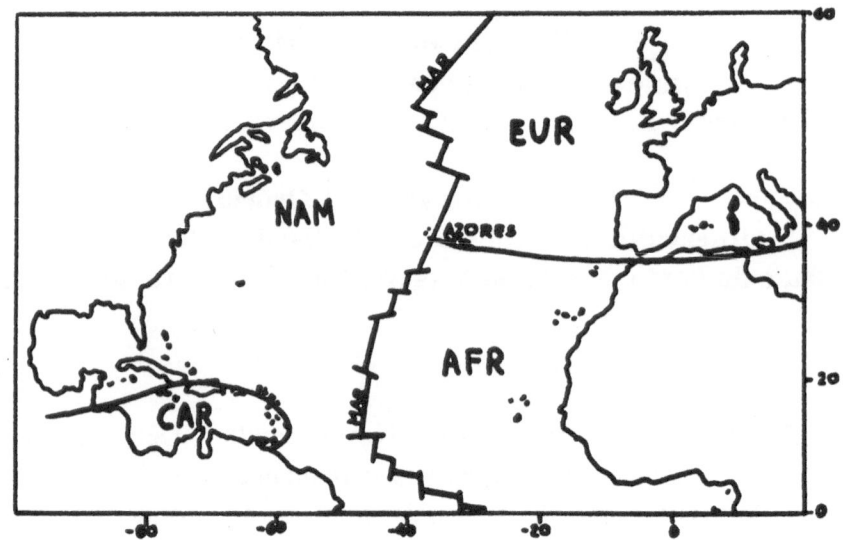

Fig. 1. Main tectonic plates in the North Atlantic area.

The Azores-Gibraltar plate boundary marks the western limit of the Alpine belt. To understand its dynamics is important for the development of an adequate model for the relative motion between Africa and Eurasia in this region of the Atlantic. Furthermore the geodynamic evolution of the Portuguese mainland and islands, where important seismic and volcanic activity occurs, is determined by the behaviour of that boundary.

Several models have been proposed (Krause et al. 1970; Machado et al. 1972; Ribeiro 1982, Hirn et al. 1980, Grimison et al. 1988) but, due to the complexity of the associated system of faults and lack of precise measurements, no conclusive solution has been reached yet.

This paper presents preliminary results of the first GPS campaign carried out from November 25 to December 5, 1988. Due to the large amount of data envolved the reduction is beeing done in steps, starting with subsets of the whole network. Here we discuss the results of the local Azores network and the connection of the Caribbean stations to Bermuda and to the American continent. In this processing only broadcast ephemeris were used but because of the high accuracy demands (0.1 to 0.01 ppm in baselines up to thousands kilometers) a GPS orbit determination will be performed based on the fiducial point concept and using the observations from the CIGNET stations.

CAMPAIGN DESIGN

The TANGO campaign was carried out within ten days in fall 1988 occupying all stations of interest on the European continent (three receivers, 7 stations with two tide gauges), on Madeira (one receiver), in the French Caribbean (one receiver, two stations), on Bermuda (one receiver) and on the Azores islands (four receivers, 12 stations including three tide gauges). Besides the mobile sites, which were occupied for about two to five days, 4 permanent sites were observing in Madrid (VLBI site), Madeira, Terceira (Azores) and Bermuda.

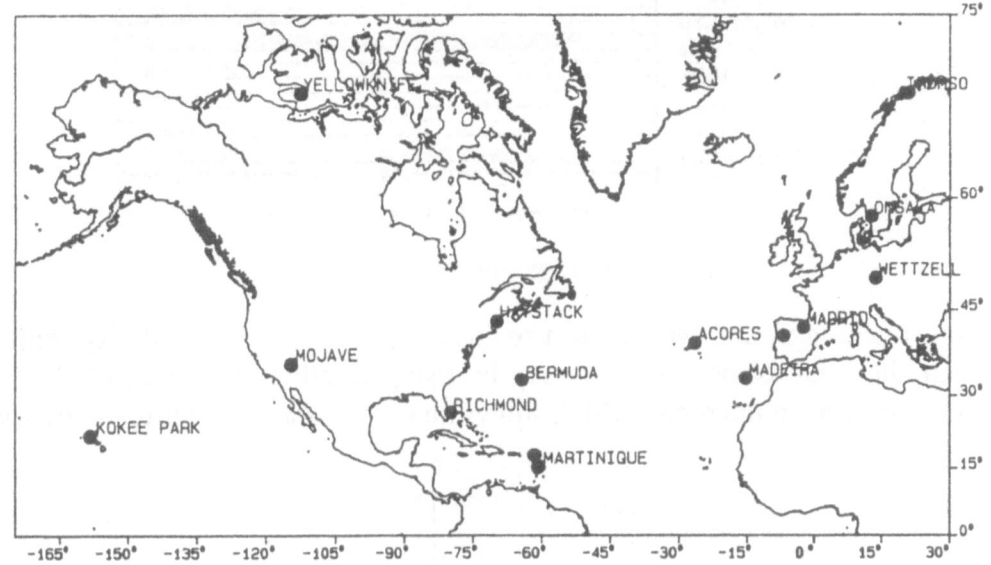

Fig. 2. Location of participating stations during TANGO campaign.

The station occupation is summarized in Fig. 3 for the 10 observation days.

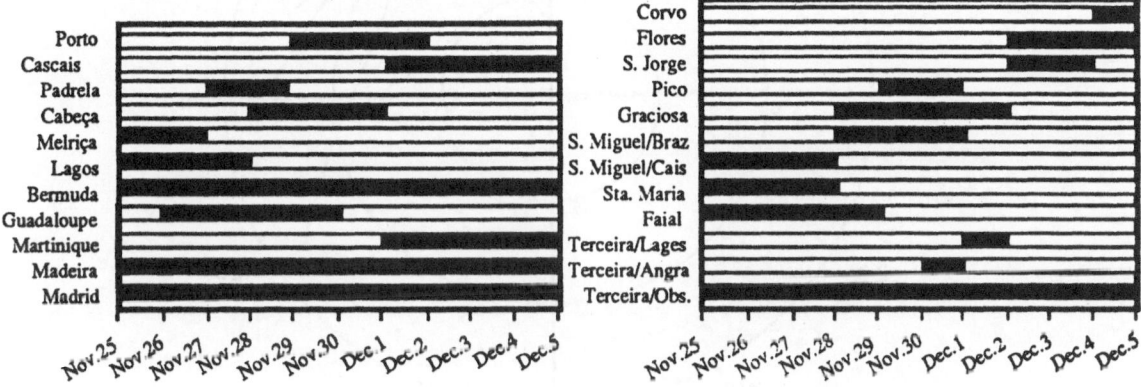

Fig. 3a,b. Station occupation of TANGO campaign.

Observation periods concerning the Azores sites for each day are given in Fig. 4 showing that two different types of observation plans were used. This was caused by the fact that two different operating systems were used (TI Navigator and GESAR for two receivers each). The Navigator software was sampling data with 3-second interval and GESAR was programmed to use a 15-second interval. Due to the inability of the Navigator software to record data with less than three satellites and the small sampling interval we decided to use these types only for 2.5 hours starting at about midnight, whereas the GESAR receivers were operating from about 2.00 pm to 2.30 am.

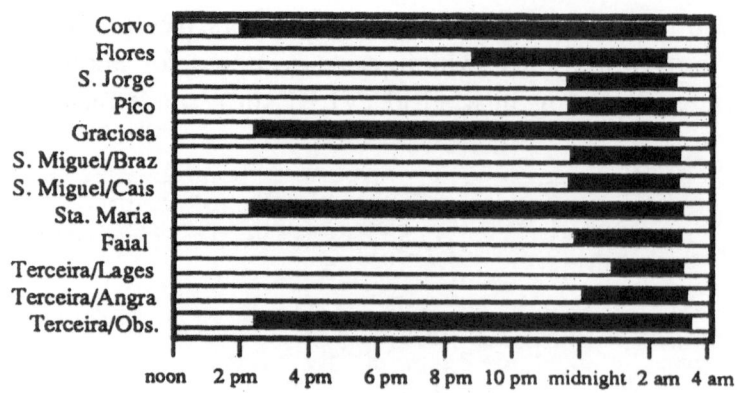

Fig. 4. Observation time spans.

This long tracking period was chosen to make use of the intervisibility with the North American sites and to increase accuracy by using satellites in a part of the hemisphere different to the one in the night, which shows an unfavourable satellite geometry (Fig. 5).

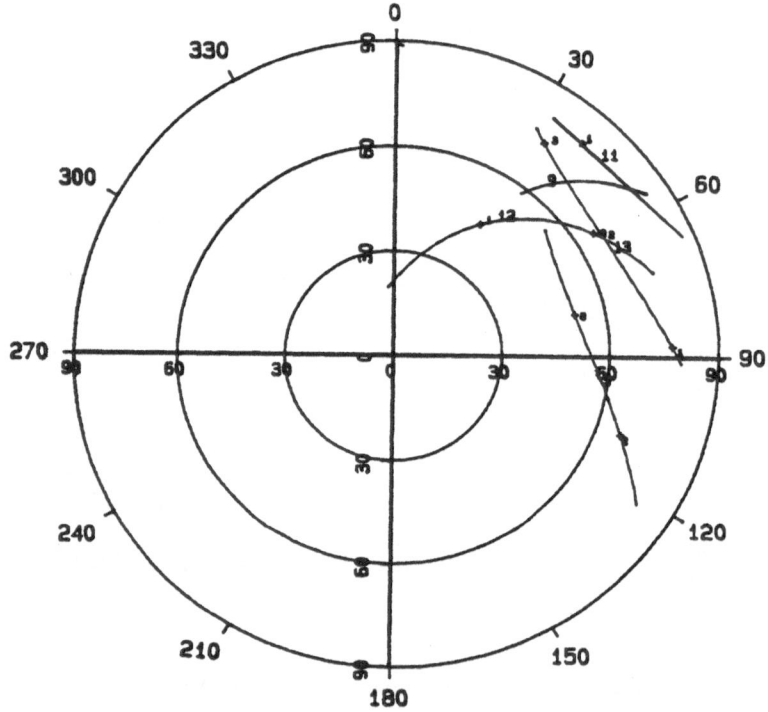

Fig. 5. Satellite geometry for 4-satellite configuration.

The GDOP value for the two four-satellite configurations were larger than 30 for the first 45 minutes (first configuration) and between 6 and 11 for the next 45 minutes (second configuration) resulting in a total of about 1.5 hours four-satellite tracking time period. The site occupation was chosen in such a way that long baselines were observed with GESAR software allowing long tracking periods. TI Navigator software was then employed for short baselines in some kind of collocation with the GESAR receivers.

The data tapes and field logs were sent to the IAPG (Institute of Astronomical and Physical

158

Geodesy, University FAF Munich), where the data were decoded, preprocessed, and archived.

DATA ANALYSIS

Estimation model

In order to reach accuracies that can support geophysical research, in a regional or global scale, with GPS, a careful modelling of all error sources is necessary. In our data reduction process we are applying the TOPAS software (Landau, 1989). TOPAS is a multistation multisession software for GPS undifferenced code and carrier phase data, using an extended kalman filtering in UD-mechanization, which allows recursive estimation of all the parameters considered (Landau, 1988).

Of special interest in the processing of the Azores network was the possibility of fixing the ambiguities. The baselines in that network range from 1 km to 300 km and because of the bad satellite geometry and the short observation time span with four satellite , it was not possible to reach the desired accuracy without fixing the ambiguities.

The technique used in the the data reduction is based on wide/narrow laning in combination with P-code data. In a first step the wide-lane ambiguity is computed from dual-frequency phase data and pseudo-range data via the relation

$$N_{LW} = \psi_{L1} - \psi_{L2} + (\Delta t_{L1} f_1 + \Delta t_{L2} f_2) \cdot \frac{f_1 - f_2}{f_1 + f_2};$$ (1)

with

N_{Lw}	... wide-lane ambiguity,
ψ_{L1}, ψ_{L2}	... carrier phase data in two frequencies,
$\Delta t_{L1}, \Delta t_{L2}$	... transit times in two frequencies from code-phases,
f_1, f_2	... carrier frequencies.

In the next step the observation equations are linearized with respect to the narrow-lane ambiguity N_{Ln} in the double difference phase introduced in the estimation model by implicit differencing. The ionosphere-free double-difference carrier-phase observable can be described by

$$\nabla \Delta \psi_{Lc} = \frac{f_1 \nabla \Delta (\psi_{Lw} + N_{Lw})}{2(f_1 - f_2)} + \frac{f_1 \nabla \Delta (\psi_{Ln} + N_{Ln})}{2(f_1 + f_2)}.$$ (2)

The partial derivative with respect to the narrow-lane ambiguity is then

$$\frac{\partial (\Delta \nabla \psi_{Lc})}{\partial (\nabla \Delta N_{Ln})} = \frac{f_1}{2(f_1 + f_2)}.$$ (3)

The determination of the narrow-lane ambiguity is performed in a sequential way in a multi-station processing mode.

A different problem we have to face in TANGO is due to the extend of the network which has no intervisibility between the European and American stations. TOPAS is well suited to overcome this problem because it uses carrier and code phase data instead of applying differencing techniques. That has also the advantage of allowing the modelling of oscillator errors which is useful for the orbit determination process.

In the preprocessing of the data a package described in Landau (1988) is used. Of particular importance at this stage is the problem of cycle slip detection and correction. This is done in an automatic mode using different levels of analysis:

- Phase-range combination
- Ionospheric residual
- Raw-phase data analysis
- Double difference

The process has proven to be efficient in most of the data analysed. Nevertheless in the data from some of the CIGNET stations we were faced with an abnormal amount of cycle slips and unsuable tracker related data. As a consequence phase continuity could not allways be restored automatically. Aditional problems in fixing cycle slips occured with data from Yellowknife due to small scale variations in the ionosphere. This difficulty has already been reported by others (Beutler et al., 1988) in stations with similar high latitude.

Preliminary results

The data is being processed in independent modules, starting with the local networks and leading to a final multi-station multi-session solution considering all data and using improved orbits computed from the observations.

We discuss here the first results obtained for the connection between the Azores islands and for the connection between Westford and Martinique and Bermuda.

Connection between the Azores islands. Ten single-day solutions were performed for the Azores network by applying the ambiguity determination technique above and using broadcast ephemeris data only. For this preliminary analysis we were restricting the observational data to the 2.5 hours time span starting around midnight. It turned out that the data on Dec. 2 - 3 on the permanent station Terceira and on Corvo were not usable. Fig. 6 shows the network which was processed.

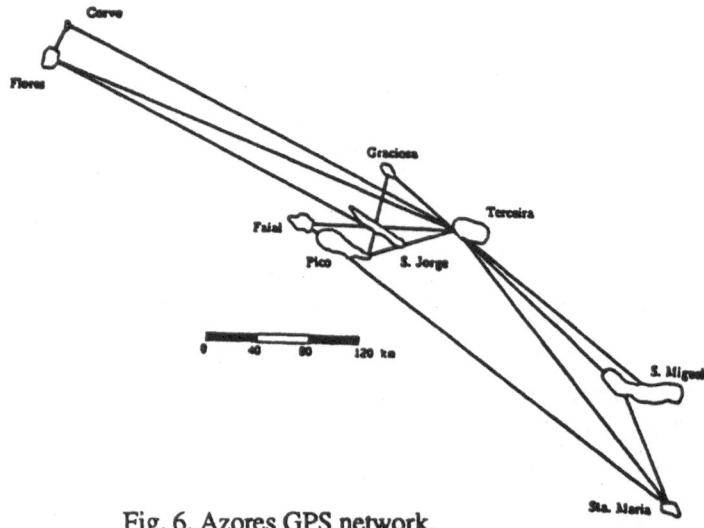

Fig. 6. Azores GPS network.

The single-day solution for ten stations were finally combined in a network adjustment. The standard deviations derived from the adjustment are tabulated in Table 1.

The results for stations S. Jorge, Flores, and Corvo are considerably worse compared to the other stations due to the failure of reference station Terceira on Dec. 2 - 3. The mean observation residuals after the adjustment are 2.1 cm in north, 4.2 cm in east, 5.5 cm in height, and 3.4 cm in length. This confirms the expected positioning accuracy derived in a simulation study (Bastos et al. 1988).

Table 1. Standard deviations for adjusted positions in centimeters (Terceira was fixed).

Station	σ_{North}	σ_{East}	σ_{Height}
Terceira / Observatory	-	-	-
Terceirs / Angra	0.4	0.9	0.9
Terceira / Lages	0.7	2.6	1.8
S. Miguel / S. Braz	2.0	5.1	5.4
Pico / Porta da Ilha	1.8	5.0	5.0
Corvo	7.3	14.8	13.1
Flores	7.2	14.4	13.1
S. Miguel / Cais	1.6	4.5	4.8
S. Jorge	7.2	15.0	14.3
Sta. Maria	1.5	4.2	4.6
Graciosa	1.1	3.4	3.3
Faial	1.4	3.9	3.7

Connection Westford-Bermuda-Martinique. For the baseline determination between Bermuda and Westford five single day solutions were performed whereas for the baseline Westford-Martinique only three days are available. In these cases the full data set was used corresponding to twelve or more observation hours.

Table 2. Preliminary results with broadcast ephemeris.

Baseline from Westford to	Day	Distance [m]	Scale Difference to day 338 [ppm]
Bermuda	332	No data in Westford	
"	333	1282210.764	-0.3432
"	334	Sat 6 unhealthy	
"	335	1282211.089	-0.0897
"	336	11.023	-0.1410
"	337	10.764	-0.3432
"	338	11.204	-
Martinique	336	3213913.809	-0.0772
"	337	12.900	-0.3581
"	338	14.051	-
After applying scale factor derived from Westford - Bermuda			
Martinique	336	3213914.256	-0.0640
"	337	14.002	-0.0152
"	338	14.051	-

As can be seen from analysis of Table 2, a repeatability of some parts to 10^{-7} was reached. The small scale differences obtained prove the good quality of the broadcast orbits during the campaign.

CONCLUSIONS AND FUTURE WORK

The first experiments with broadcast orbit show promising results and we can expect accuracies in the range of some parts to 10^{-8} on baselines reaching 3000 km, using an improved orbit.

In the specific case of the Azores network it was shown that, with most unfavourable conditions, a repeatability of 3 cm in baseline length can be obtained in baselines up to 300 km, with ambiguity determination. This can still be improved by using the full data set for the stations operated with the GESAR software. We also expect better results for stations S. Jorge, Flores and Corvo by using the new measurements obtained for those stations in June 1989.

The next step will be to perform an orbit determination using the full data set which includes all the CIGNET stations in Europe and North America.

Local networks with stations in the Portuguese mainland and islands will be process for the purpose of datum unification.

The repeatability of Bermuda, Caribbean, Azores and Madeira will be analysed within a global framework.

The reobservation of the whole network is expected to be accomplished within the next two years, whereas a gravimetric campaingn is already scheduled for Spring 1990 to establish absolute stations in the Portuguese mainland and islands. This information together with satellite altimetry will be combined in the future adjustment of the TANGO network.

Acknowledgment. The realization of this project was possible through the efforts of international cooperation envolving the following organizations:

- Alfred-Wegener-Institut für Polar-und Meeresforschung, Bremerhaven,
- Azores University,
- Danish Geodetic Institute, Copenhagen,
- Institut für Angewandte Geodäsie, Frankfurt,
- Institut für Weltraumforschung, Graz,
- Institut Géographique National, Paris,
- Instituto Geográfico e Cadastral, Lisboa,
- Instituto de Investigação Científica Tropical, Lisboa,
- Instituto Nacional de Meteorologia e Geofísica, Ponta Delgada,
- Technical University Munich, Institut für Astronomische und Physikalische Geodäsie,
- University FAF, Institute of Astronomical and Physical Geodesy,
- University FAF, Institut für Geodäsie,
- University of Hannover, Geodetic Institute,
- University of Madrid, Instituto de Astronomia e Geodesia,
- University of Maine, Department of Surveying Engineering,

- University of Porto, Observatório Astronómico e Centro de Astronomia,
- U. S. National Geodetic Survey, Rockville MD.

The authors would like to express their warmest thanks to all this agencies and institutions who have kindly supported our campaingn by making available equipment and/or staff or were engaged with the logistics.

We further thank the efficient collaboration of the Portuguese Air Force in making the connecting flights between the Azores islands.

This project was possible due to a grant from the Portuguese Junta Nacional de Investigação Científica e Tecnológica.

REFERENCES

Beutler, G., Gurtner, W., Hugentobler, U., Rothacher, M., Schildknecht, T. and Wild, U. (1988). Ionosphere and GPS Processing Techniques, Presented at the *Chapman Conference on GPS Measurements for Geodynamics*, Ft. Lauderdale, FL, 19-23 September.

Bastos, L., Osorio, J., Landau, H. and Hein, G. (1988). Transoceanic Connection between Azores Island, Madeira, and the European Continent for Geodetic and Geodynamic Purposes - A Covariance Analysis Study, Presented at the *Chapman Conference on GPS Measurements for Geodynamics*, Ft. Lauderdale, FL, 19-23 September.

Grimison, N., and Chen, W., (1988). Source mechanisms of four recent earthquakes along the Azores-Gibraltar plate boundary, *Geophysical Journal* 29, p. 391-401.

Hirn, A., Haessier, H., Trong, P., Wittlinger, G., Victor, L. (1980). Aftershock sequence of the January 1st, 1980, earthquake and present-day tectonic in the Azores, *Geophysical Research Letters* 7, p. 501-504.

Krause, D. and Watkins, N. (1970). North Atlantic Crustal Genesis in the Vicinity of the Azores, *Geophysical Journal* 19, p. 261-283.

Landau, H. (1988). Zur Nutzung des Global Positioning Systems in Geodäsie und Geodynamik: Modellbildung, Software-Entwicklung und Analyse, Schriftenreihe, *Studienermessunqswesen*, 36, Universität der Bundeswehr München, Neubiberg.

Landau, H. (1989). An Analysis of Alternatives in Modelling GPS Observational Data and Satellite Orbits, Presented at the *Fifth Symposium on Satellite Positioning*, Las Cruces, 13--17 March.

Machado, F., Quintino, J. and Monteiro, J. (1972). Geology of the Azores and the Mid--Atlantic Rift, *Proceedings of the 24th International Geology Congress, Montreal*, p. 134--142.

Ribeiro, A. (1982). Tectónica de Placas: Aplicação à Sismotectónica e à Evolução da Fronteiras de Placas Açores - Gibraltar, *Geonovas*, 4, p.87-98.

THE AUSTRALIAN GPS ORBIT DETERMINATION PILOT PROJECT: A STATUS REPORT

Chris Rizos
School of Surveying, University of N.S.W.,
Sydney, N.S.W. 2033, AUSTRALIA

Ramesh Govind
Australian Survey and Land Information Group,
Dept. of Administrative Services,
A.C.T. 2616, AUSTRALIA

Art Stolz
School of Surveying, University of N.S.W.,
Sydney, N.S.W. 2033, AUSTRALIA

BACKGROUND

In mid 1987 the Division of National Mapping (now amalgamated into the Australian Survey and Land Information Group (AUSLIG) of the Department of Administrative Services), and the School of Surveying of the University of New South Wales (UNSW) jointly initiated and embarked upon an Australian GPS Orbit Determination Pilot Project (Rizos et al., 1987). An observation campaign involving 5 temporary tracking stations was commenced on the 13th August, 1987, and dual-frequency TI4100 receivers were used to track 6 GPS satellites, for 7 days.

Aims of the Project

Errors in the GPS satellite ephemeris information can be minimised through the use of differential positioning techniques. An adequate rule-of-thumb for estimating the effect of orbit error on the relative position of two receivers is that a given error in the satellite ephemeris introduces an error in the baseline reduced by the ratio of the baseline length to the satellite's altitude (20,000km for GPS). For example, a 20m orbit error will result in a baseline accuracy of approximately 1 part per million, or 10cm in 100km, while a 2m orbit error results in a part in 10^7 baseline accuracy, or 1cm in 100km, and so on. Consequently an important limitation on the precision of GPS surveys is the accuracy of the ephemerides available for the data reduction.

Up to the present, the accuracy of the Broadcast Ephemeris has been sufficient for positioning applications demanding accuracies of up to a few parts per million. However this ephemeris data is likely to be degraded when the policy of "Selective Availability" is enforced with the launch of the Block II satellites. Furthermore, for higher precision applications such as crustal motion surveys, an orbit adjustment capability is essential for high precision GPS data reduction software. The Australian GPS Orbit Determination Pilot Project was initiated in order to acquire the data necessary to test the feasibility of establishing an independent tracking <u>and</u> orbit determination capability for the Australasian region. In particular, the benefits of this project to both AUSLIG and UNSW were seen to be:

1. Development of the infrastructure necessary for the planning, establishment and

maintenance of a regional satellite tracking network.

2. Develop the expertise, procedures and software to decode, transcribe, preprocess and archive GPS tracking data.

3. Develop, test and refine state-of-the-art multi-station and orbit determination software capable of supporting GPS surveying at the few parts in 10^7 or better.

4. Develop the expertise and strategies necessary for processing GPS data in order to derive the highest possible quality orbits.

The Experiment

A temporary tracking station network well-spaced across the Australian continent, and including a station in New Zealand, was established (Fig. 1, Table 1).

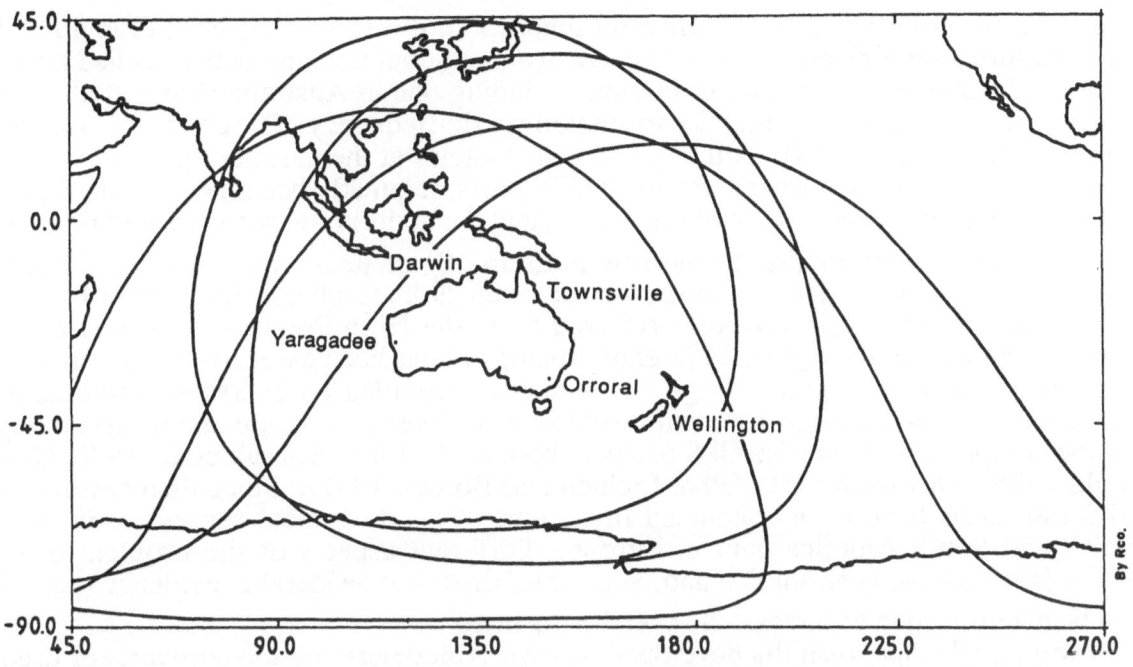

Fig. 1. Locations of tracking stations involved in the GPS Orbit Determination Pilot Project (20° visibility circles are also shown).

Table 1. Distances in kilometres between tracking sites.

	Yaragadee	Townsville	Darwin	Wellington
Orroral	3195.1	1820.6	3114.2	2317.4
Yaragadee		3321.1	2423.4	5306.1
Townsville			1856.3	3557.2
Darwin				5168.6

Each of the five stations were equipped with a dual-frequency TI4100 GPS receiver connected to an external atomic frequency standard. Observations were commenced as soon as four satellites were visible above 20° elevation and continued for as long as at least two satellites were above the minimum elevation of 20°. About 7 hours of continuous tracking data was obtained each day, with a total of approximately 244 hours of data for the entire campaign: 45h28m at Orroral, 49h10m at Yaragadee, 52h48m at Townsville, 49h25m at Darwin, and 47h10m at Wellington. The total observation window was broken up into sessions during which the same constellation of satellites were tracked by all sites. The selection of satellites to be tracked during any one session was influenced by the desire to obtain the best possible overall tracking coverage for the 6 satellites (PRN 3, 6, 8, 11, 12, 13) as they rose and set, within the constraints imposed by the TI4100's capabilities.

With the successful completion of the tracking campaign, the analyses began.

Open Questions in the Analyses

The progress of GPS orbit determination schemes in the 1980's has been dramatic. The idea of estimating high precision GPS orbits from regional tracking networks had been the subject of a number of simulation studies, including one in Australia (Stolz et al., 1984), however it was first put into practice to support single-frequency Macrometer V–1000 GPS surveys (Bock et al., 1984). Although orbit accuracies at the part per million level (20m orbit uncertainty) were adequate for most GPS users, it quickly became apparent that with the use of dual-frequency phase observations, and the appropriate software and processing strategy, orbit determination at the few parts in 10^7 or better was possible. The first contintental-scale GPS field test took place between 29th March and 5th April, 1985, in the U.S. This experiment is variously referred to as the High Precision Baseline test, the March 1985 test, or the Spring 1985 experiment, and has been extensively referenced in a number of conference proceedings. This dataset provided an enormous impetus for a number of academic and government institutions to develop software and refine strategies for the computation of precise GPS orbits (Abbot et al., 1986; Abusali et al., 1986; Beutler et al., 1986a; Davidson et al., 1986; Lichten and Border, 1987). Since then several other GPS field tests have been conducted in support of geodynamical surveys, primarily in North and South America, and in Europe. Different aspects of the problem of orbit determination have been studied and, at present, there is considerable evidence that orbits can be determined to accuracies of several parts in 10^8.

However, whether Australia developed its own orbit determination software, or used an already developed software package, it would still be necessary to test, verify and, perhaps, modify the accepted processing strategies (developed overseas during the last few years) using the tracking dataset acquired during the GPS Orbit Determination Pilot Project. The following factors were identified, at the time, as requiring some investigation (Rizos et al., 1987):

1. The **Fiducial Concept** is fundamental to most orbit computation experiments (e.g. Kellogg et al., 1989). In such a strategy, a network of tracking stations whose locations are accurately known from VLBI define a self-consistent coordinate frame to which all GPS orbits (and baseline) solutions are referred. An alternative to the fiducial approach – the **Free Network** approach – described in Beutler et al. (1986b) is however more appropriate to the Australian experiment.

2. **Clock modelling**: There is little doubt now that the double differenced phase observable, or its equivalent (King et al., 1987), from which the satellite and receiver clock errors are eliminated, is capable of high precision solutions for orbits and baselines (e.g Kellogg et al., 1989; Dong and Bock, 1989). Hence investigating

alternative (explicit) clock error modelling is unlikely to improve solutions.

3. **Orbit force modelling** and **variational computations**: The sensitivity of the orbit solution to variations in the perturbing force models although proposed for study, is now unlikely to bring much benefit. GPS is a relatively "robust" technique, and orbit force modelling to the level of sophistication required for LAGEOS and low-flying satellites appears not to be warranted.

4. **Orbit element modelling**: It appears sufficient to adjust the initial satellite positions and velocities, as well as one or two solar radiation scaling parameters per satellite, depending upon the length of arc adjusted.

5. The analysis of **pseudo-range** and **phase data** together has been advocated, following simulation studies carried out by Wu et al. (1986). Blewitt (1989) has shown that the combination of data is particularly useful for cycle ambiguity (though it is unlikely that ambiguities could be resolved for the long baselines involved in this experiment).

6. **Tropospheric modelling** is an important factor influencing orbit and baseline computations. Almost all investigators estimate a constant zenith tropospheric delay parameter per session per site, though the trend towards dynamic modelling is growing.

7. **Variations in tracking network configuration.**

8. **Variations in arc length**: The original intent was to test satellite arc lengths varying from 1 to 7 days. The evidence suggests that multi-day arc lengths are the most appropriate for highest precision applications.

9. **Variations in data rate**: Data collected during this experiment using the TI4100 receivers consisted of phase and pseudo-range measurements on two L-band frequencies every three seconds. This was a very large quantity of data, and it was decided to immediately cull the data to obtain 30sec sampling.

To study the impact of different data reduction strategies on the accuracy of the GPS orbits it was decided to first perform a covariance analysis study.

COVARIANCE ANALYSIS AND SYSTEMATIC ERRORS

The study of systematic errors in simulated least squares adjustments is known variously as **covariance analysis, sensitivity analysis** or **consider analysis** (Grant, 1989). These names arise because the **cross covariance** matrix between systematic biases and the estimated parameters is used to determine the **sensitivity** of the solution to unadjusted parameters that are **considered** to be biased.

In the case of GPS phase adjustment, systematic biases have origins in the satellite orbits (force models), atmospheric refraction (troposphere and ionosphere) and the reference system (fixed station coordinates, reference frame orientation parameters, etc.). In this report we summarise the results of a covariance analysis in which the effects of reference system uncertainties and length of satellite arc on the initial (adjusted) elements of the GPS satellites are studied. The DASH software, developed at UNSW, was used to simulate a simultaneous satellite orbit and tracking station adjustment using GPS phase data (Grant, 1989).

A Priori Uncertainties of Observations and Adjusted Pararmeters

Tracking Station Coordinates. None of the tracking stations were located at VLBI sites, although the relative position of the Orroral and Yaragadee sites are known to the subdecimetre level from Satellite Laser Ranging (Rizos et al., 1987). The other sites have coordinates derived from TRANSIT Doppler observations, and their relative coordinates are only known at the few metre level. It was decided to initially hold the coordinates of Orroral and Yaragadee fixed and attempt to solve for the coordinates of the remaining tracking stations, in addition to the satellite, GPS bias and atmospheric parameters. We refer to this as Option 1 - the Free Network Option.

Satellite Parameters. The initial elements (position and velocity) of the GPS satellites were adjusted, for various lengths of arc ranging from 1 day to 7 days. The initial elements refer to the starting epoch of the arc.

GPS Biases. As DASH simulates a GPS phase adjustment, cycle ambiguities and explicit satellite and receiver clock errors are included in the solve-for parameter set. No attempt was made to resolve ambiguities to their integer values.

Atmospheric Refraction. Ionospheric delay was assumed to be known without error from dual-frequency observations. Residual tropospheric delay error after the observations have been corrected *a priori* using surface met data, a tropospheric refraction model, or both, was assumed to have a constant zenith magnitude of 0.048m (approximately 2% of full effect).

Measurement Noise. The receiver measurement error (all receiver and antenna errors) was assumed to be 0.004m for 4 minute data.

The Systematic Error Models

In addition to defining observation scenario and specifying the *a priori* uncertainties of the observations and adjusted parameters, *a priori* uncertainties of the unadjusted parameters need also to be specified.

Tracking Station Coordinates. The coordinates of Orroral were held fixed, and considered free from error, hence they define the reference frame origin. The uncertainty of the coordinates of Yaragadee with respect to this origin was considered to be a systematic error. Simulation studies were also carried out in which the tracking station coordinates of Darwin, Townsville and Wellington were assumed known at the 2cm level. This is referred to as Option 2 - the Fiducial Network Option.

Reference Frame. The uncertainty in the coordinates of the geocentre (about which the satellites orbit) in the reference frame defined by the fixed tracking station coordinates was assumed to be an additional source of systematic error. Uncertainties in polar motion, UT1 and GM were also included.

Satellite Perturbation Model. Uncertainties in the gravity field model and from residual solar radiation pressure (it was assumed to be modelled with an uncertainty of 5% of its full effect) were also considered.

Table 2. *A priori* uncertainties of observations, adjusted and unadjusted parameters in the simulation of a combined orbit / network adjustment.

OBSERVATIONS
    4 minute data   0.004m

ADJUSTED PARAMETERS
Receiver Coordinates
    Tracking stations (Option 1)   3.0m (east, north, height)
Orbits
    Initial elements   20m (5m, 12m, 15m HCL)

UNADJUSTED PARAMETERS
Receiver Coordinates
    Origin (Orroral)   0m by definition
    Yaragadee (Option 1)   0.1m (east, north, height)
    Tracking stations (Option 2)   0.02m (east, north, height)
Reference Frame
    Centre of mass   0.5m (x, y, z)
    Polar motion, UT1   0.01 arcsec (0.3m)
    GM   0.02 ppm
Force model
    Gravity field   100% of GEM-9 to (8,8)
    Solar radiation pressure   5% of effect
Troposphere
    Constant zenith delay   0.048m (2% of effect)

The Results

The accuracies of the estimated satellite <u>initial</u> positions assuming tracking Option 1 (that is, adjust coordinates of Darwin, Townsville and Wellington) are shown in Fig. 2, for various arc lengths. The total position error is the RSS ("root sum squared") of the individual errors, for each position component.

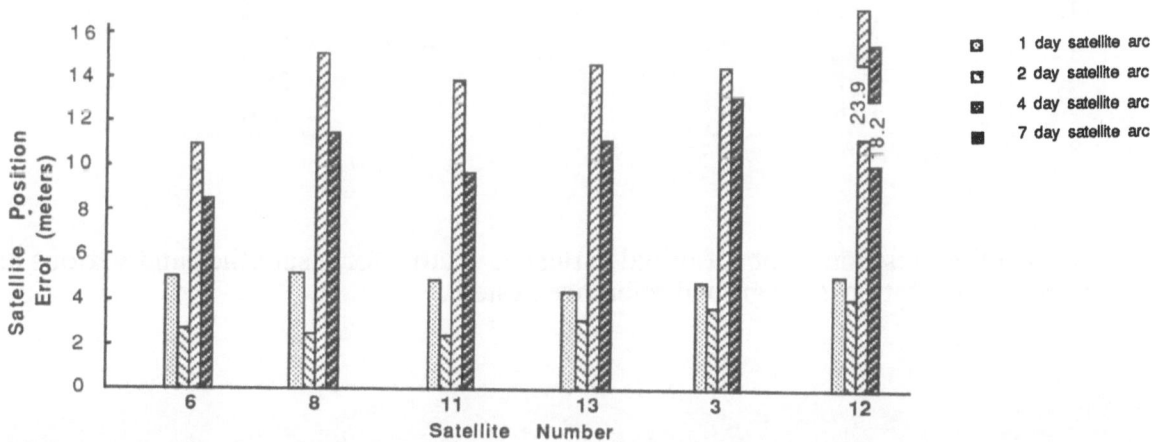

Fig.2. Satellite position error (of initial cartesian vector) for 6 satellites and various arc lengths assuming Option 1 for ground reference system.

A breakdown of the error sources is given in Fig. 3 for satellite 13 (a typical example), for various arc lengths. The "Fiducial Station" contribution is from Yaragadee's position uncertainty relative to Orroral. The "Origin" contribution is the centre-of-mass uncertainty.

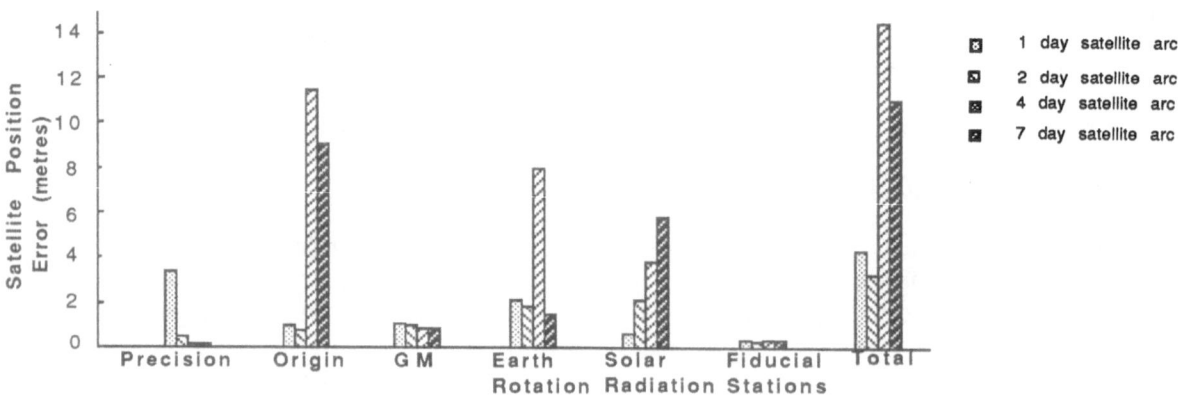

Fig. 3. Components of position error for satellite PRN13 assuming Option 1 for ground reference system.

The accuracies of the estimated satellite initial positions assuming tracking Option 2 (that is, all tracking stations held fixed) are shown in Fig. 4, for various arc lengths. A breakdown of the error sources is given in Fig. 5 for satellite 13, for various arc lengths.

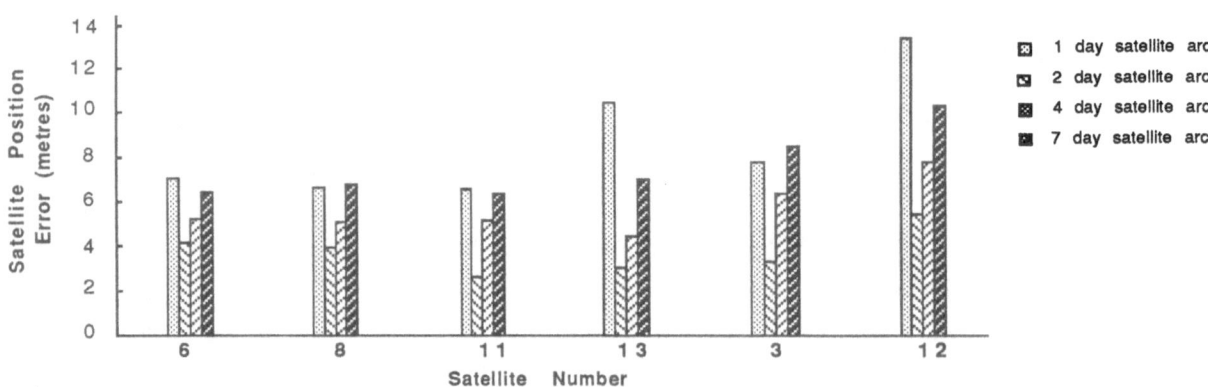

Fig.4. Satellite position error (of initial cartesian vector) for 6 satellites and various arc lengths assuming Option 2 for ground reference system.

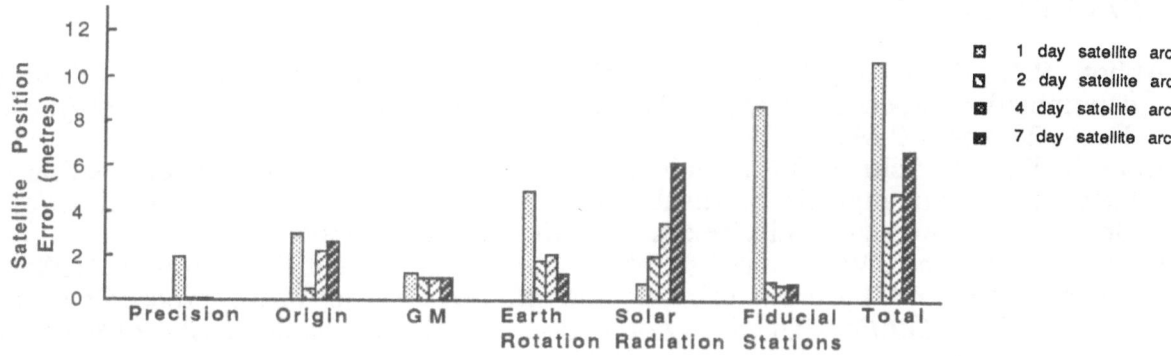

Fig. 5. Components of position error for satellite PRN13 assuming Option 2 for ground reference system.

Comment. The effect of gravity field model errors and residual tropospheric errors were relatively minor. The reference frame effects are not necessarily "errors" if the orbits are computed consistently within this frame. The magnitude of the uncertainty of each individual systematic error source given in Table 2 can be scaled up or down and a new RSS total error obtained. For example, the 0.5m uncertainty in the geocentric coordinates of Orroral may be too pessimistic, and can be changed to, for example, 0.1m, and new RSS values obtained for Figs. 2 to 5. Note that the accuracy of the <u>initial</u> satellite position is obtained, not the average accuracy through the arc (some error would tend to grow with time since start of arc, while others may be constant or even diminish).

CONCLUSIONS

As expected, the satellite position accuracy appears to improve with increasing length of arc. Analysis of data over 7 days appears to satisfy the requirements for orbit accuracy at the several parts in 10^7 level. The dominant error source appears to be the uncertainty in the fixed tracking station coordinates. Certainly 2cm accuracy with respect to Orroral is optimistic at present. However with an increase in the frequency of high precision GPS campaigns in the S.W. Pacific and Asian region, it should be possible in the near future to connect the 5 tracking stations to each other, and to the already well established Fiducial Network in North America and Europe, at the several centimetre level.

The analysis of the tracking data has been completed using the University of Berne Secong Generation Software System (Gurtner et al., 1985). In early 1990 these results will be verified using the GPS multi-station/orbit determination software developed at UNSW (Rizos and Stolz, 1988). The completion of this phase would be a significant milestone, and would have demonstrated the feasibility of a satellite tracking and orbit determination capability in which a major component could be provided by Australian institutions.

REFERENCES

Abbot, R.I., Bock, Y., Counselman, C.C. and King, R.W. (1986). GPS orbit determination, proceedings *4th Int.Geod.Symp.Sat.Pos.*, University of Texas at Austin, Texas, 28 April - 2 May, 271-274.

Abusali, P.A.M., Schutz, B.E., Tapley, B.D. and Ho, C.S. (1986). Determination of GPS orbits and analysis of results, proceedings, *4th Int.Geod.Symp.Sat.Pos.*, University of Texas at Austin, Texas, 28 April - 2 May, 355-364.

Beutler, G., Gurtner, W., Rothacher, M., Schildknecht, T. and Bauersima, I. (1986a). Determination of GPS orbits using double difference carrier phase, proceedings, *4th Int.Geod.Symp.Sat.Pos.*, University of Texas at Austin, Texas, 28 April - 2 May, 319-336.

Beutler, G., Gurtner, W., Rothacher, M., Schildknecht, T. and Bauersima, I. (1986b). Evaluation of the March 1985 high precision baseline (HPBL) test: fiducial point concept versus free network solutions, *EOS Trans. AGU*, **67**, 911.

Blewitt, G. (1989). Carrier phase ambiguity resolution for the Global Positioning System applied to geodetic baselines up to 2000km, *J. Geophys. Res.*, in press.

Bock, Y., Abbot, R.I., Counselman, C.C., Gourevitch, S.A., King, R.W. and Paradis, A.R. (1984). Geodetic accuracy of the MACROMETER model V-1000, *Bull.Géod.*, **58**, 211-221.

Davidson, J.M., Thornton, C.L., Stephens, S.A., Border, J.S., Sovers, O.J. and Lichten, S.M. (1986). Improved application of the fiducial concept for GPS-based geodesy, *EOS Trans. AGU*, **67**, 262.

Dong, D. and Bock, Y. (1989). GPS network analysis with phase ambiguity resolution applied to crustal deformation studies in California, *J. Geophys. Res.*, **94(B4)**, 3949-3966.

Grant, D.B. (1989). Combination of terrestrial and GPS data for earth deformation studies in New Zealand, UNISURV **S-32**, School of Surveying, University of New South Wales, Australia, in press.

Gurtner, W., Beutler, G., Bauersima, I. and Schildknecht, T. (1985). Evaluation of GPS carrier difference observations: the Bernese second generation software package, proceedings, *1st Int. Symp. on Precise Pos. with GPS*, Rockville, Md., 15-19 April, 363-372.

Kellogg, J.N., Dixon, T.H. and Neilan, R.E. (1989). CASA Central and South America GPS geodesy, *EOS Trans. AGU*, **70**, 649.

King, R.W., Masters, E.G., Rizos, C., Stolz, A. and Collins, J. (1987). *Surveying with GPS*, Dümmler Verlag, Berlin.

Lichten, S.M. and Border, J.S. (1987). Strategies for high-precision GPS orbit determination, *J.Geophys.Res.*, **92**, 12751–12762.

Rizos, C., Govind, R., Stolz, A. and Luck, J.M. (1987). The Australian GPS orbit determination pilot project, *Aust.J.Geod.Photogram.Surv.*, **46 & 47**, 17-40.

Rizos, C. and Stolz, A. (1988). The UNSW satellite measurement analysis software system (USMASS), *Aust.J.Geod.Photogram.Surv.*, **48**, 1-28.

Stolz, A., Masters, E.G. and Rizos, C., (1984). Determination of GPS satellite orbits for geodesy in Australia, *Aust.J.Geod.Photogram.Surv.*, **40**, 41-52.

Wu, S.C., Lichten, S.M., Bertiger, W.I., Wu, J.T., Border, J.S., Williams, B.G. and Yunck, T.P. (1986). Precise orbit determination of GPS and LANDSAT-5, proceedings, *4th Int.Geod.Symp.Sat.Pos.*, University of Texas at Austin, Texas, 28 April - 2 May, 275-288.

A GPS SURVEY
IN THE NORTH–EAST VOLCANIC ZONE OF ICELAND 1987
FIRST RESULTS

Cord - Hinrich Jahn
Günter Seeber
Institut für Erdmessung (IfE)
D-3000 Hannover 1, Nienburger Str.6,   Federal Republic of Germany

Gillian Foulger
Dept. of Geological Sciences
Durham DH1 3LE,   United Kingdom

Axel Björnsson
ORKUSTOFNUN
Reykjavik,   Iceland

INTRODUCTION

Iceland lies on the mid - Atlantic ridge at a latitude where the full spreading rate is 2.2 cm / year, which was estimated from seafloor anomalies south of Iceland (Talwani and Eldholm, 1977). The ridge on land is comprised of transform zones and the so-called neovolcanic zone that is made up of numerous volcanic systems. The first geodetic measurements in the neovolcanic zone were carried out in 1938 by Niemczyk (Möller, 1989) in north-east Iceland. Subsequently a classical geodetic network with an extent of 110 km in the east-west direction has been developed, spanning the main part of the rift zone.

Since 1964 new points have been added to this network and the configuration substantially improved. Regular remeasurements of this network have been made (Möller, 1989). After 2 centuries of inactivity, a major rifting episode started in 1975 in the centre of this network. Displacements of 1 - 8 m occurred in the Krafla volcanic system, and of a few dm at distances of 50 km from the plate boundary.

In 1986 the first GPS campaign of Iceland was carried out in the south - west and north part of the island by a large, international consortium (Foulger, 1987a). Points spaced at 50 km on average were observed in two short sessions (morning and evening window).

In 1987 a further international GPS project was carried out in the north - east and south - east of Iceland. This survey covered the rift zone, included many points of the pre-existing terrestrial control network, and was tied to points distant from the rift zone, in tertiary rocks to the east and west. The distances varied from a few km to more than 100 km in the stable plate interior. This contribution documents the 1987 GPS project.

GROUPS PARTICIPATING IN THE 1987 GPS PROJECT

In 1986 several independant groups from different countries planned GPS measurements in Iceland. In order to coordinate all groups, and guarantee optimal results of GPS measurements the *Icelandic GPS Coordinating Committee* was established by the

- Iceland Geodetic Survey, Reykjavik, Iceland,
- National Energy Authority, Reykjavik, Iceland,
- Nordic Volcanological Institute, Reykjavik, Iceland,
- Science Institute, University of Iceland.

In addition to these institutions, the participants of the 1987 campaign included the *Dept. Geological Sciences*, University of Durham, U.K. (Foulger, 1987b) and the *Institut für Erdmessung*, University of Hannover, Fed. Rep. of Germany (Jahn and Seeber, 1989).

THE GOALS OF THE ICELAND 1987 SURVEY

The principal objective of the survey was to study crustal deformation along a substantial portion of the subareal accretionary plate boundary, and to relate the results to tectonic processes. The 1987 survey established a first epoch measurement of the network and created a precise three dimensional survey. After remeasurement of this network, the combination of the two epochs in a three dimensional deformation analysis will yield three dimensional crustal strainfield data.

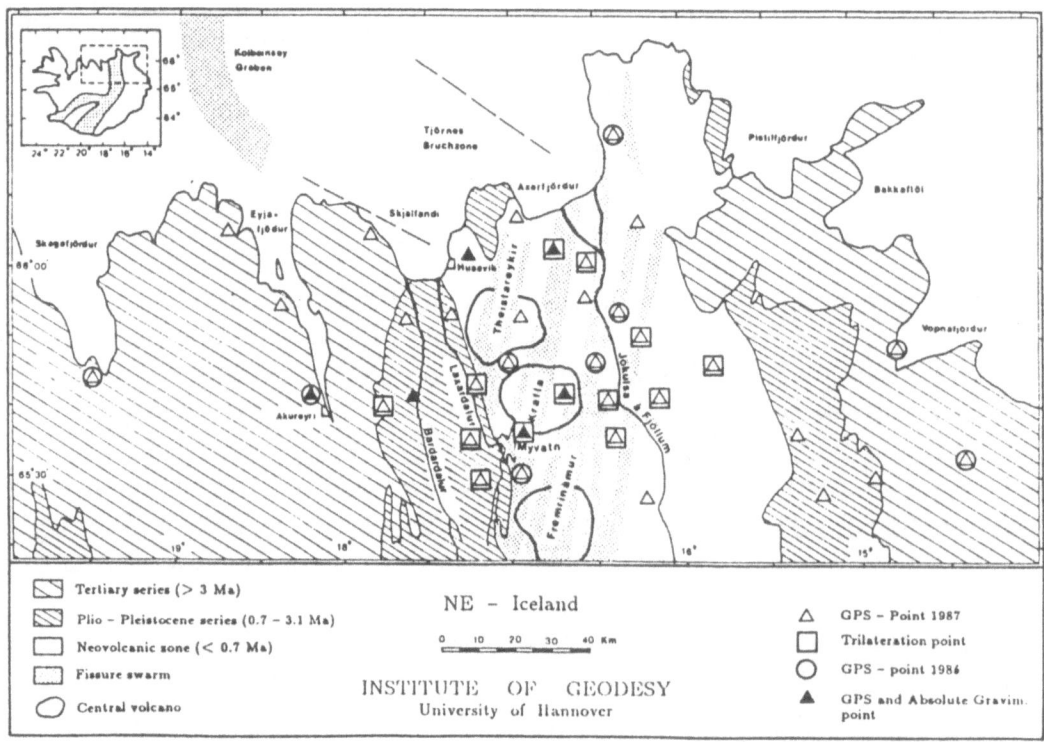

Fig. 1. Tectonic structure and geodetic points in North-East Iceland

The network covers an area of 300 * 280 km and contains 63 points. It is structured such that the accretionary zone is densely covered, with particular emphasis on the Krafla and Askja central volcanoes. More widely spaced points to the east and west of the accretionary zone extend the network up to about 150 km into intraplate regions.

Figure 1 shows the northern part of the network and illustrates the structure of the survey. Twelve points of the pre-existing terrestrial network of the University of Braunschweig, FRG were included in the GPS network. That network encompasses the Krafla volcanic system with point spacings of 15 to 40 km. Parts of it were repeatedly remeasured during the 1975 - 1985 spreading episode in the Krafla region (Wendt et al., 1985).

Twelve points of the 1986 GPS survey were included in the northern part of the 1987 GPS network. These points were distributed throughout the network and can be used for control, and a preliminary deformation analysis. In the southern part, west of the Vatnajokull icecap, two of the 1986 GPS points were reobserved.

Six of the GPS points coincide with absolute gravity stations which were measured with a transportable absolute gravity-meter immediately prior to the GPS measurements (Torge, 1989). The GPS results will provide high accuracy coordinates for these stations. Within and flanking the Askja volcanic system a dense network of 17 GPS points was measured in four sessions.

Two measurement sessions around the Vatnajokull glacier were performed in order to connect the 1986 GPS campaign in south Iceland to the 1987 survey. Several points on the east and south-east coast were included. These stations belong to the first order GPS network, planned for the whole country.

THE FIELDWORK

The network was measured during the period 2nd - 24th August 1987, using 7 TI4100 receivers operating TI4100 software. Four of these were provided by the Universität Hannover, and three by the University of Durham. The recording medium used was Verbatim tape cassettes.

The observation plan was split into three parts. The first part comprised the Northern Volcanic Zone, including the Krafla area. The second part comprised a two day session around the Vatnajokull icecap, and the third part comprised the network around the Askja central volcano.

A single, early morning observation window of 3 hrs 20 mins, involving 6 satellites, was used every day. Figure 2 shows the network configuration. Point No. 17 was the permanent station where measurements were made throughout the northern survey (2nd - 16th). This station is near Skutustadir, at lake Myvatn.

During the 17th - 18th the two Vatnajokull sessions were observed. Because of the long drive times around the icecap no observations were planned for the 19th, and all groups returned to the base. During the period 20th to 23rd the Askja network was measured in four sessions. On the last observation day, the 24th, three receivers reoccupied a small triangle in the Askja area while four receivers reoccupied a quadrilateral in the northern network.

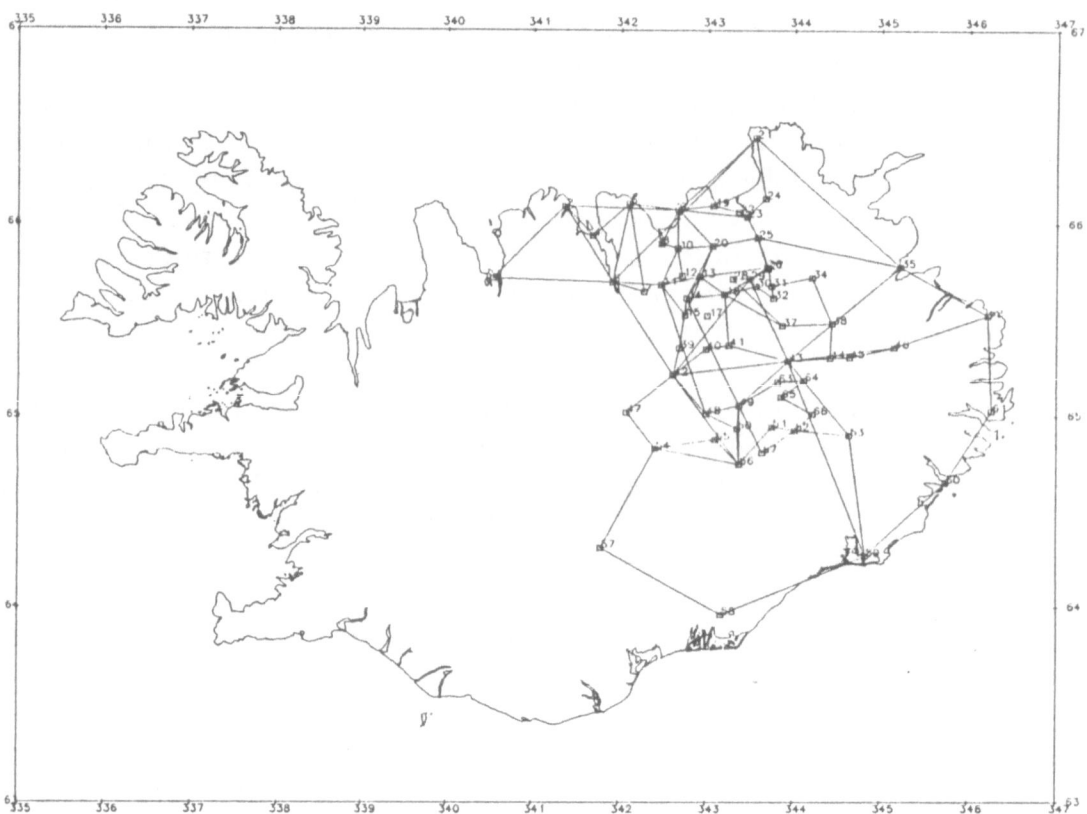

Fig. 2. The 1987 GPS survey network.

Field vehicles and auxiliary equipment were provided by the participating groups. The German group transported three VW buses via ferry to Iceland. One vehicle was provided by University of Durham, and all other vehicles were either rented by the participating groups or provided by the Iceland GPS committee.

The IfE provided a HP 200 field processing center with capabilities for copying and checking raw data and computing satellite predictions. The Durham group used the $MAGNET^{TM}$ software for preprocessing on an IBM PC, and compared these preliminary results with geodimeter measurements available for some lines of the network. The main data processing was done in Hannover with the software package GEONAP.

The individual polygons measured each day were joined to each other via two common field stations and the permanent station. Thus in total each polygon was connected to the next via three points. This arrangement was used in order to achieve sufficient redundancy. Independent control through terrestrial measurements are available for a few lines in the inner part of the network. Because of the large number of new GPS points it is advisable to have a high degree of redundancy in order to recover ambiguities, and as a safety net in case of bad data.

176

TECTONIC SETTING OF ICELAND AND THE NORTHERN VOLCANIC ZONE

Iceland is the 100,000 square km subaerial crown of a massive basalt dome that rises from the seafloor in the north Atlantic. It lies astride the mid-Atlantic ridge, and it's existence is due to the presence of a large hotspot beneath the spreading plate boundary. This has resulted in the production of basaltic magma far in excess of that required to enable the 2 cm per year of crustal spreading that occurs at these latitudes (Talwani and Eldholm, 1977). The surplus magma has been erupted onto the surface of the seafloor over the last 25 Ma or so, and built up a massive basalt dome that projects up above sea level.

Structurally, Iceland is characterized by a crust that is greatly thicker than that of oceanic areas, and a geophysically anomalous mantle. Not surprisingly, the morphology of the plate boundary on land in Iceland contrasts with that of the sea floor. The accretionary plate boundary is identified as the so-called neovolcanic zone, which is the zone where post-glacial lavas are seen (younger than 10,000 years). This zone crosses Iceland from the SW to the NE, and is up to 50 km wide. A closer look at its detailed structure shows that it may be subdivided into discrete volcanic systems, most of which are comprised of a fissure/rift system, a central volcano possibly with a shallow magmatic system, and high temperature geothermal resources. In addition to this accretionary zone, two fracture zones occur, which are analogous to sea-floor transform zones, but much more complex in their sub-aerial form.

The Northern Volcanic Zone, which is the primary target of the work described in this contribution, is a 200 km long portion of the accretionary plate boundary. It is comprised of 5 en-echelon volcanic systems, three of which, Krafla, Askja and Kverkfjoll, contain very active central volcanoes (Saemundsson, 1978). It terminates in the north where it joins with the Tjornes Fracture Zone. Recently, volcanic and tectonic activity has been very intense in the Krafla and Askja systems, which were particularly densely surveyed in 1987.

During the period 1975-1985 a magma chamber beneath the Krafla volcano continually inflated at a rate of approximately 5 cu m per sec., and at intervals of a few months, magma burst from it and flowed along the rift zone to the north and south of the volcano. During the first few years, the magma formed dykes in the crust, and the fissure zone widened by up to 8 m in its central zone. Differential vertical movements were up to 1.5 m. During the later years of activity, the fissure swarm could no longer accomodate further dykes, and the magma formed surface lava flows (Björnsson, 1985 and 1989).

The Askja volcanic system lies parallel to the Krafla system, and contains a 150 km long fissure zone, and a spectacular central volcano comprised of at least 4 calderas. It lies only 30 km from Kverkfjoll, the proposed centre of the Iceland hotspot. Askja was active in 1875, when a 500 m wide explosion crater was formed, spreading volcanic ash as far afield as Scandinavia. Intense subsidence occurred in the main caldera, and a new crater lake 100 m deep formed, which indicates the minimum vertical movements that must have occurred in this episode. Lava was also erupted in the fissure swarm, indicating that the tectonic mechanism may have been similar to that of Krafla 1975-1985 (Sigurdsson and Sparks, 1978). Activity also ocurred in the Askja volcano in 1921-1929 and 1961. More recently crustal movements have been detected by repeated levelling (E. Tryggvason, unpublished data).

SOFTWARE AND DATA PROCESSING

Different software packages for static positioning with GPS have been developed at the IfE since 1983. The software package being used for the 1987 Iceland project is the GEONAP (**GEO**detic **NAVSTAR P**ositioning) software. The basic concept of this package is the simultaneous adjustment of undifferenced GPS observables in a multi-station, multi-session, multi-receiver and multi-frequency mode. The parameter estimation concept is used rather than the parameter elimination approach. Selectable parameters are, for example, the tropospheric scaling parameter, satellite and receiver clock parameters, satellite orbits and coordinates.

The solution of ambiguities is done by code- and carrier phase combinations, by ionospheric modelling and/or by "narrow-", "wide-" or "extra wide laning" techniques (Wübbena, 1989).

After copying the raw data, preprocessing began with the decoding of the data. The original receiver measurements were compressed to normal points. This procedure has the advantages of facilitating the detection of bad data sets, and saving computer storage space. These compressed data files were used as input to the main program.

The network computations were done session by session. After finishing processing all sessions, the whole network was processed in a free network adjustment. The final output of the GEONAP software package is a set of three dimensional coordinates and a complete covariance matrix.

One of the problems of the Iceland 1987 project was created by SV 9. This satellite was set unhealthy three days before starting the project and set healthy again only two days before finishing the survey (!). It was therefore possible to use SV 9 only for the last two days of the project. On all other days SV 8 was used instead, which had an inaccurate clock. However, with appropriate clock modelling, this satellite was used without any problems.

FIRST RESULTS OF PROCESSING

Figure 2 shows the entire 1987 network (independent lines are drawn for illustation). The first data processing was performed in three parts (Krafla-, Vatnajokull- and Askja network). Table 1 shows the standard deviation of coordinate differences and the repeatability for some lines of the Krafla network, resulting from the free network adjustment. Most standard deviations are of the order of 0 to 10 mm for the north and the east components and from 0 to 20 mm for the height component. The values for distances correspond to those of the north components. The standard deviations are mostly independent of the distances thus given very homogeneous results over the whole area. For all parts of processing standard GEONAP routines were used (broadcast ephemeris without orbit modelling; no estimation of troposperic parameters).

Table 1. Standard deviation and repeatability of coordinate differences of the Krafla network

Line	S [km]	North [m]	East [m]	Up [m]	Distance [m]	North [m]	East [m]	Up [m]	Distance [m]
17⇒09	64	0.005	0.006	0.012	0.004	-0.003	0.011	-0.018	-0.007
17⇒11	30	0.006	0.006	0.011	0.006	-0.001	-0.007	-0.013	0.005
17⇒13	23	0.005	0.005	0.009	0.003	-0.004	-0.004	-0.001	-0.003
17⇒18	15	0.004	0.004	0.007	0.004	0.003	0.003	0.019	0.001
17⇒29	31	0.004	0.005	0.009	0.005	-0.001	0.002	0.017	0.002
17⇒31	38	0.005	0.005	0.010	0.006	-0.005	0.000	-0.008	-0.002
17⇒35	106	0.006	0.006	0.012	0.007	0.009	-0.011	-0.021	-0.009
17⇒42	39	0.005	0.006	0.011	0.005	0.002	0.003	0.023	-0.003
17⇒43	50	0.008	0.008	0.014	0.008	-0.007	-0.016	0.012	-0.009
05⇒06	47	0.007	0.008	0.016	0.006	0.011	-0.007	-0.022	0.018
09⇒23	36	0.009	0.011	0.016	0.010	0.017	0.002	-0.004	-0.003
27⇒31	10	0.005	0.006	0.011	0.004	-0.008	-0.004	-0.002	-0.007
35⇒38	48	0.006	0.007	0.013	0.007	-0.001	-0.001	-0.010	-0.004
01⇒35	215	0.011	0.011	0.022	0.012	—	—	—	—
01⇒62	263	0.014	0.015	0.026	0.016	—	—	—	—
21⇒42	144	0.018	0.014	0.019	0.020	—	—	—	—

The right-hand column of Table 1 show the repeatability of coordinate differences processed by overestimated lines. In most cases two observations are used, but in the case of line 17⇒18 for example the repeatability comes from five observations. In the

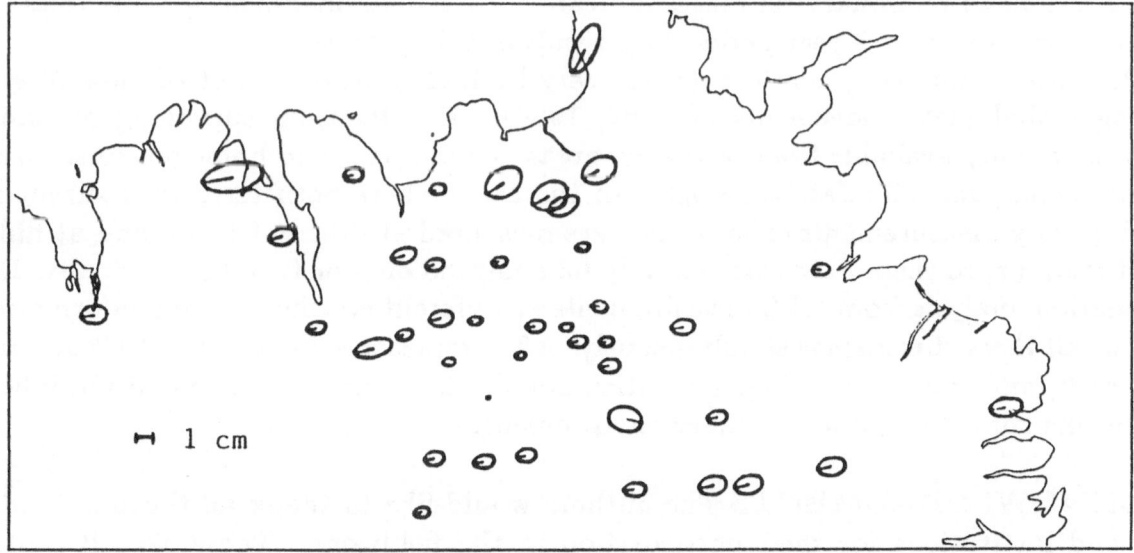

Fig. 3. Error ellipses for the 1987 Iceland north part network.

second part of the network, because of a bad data set, two sessions (day 229 and 230) yielded a larger standard deviation than in the northern part. However, the accuracy requirement for this part of the network was only of the order of a few cm (it was measured to create a GPS first order network) and this was achieved. The third part of the network, the Askja network, yielded similar results to the first part. All survey precision goals were therefore met.

A good overview of the accuracy of the coordinate differences and the stability of the whole network can be shown by error ellipses. Different ways of looking are possible, i.e. absolute or relative ellipses. Figure 3 contains the relative error ellipses of the GPS points in the Krafla network. The diameters of the ellipses for points in the inner part of the network and also for the points in tertiary rocks are in the order of 1 - 2 cm. The orientation of the ellipses is mainly in the east / west direction. This reflects the satellite geometry and the high stability in the north / south direction.

FUTURE GPS MEASUREMENTS IN ICELAND AND CONCLUSIONS

The primary goal of this work is neotectonic studies. This is only possible when a deformation field is determined by remeasurement. A reoccupation of the 1987 GPS network is planned for 1990, again within the framework of international cooperation. What is the outlook today for a significant deformation signal after a period of 3 years? With spreading rates of 2.2 cm/year and a GPS network accuracy of $\pm$ 1 - 2 cm, deformations should be detectable after 3 years if spreading within the rift zone equals the time averaged rate. Short term deformation signals could also result from earthquakes. The earthquake activity in the Northern Volcanic Zone has been at a low level since the 1987 survey and only a few events greater than magnitude 3 have occured (Foulger, 1989). Significant deformation from this source is therefore unlikely at the time of writing. Crustal movements may also have ocurred as a result of viscous rebound, resulting from the very fast, 8 m extension that occurred within the rift zone. Preliminary estimates indicate that displacements of up to several cm could occur within a 3 year period as a result of this process.

Presently available geodetic data are very limited from the point of view of studying global plate movements (Jacoby, 1988). Results from surveying measurements are only available from restricted areas. Furthermore no homogeneous, three-dimensional, geodetic data set exists. In the cases where both terrestrial surveying and gravity measurements exist, data were measured at different times and at different stations, so these data sets can only be analyzed one- or two-dimensionally. Deformation analysis from GPS measurements at different epochs can give information about all three dimensions simultaneously. After remeasurement of the 1987 Iceland network, and data processing, we anticipate the first three-dimensional strainfield from this spreading plate boundary environment.

ACKNOWLEDGEMENTS The authors would like to thank all those individuals and institutions for their participation in the fieldwork. These are all contributers from the Icelandic GPS Committee, staff and students of the participating organisations. Jon Hemingway, Geophysical Services Inc., did a first class job in the field. The University of Durham contribution was funded by NERC grant GR3/6676, NERC/SERC grant GST/02/151 and seed money from University of Durham. The IfE contribution was funded by the Deutsche Forschungsgemeinschaft (DFG, German Research Council grant Se 313/8-1).

REFERENCES

Björnsson, A. (1985). Dynamics of crustal rifting in NE Iceland *J. Geophys. Res.*, **90**, 10,151 - 10,162, 1985.

Björnsson, A. (1989). Crustal rifting in NE Iceland. *Zeitschrift für Vermessungswesen* **114**, 2-9, 1989.

Foulger, G.R. (1987a). The Iceland GPS Geodetic Field Campaign 1986. *EOS Trans. AGU* **68**, 1809-1818, 1987.

Foulger, G.R. (1987b). A GPS geodetic survey of the Northern Volcanic Zone of Iceland, 1987.(Abstract), *EOS Trans. AGU*, **68**, 1236, 1987.

Foulger, G.R. (1989). Neotectonics of the North Iceland Accretionary Plate Boundary. Interim report to NERC on Grant GR3/6676, 1989.

Jacoby, W.R. (1988). Geodynamics of Iceland studies with the aid of terrestrial geodetic and GPS experiments. In: *GPS-Techniques Applied to Geodesy and Surveying*, Proceedings of the International GPS-Workshop, Darmstadt (FRG), April 10-13, 1988.

Jahn, C.-H. and Seeber, G. (1989). GPS measurements in the Northern Volcanic Zone of Iceland - preliminary results.(Abstract), 302. *Annales Geophysicae*. Special Issue to the XIV General Assembly of the Europaen Geophysical Society, Barcelona, March 13-17, 1989.

Möller, D. (1989). Terrestrische geodätische Arbeiten zur Erfassung horizontaler rezenter Oberflächenbewegungen. *Zeitschrift für Vermessungswesen* **114**, 10-25, 1989.

Saemundsson, K. (1978). Fissure swarms and central volcanoes of the neovolcanic zones of Iceland *Geol. J. Spec. Issue*, **10**, 415-432, 1978.

Seeber, G. (1989). GPS-Messungen in der jungvulkanischen Zone NO-Islands. *Zeitschrift für Vermessungswesen* **114** 39-43, 1989.

Sigurdsson, H. and Sparks, R. S. J. (1978). Rifting episode in north Iceland, 1874-1875 and the eruption of Askja and Sveinagja. *Bull. Volc.*, **41**, 149-167, 1978.

Talwani, M. and Eldholm, O. (1977). Evolution of the Norwegian-Greenland sea. *Bull. Geol. Soc. Am.*, **88**, 969-999, 1977.

Torge, W. (1989). Schweremessungen in Nordostisland 1938-1987. *Zeitschrift für Vermessungswesen* **114**, 44-55, 1989.

Wendt, K. and Möller, D. and Ritter, B. (1985). Geodetic measurement of surface deformations during the present rifting episode in NE-Iceland. *J.Geophys.Res.* **90**, 10163-10173.

Wübbena, G. (1989). The GPS Adjustment Software Package - GEONAP - Concepts and models. Fifth International Symposium on Satellite Positioning, Las Cruces, New Mexico, March 13-17, 1989.

MODELLING OF GROUND SUBSIDENCE FROM A COMBINATION OF GPS AND LEVELLING SURVEYS

Adam Chrzanowski
Department of Surveying Engineering
University of New Brunswick
P.O. Box 4400, Fredericton, N.B., E3B 5A3,Canada

Yong-qi Chen
Wuhan Technical University of Surveying and Mapping
39 Lo-Yu Road, Wuhan, People's Republic of China

Julio Leal
Maraven, S.A., Lagunillas, Venezuela

INTRODUCTION

Since 1984, the authors have been involved in ground deformation studies in oil fields along the east coast of Lake Maracaibo in Venezuela. Due to oil extraction, an area of about 50 km x 50 km has been subjected to subsidence which in some places reaches 20 cm/year. Subsidence in the whole area has been monitored since 1929 using conventional geodetic levelling with over 1600 benchmarks connected to some points considered to be outside of the subsidence influence. The levelling network consists of 800 km of primary (main network) and 600 km of second-order densification surveys. The complete survey has been repeated every two years with a portion of the network (about a third of the whole area) remeasured every six months for the purposes of upgrading protective coastal dykes, updating the irrigation system, and controlling the stability of offshore platforms and plants. An evaluation (Leal, 1989) of the last few survey campaigns indicates that the standard deviation of the levelling surveys is about $2 \text{ mm}/\sqrt{\text{km}}$ for the main network and $4 \text{ mm}/\sqrt{\text{km}}$ for the densification survey. The average accuracy of the subsidence determination is about 10 mm at a one sigma level. The levelling survey is a slow, expensive, and labour intensive operation. Several survey crews need about three months to measure the whole subsidence area at a total cost of about US$200 000 per campaign.

Early in 1986, in a search for a more economical monitoring method, the authors suggested replacing the main levelling network and the long connecting lines with a network of GPS measurements combined with the second-order levelling surveys which could bring about 30% savings in time and money. Figure 1 shows the suggested integrated monitoring scheme. A pre-analysis of the combined survey (Leal, 1989) has indicated that in order to achieve the same average accuracy of the subsidence determination as previously, the average standard deviations of the GPS-determined height differences should be smaller than 15 mm.

Before implementing the use of GPS, extensive tests on the achievable accuracy had to be performed, and a mathematical model for the integration of GPS with geodetic levelling had to be developed. The first test measurements were performed in 1986 and 1987 on a test

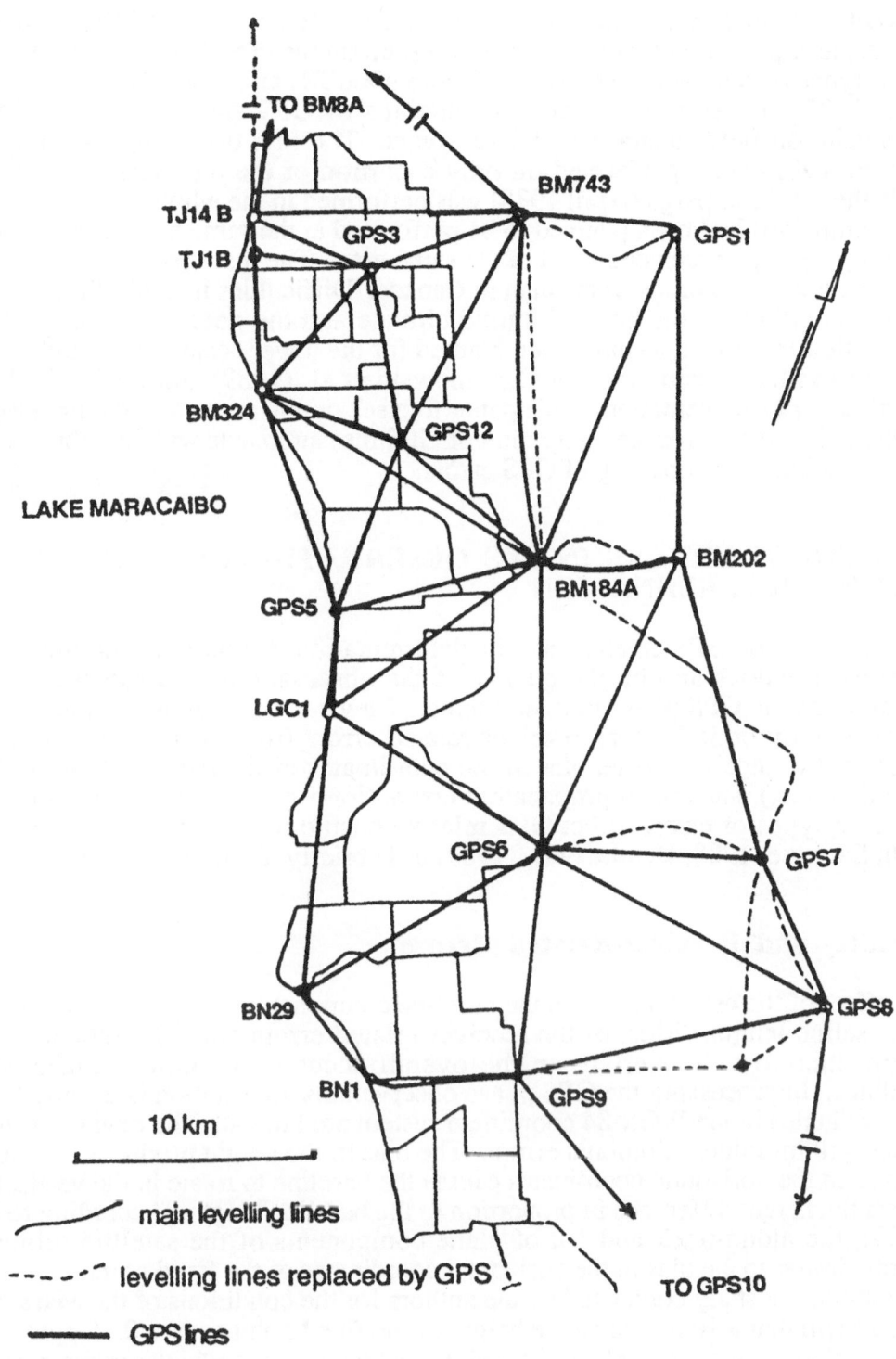

Figure 1. Main levelling and GPS aided subsidence monitoring network

network established by the University of New Brunswick (UNB) near Fredericton, Canada, as a part of a broader research program on the use of GPS in engineering surveys. Three types of receivers (TI 4100, Trimble 4000SX, and WM101) were used. Between April 1987 and April 1988, three campaigns of GPS surveys were performed in the Maracaibo oil fields using WM101 receivers. The first two campaigns (April 1987 and October 1987) were performed on only a portion of the monitoring levelling network, while the third campaign (April 1988) was performed in the whole subsidence area. All the GPS campaigns had been planned to be performed at the same time when the conventional levelling surveys were conducted in the same area. The test surveys in Venezuela gave an accuracy about two times worse than in Canada. Difficulties in controlling the tropospheric refraction effects in the hot and humid climate and the poorer geometry of the satellite distribution in Venezuela have been blamed for the lower accuracy. Detailed results of the test surveys have been given in Chrzanowski et al. (1989) and Leal (1989) and are not repeated in this presentation. This paper focuses on the aspects of the development of the mathematical model for the integration of GPS measurements with levelling surveys and on the evaluation and modelling of GPS errors.

EFFECTS OF ERRORS IN GPS OBSERVATIONS ON RELATIVE HEIGHT DETERMINATION

The accuracy of GPS relative height determination depends on the distribution of the observed satellites and on the quality of the observations. Different sources of errors contaminate the GPS phase measurements. They can be categorized into satellite-related errors (mainly orbital error), receiver-related errors (multipath, variation in the antenna phase centre, receiver noise, bias in the coordinates of the receiver holding fixed in data reduction, etc.), and signal propagation errors (ionospheric and tropospheric refractions). Different types of errors affect GPS relative positioning in different ways (Wells et al., 1986; Santerre, 1988; Beutler et al., 1989) as is briefly discussed below.

Satellite- and Receiver-Related Errors

The effect of the orbital errors on the baseline components is approximately proportional to the baseline length. Most of the receiver-related errors are independent of the baseline length, therefore their effect on the overall accuracy is more significant over short baselines. In processing the GPS phase observations, one station is usually held fixed and its coordinates in the WGS-84 coordinate system are known. The deviation from their true values will introduce additional errors. The bias in the height produces a scale error, while the bias in the horizontal coordinates causes the baseline to rotate in the vertical plane. This affects the height difference in proportion to the baseline length. According to Beutler et al. (1989), the along-track and out of plane components of the satellite orbital error have effects similar to the bias in the horizontal coordinates of the fixed point.

A simulation study conducted by the authors for the conditions of the Maracaibo network have shown that a 10 m bias in the height of the fixed point causes 0.15 ppm rotation in the vertical plane around the axis of azimuth 67° which is perpendicular to the symmetrical axis of the satellite distribution (see plot of the satellite distribution in Fig. 2); and a 10 m bias in the x coordinate introduces another 0.3 ppm rotation in the vertical plane around the y axis.

Refraction Errors

In an ideal situation when the observed satellites are evenly distributed in the sky, the absolute refraction error, defined as that common to both stations of the observed baseline, introduces a scale error which does not affect the height component. Its uncommon part

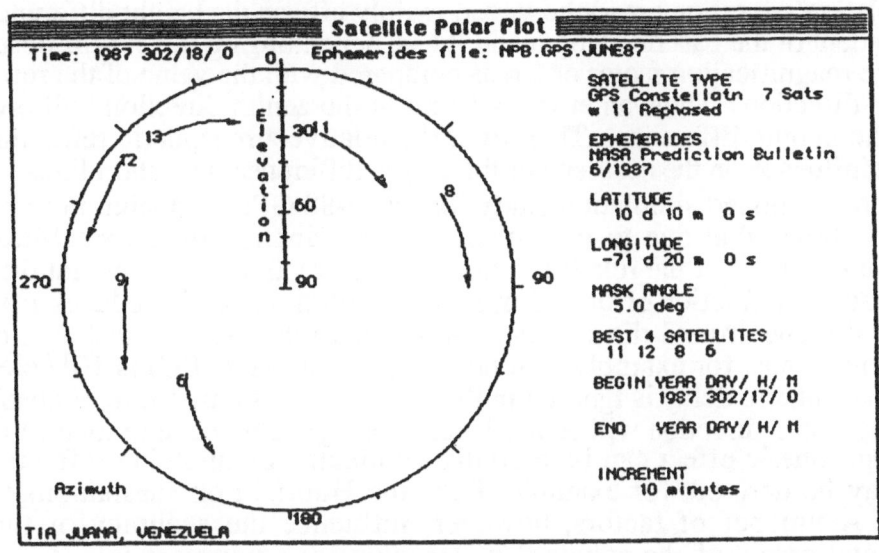

Figure 2. Typical polar plot of satellite distribution during GPS measurements in Venezuela

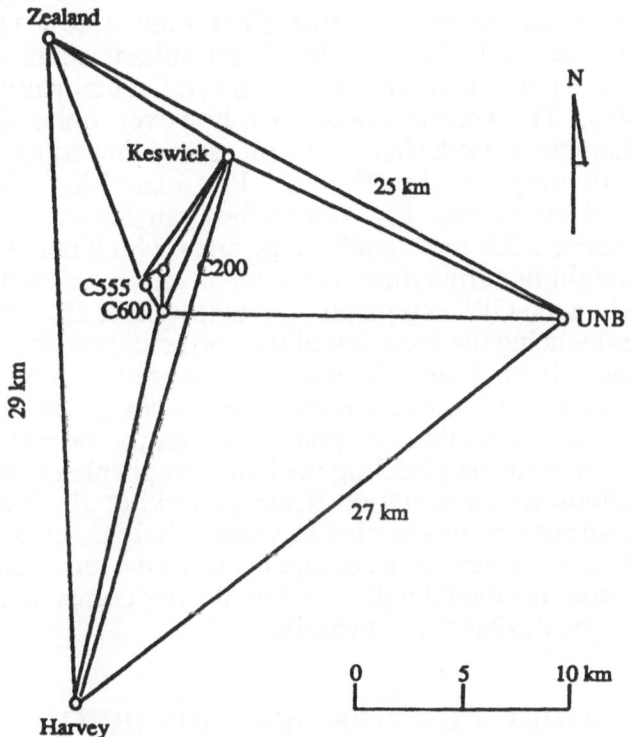

Figure 3. UNB test network

(called relative refraction error), however, mainly affects the height difference. This effect is independent of the baseline length and it has, depending on the cut-off observation angle, an average magnification factor of 2.6 as compared with the value of the refraction error in the zenith direction. Thus, an error of 1 mm in the zenith direction will produce 2.6 mm error in the height difference. Therefore, the relative atmospheric refraction may have a dominant influence on the accuracy of the height difference over short lines.

The aforementioned simulation study for the 1988 GPS campaign in the Maracaibo oil fields has shown that due to the rather poor distribution of the satellites (Fig. 2), the magnification factor of the relative atmospheric refraction for the height difference is 4.1, and an absolute refraction error of 1 m in the zenith direction introduces 1 ppm rotation in the vertical plane around the axis perpendicular to the symmetrical axis of the satellite distribution. Thus, for example, if total electron content (TEC) is 10^{17} electrons per m^2 and the ionospheric delay is ignored in the GPS observations, 1.6 m of absolute refraction error in the zenith direction will be produced causing the baseline to rotate 1.6 ppm.

The tropospheric effect can be partially eliminated by applying refraction corrections which may be derived, for example, from the Hopfield or Saastamoinen tropospheric models. A number of factors, however, influence the accuracy of the corrections. Instrumental errors of the standard meteorological equipment introduce biases in the measured temperature, pressure, and humidity. A more severe problem is that the meteorological conditions measured near the ground surface at the receiver sites are not representative of the atmospheric model above the site due to local micro-climatic effects. In practice, in small diameter networks (less than a few tens of kilometres) it is recommended that some kind of average local atmospheric conditions be used, the same for the entire network, rather than use the observed meteorological values (Chen and Chrzanowski, 1989). The average conditions, however, quite often do not represent the meteorological situations at both sites of the measured line equally well, hence the relative tropospheric errors may be significant. For example, when T=30°C, H=100%, P=1000 mb (typical conditions for Maracaibo), an error of only 1% in the relative humidity will introduce a 3.8 mm zenith range error which may be translated into an error of 11 mm in the height determination in the ideal satellite distribution and 16 mm in the case of the 1988 Maraven GPS campaign (Chrzanowski et al., 1989).

Another factor influencing the accuracy of the corrections is an uncertainty of the accepted tropospheric model. Both Hopfield and Saastamoinen models are based on some assumptions of the vertical profile of the atmospheric conditions. These assumptions prove to be justified for the dry refraction component but not for the wet component. The authors have performed experiments in profiling the lower tropospheric conditions in the Maraven oil fields using balloon instrumentation (Chrzanowski et al., 1989; Pedroza, 1988). The meteorological conditions were sampled at various heights up to 1000 m. A comparison between the refractivity distribution as computed from the recorded meteorological data and the values derived from the model indicates that the dry component matches the model very well, while the wet one deviates significantly.

STOCHASTIC MODEL FOR GPS-DERIVED HEIGHT DIFFERENCES

As one can see, several errors and biases have a significant effect on the GPS-derived height difference, especially when visible satellite distribution is poor. Some effects may be systematic for one baseline but they are randomized over the whole network, while other effects (e.g., bias in the coordinates of the fixed station, long-term orbital errors) have a systematic influence on the whole network.

In modelling ground subsidence, the stochastic model for the observations must be established. Based on the above discussion, the accuracy of the GPS-derived height differences can be generally expressed as:

$$\sigma^2_{\Delta h} = a^2 + b^2 s^2 , \tag{1}$$

where s is the baseline length. The values of a and b depend on the type of receiver, observation environment, satellite geometry, observation time, processing data technique, and so on. They should be estimated from the observations rather than taken directly from the values claimed by manufacturers. The technique of Minimum Norm Quadratic Estimation (MINQE) can be employed if there is sufficient redundancy available.

Let $(l, \mathbf{Ax}, \Sigma)$ be the Gauss-Markoff model for the adjustment of the network of GPS-derived height differences. According to the error model, eqn. (1), the variance matrix of the observations l is constructed as

$$\Sigma = a^2 \mathbf{I} + b^2 \text{diag}\{s_1^2, s_2^2, \ldots, s_n^2\} := \theta_1 \mathbf{T}_1 + \theta_2 \mathbf{T}_2 \tag{2}$$

where s_i is the ith baseline length. Using the MINQE technique (Chen and Chrzanowski, 1985), the unknown parameters θ_1 and θ_2 are estimated from

$$\hat{\underline{\theta}} = (\hat{\theta}_1 \; \hat{\theta}_2)^T = \mathbf{S}^{-1} \mathbf{q} , \tag{3}$$

where the $(i,j)^{th}$ element of matrix $\mathbf{S}$ is

$$s_{ij} = \text{Tr}\{\mathbf{R}(\underline{\theta}^0)\mathbf{T}_i\mathbf{R}(\underline{\theta}^0)\mathbf{T}_j\}, \quad i,j = 1,2$$

and the ith component of vector $\mathbf{q}$ is

$$q_i = l^T \mathbf{R}(\underline{\theta}^0)\mathbf{T}_i\mathbf{R}(\underline{\theta}^0)l$$

and

$$\mathbf{R}(\underline{\theta}^0) = \Sigma(\underline{\theta}^0)^{-1}(\mathbf{I} - \mathbf{A}(\mathbf{A}^T\Sigma(\underline{\theta}^0)^{-1}\mathbf{A})^{-1}\mathbf{A}^T\Sigma(\underline{\theta}^0)^{-1}) ,$$

where $\Sigma(\underline{\theta}^0)$ is evaluated from eqn. (2) with approximate values of $\underline{\theta}^0$. It is clear from eqn. (3) that the parameters $\hat{\underline{\theta}}$ so obtained are locally best estimates. Thus, numerical iterations are performed to ensure uniformly best estimates. Let the estimators from eqn. (3), denoted by $\hat{\underline{\theta}}^{(k)}$ be chosen as the approximate values of $\underline{\theta}$ from the kth iteration, and the MINQE is recomputed. Then, the estimators from the (k+1) iteration are

$$\hat{\underline{\theta}}^{(k+1)} = [\mathbf{S}(\hat{\underline{\theta}}^{(k)})]^{-1} \mathbf{q}(\hat{\underline{\theta}}^{(k)}) . \tag{4}$$

The iteration process continues until the solution for $\underline{\theta}$ converges. The covariance matrix of the final estimated parameters is computed from

$$\mathbf{C}(\hat{\underline{\theta}}) = 2[\mathbf{S}(\hat{\underline{\theta}})]^{-1} , \tag{5}$$

which can be used for testing the significance of the estimated parameters and adopted error model.

As discussed in the previous section, some biases in the GPS observations cause the GPS network to rotate in the vertical plane. One should be aware that this systematic effect will not show in misclosures, and therefore cannot be estimated from the MINQE. As will be shown in the next section, however, this type of systematic error can be modelled and eventually cancelled out in the subsidence studies. Thus, in this sense the values obtained from the MINQE are appropriate.

The MINQE technique has been applied to study the accuracy and the error model of the GPS-derived height differences in the aforementioned test surveys at the UNB and Maracaibo networks. The UNB network (Fig. 3) consists of 7 points with interstation distances ranging from 0.7 km to 30 km. Fifteen baselines were observed with WM101 receivers with the average observation time for each baseline being 1.5 hours. The average accuracy of the GPS-derived height differences for the 1986 and 1987 campaigns was estimated as

$$\sigma^2_{\Delta h} = (7 \text{ mm})^2 + (1.4 \times 10^{-6} \text{ S})^2 ,$$

which, over a distance of 10 km (typical for the Maracaibo network), would correspond to an error of 16 mm.

The MINQE evaluation of the 1988 Maracaibo GPS campaign has led to an unexpected conclusion: the estimated value of the b component of the error model was found to be insignificant. This limited the error model to $\sigma_{\Delta h} = a$, with the estimated value of $a = 29$ mm. The distance independent error model in the Maracaibo network may be explained by the very hot and humid climate in the area of the measurements and by the fact that the network is located along the large body of water of lake Maracaibo with some stations located along the coast and others about 20 km inland. As a result, large horizontal gradients of humidity may be expected. Therefore, the relative tropospheric effect becomes so dominant that the accuracy of the height difference determination becomes independent of the baseline length.

MODELLING GROUND SUBSIDENCE WITH COMBINATION OF GPS AND LEVELLING

The subsidence w of a point $P(x,y)$ at any time t with respect to reference time t_0 can be modelled by

$$w(x,y; t) = \mathbf{b}(x,y;t-t_0)\mathbf{c} , \tag{6}$$

where $\mathbf{b}$ is a raw vector of some selected base functions, and $\mathbf{c}$ is a vector of unknown coefficients to be estimated. The GPS-derived height difference $\Delta h_{ij}(t)$ at time t and levelled height difference $\Delta H_{kl}(t)$, for all k, l, i, j, and t, are related to the deformation model (6) by

$$\Delta h_{ij}(t) + v_{hij}(t) = H_j(t_0) - H_i(t_0) + (N_j - N_i) + [\mathbf{b}(x_j,y_j; t - t_0) - \mathbf{b}(x_i,y_i; t - t_0)]\,\mathbf{c} , \tag{7a}$$

$$\Delta H_{kl}(t) + v_{Hkl}(t) = H_l(t_0) - H_k(t_0) + [\mathbf{b}(x_l,y_l; t - t_0) - \mathbf{b}(x_k,y_k; t - t_0)]\,\mathbf{c} , \tag{7b}$$

where v corresponds to the observation error, $H_i(t_0)$ is the expected orthometric height of point P_i at the reference epoch, and N_i is the geoidal height of the point. Geoidal heights can be calculated from gravity measurements or can be modelled. Let us assume that the geoidal heights N in eqn. (7a) do not significantly change with time and are modelled using a polynomial of a general form:

$$N(x,y) = \mathbf{g}(x,y)\,\mathbf{e} , \tag{8}$$

with $\mathbf{g}$ being a raw vector of selected base functions, and $\mathbf{e}$ the vector of unknown coefficients. For example, in a case of plane fitting,

$$N(x,y) = a_0 + a_1 x + a_2 y .$$

Substituting eqn. (8) into eqn. (7a), one obtains:

$$\Delta h_{ij}(t) + v_{hij}(t) = H_j(t_0) - H_i(t_0) + [\mathbf{g}(x_j,y_j) - \mathbf{g}(x_i,y_i)]\,\mathbf{e} +$$

$$+ [\mathbf{b}(x_j,y_j; t - t_0) - \mathbf{b}(x_i,y_i; t - t_0)]\,\mathbf{c} . \tag{7a'}$$

If the field surveys, GPS measurements or levelling observations, in the area can be conducted in a short time interval (not affected by the subsidence), the network adjustment of the observations should be performed for the purpose of outlier detection and accuracy evaluation. In this case, the adjusted heights and their variance-covariance matrix are entered into the integrated analysis of deformations. The functional relation between the adjusted heights $\hat{h}_i(t)$, for example, ellipsoidal height, and the deformation model is as follows:

$$\hat{h}_i(t) + v_{hi}(t) = H_i(t_0) + N_i + \mathbf{b}(x_i,y_i; t - t_0)\,\mathbf{c} , \tag{9}$$

where N_i can also be modelled with eqn. (8).

If one is interested in single point displacements rather than in the determination of the surface of subsidence and if the vertical movements of the point may be considered as linear in time, then the deformation model for point P_i becomes

$$w(x_i, y_i; t - t_0) = (t - t_0) \dot{H}_i \; , \tag{10}$$

where $\dot{H}_i$ is the subsidence rate of point P_i. Then the deformation model in eqn. (7) is expressed in terms of the rates of subsidence of individual points:

$$[\mathbf{b}(x_j, y_j; t - t_0) - \mathbf{b}(x_i, y_i; t - t_0)]\mathbf{c} = (t - t_0)(\dot{H}_j - \dot{H}_i) \; . \tag{11}$$

As discussed earlier, some systematic errors cause the GPS network to rotate in the vertical plane. The magnitude of the effect depends on the visible satellite distribution and the error nature. Change in the visible satellite distribution and the systematic errors between two GPS campaigns yields additional errors to the subsidences measured with GPS. To accommodate this type of error, one could introduce into eqn. (7a) an additional term as

$$(x_j - x_i)\delta a(t) + (y_j - y_i)\delta b(t) \; , \tag{12}$$

where the unknown coefficients $\delta a(t)$ and $\delta b(t)$ are related to the GPS campaign at epoch t with respect to the initial one.

Summarizing the above discussion and combining all the measurements, one can write the observation equations in matrix form as

$$l + \mathbf{v} = \mathbf{Ax} + \mathbf{Bc} + \mathbf{Ge} + \mathbf{D}\delta\mathbf{a} \; . \tag{13}$$

where $\mathbf{x}$ is the vector of orthometric heights at reference epoch t_0, $\delta\mathbf{a}$ is the vector of unknown coefficients which model the systematic errors (see eqn. (12)), $\mathbf{c}$ and $\mathbf{e}$ have been defined before, and $\mathbf{A}$, $\mathbf{B}$, $\mathbf{G}$, and $\mathbf{D}$ are the corresponding design matrices. Applying the least-squares criterion to eqn. (13), the coefficients $\mathbf{c}$ and $\mathbf{e}$ and their variance-covariance matrix are estimated. The significance of the deformation model and the estimated parameters is then tested. For detailed procedures and method of deformation analysis, the reader is referred to Chrzanowski et al. (1986).

The system (13) cannot always be solved. The problems of solvability need the following additional discussion.

The orthometric heights. To solve for the orthometric heights of the points for the reference epoch in a monitoring network without configuration defects, the height of one point should be given to define the datum. If one is not interested in the heights of the points, they can be eliminated from the normal equations formulated from eqn. (13).

The deformation parameters. Consider the deformation model of the subsidence surface with n_c unknown parameters. It is required that the number of height differences surveyed with GPS or levelling be larger than n_c. They should be surveyed in at least two epochs if the model is linear in time, or in at least $(m+1)$ epochs if the model has a time function of power m. It should be emphasised that the height difference of a line measured with GPS at one epoch and with levelling at another epoch will not contribute to the determination of deformation parameters if the geoidal heights are unknown. If the geoidal heights are solvable (see below), however, the restriction that height differences be measured with the same technique in both epochs can be eliminated. In addition, a constant term should be excluded from the deformation model if no stable reference point exists. For the point velocity model (eqn. (10)), all the points must be connected with GPS or levelling, and one point with zero velocity must be assigned.

The geoidal height. Consider the geoid to be fitted by a model with n_e unknown coefficients. The number of the height differences surveyed with both GPS and levelling should not be smaller than n_e. Both types of measurements can be done at the same epoch or at different epochs. For the latter case, the deformation model should be estimable. One should note that the constant term in eqn. (8) is unsolvable. If the geoidal height of each surveyed point is required, all the points must be connected with both GPS and levelling, and the value N of one point has to be given.

Modelling the systematic error. In order to solve for rotation parameters, the changes in the height differences of at least two lines not lying in the same direction should be monitored with both GPS and levelling.

CONCLUDING REMARKS

The developed methodology has several merits. It takes full advantage of all the survey data available. In addition, the biases in GPS-derived height differences can be filtered out providing an unbiased contribution to the determination of deformation model. The method is very flexible. The parameters for modelling geoidal height and possible systematic changes in the GPS-derived height differences can be statistically tested for their significance.

The technique of MINQE is useful in formulating the stochastic model for GPS observations if there is sufficient redundancy in the observations. Other error models, not only the one given in eqn. (1), also can be evaluated using the technique. One should be aware that the accuracy of GPS-derived height differences depends on many factors and should estimate it, if possible, rather than blindly use a model suggested by others.

The above methodology for modelling the ground subsidence from the combined observations has been adopted by Maraven, S.A., a petroleum company in Venezuela, which is in charge of the subsidence monitoring. Software for the implementation of the method has been developed and implemented (Leal, 1989) using the point velocity model. There are two options for modelling the geoidal heights. One is a polynomial function model and the other is a discrete point model. The orthometric heights of all the benchmarks at a specified reference epoch can also be obtained. Statistical tests on the significance of the subsidences and on the geoidal model are introduced in the software. Further improvements are under way to make the software more universal.

Acknowledgment. The research for this project has been sponsored by Maraven, S.A., Venezuela, by the Natural Sciences and Engineering Research Council of Canada, and partially by the Industrial Research Aided Program (IRAP) of the Canadian National Research Council in cooperation with Usher Canada Limited.

REFERENCES

Beutler, B., I. Bauersima, S. Botton, W. Gurtner, M. Rothacher, and T. Schildknecht (1989). Accuracy and biases in the geodetic application of the Global Positioning System. *Manuscripta Geodaetica*, **14**, No. 1, 28-35.

Chen, Y.Q., and A. Chrzanowski (1988). Assessment of levelling measurements using the theory of MINQE. *Proceedings of NAVD Symposium '88*, 389-400.

Chen, Y.Q., and A. Chrzanowski (1989). Experimental study on the methods for processing GPS observation data. *Journal of Wuhan Technical University of Surveying and Mapping*, No. 1, 1-10.

Chrzanowski, A., Y.Q. Chen, and J. Secord (1986). Geometrical analysis of deformation surveys. *Proceedings of Deformation Measurements Workshop*, MIT, Cambridge, Mass., 170-206.

Chrzanowski, A., Y.Q. Chen, R. Leeman, and J. Leal (1989). Integration of the Global Positioning System with geodetic levelling surveys in ground subsidence studies. CISM Journal ACSGC, winter (in press).

Leal, J. (1989). Integration of satellite Global Positioning System and levelling for the subsidence monitoring studies at the Costa Bolivar oil fields in Venezuela. M.Sc. thesis, Department of Surveying Engineering, University of New Brunswick, Fredericton, N.B., April.

Pedroza, M. (1988). Preliminary analysis of the effects of troposphere and geometry satellite distribution in GPS applied to ground subsidence studies. M.Eng. thesis, Department of Surveying Engineering, University of New Brunswick, Fredericton, N.B.

Santerre, R. (1988). The impact of the sky distribution of GPS satellites on precise positioning. Presented at the Chapman Conference on GPS Measurements for Geodynamics, Florida.

Wells, D., N. Beck, D. Delikaraoglou, A. Kleusberg, E. Krakiwsky, G. Lachapelle, R. Langley, M. Nakiboglu, K.P. Schwarz, J. Tranquilla, and P. Vaníček (1986). *Guide to GPS Positioning*. Canadian GPS Associates, Fredericton, N.B.

A CRUSTAL DEFORMATION NETWORK USING GPS

Yola Georgiadou
Geodetic Research Laboratory
University of New Brunswick, POB 4400
Fredericton, N.B., E3B 5A3, CANADA

ABSTRACT

This paper describes the analysis of Global Positioning System (GPS) measurements of the Port Alberni network, established to monitor crustal deformation on the west coast of Canada. A total of 28 independent baselines varying in length between 18 km and 116 km were observed with Texas Instruments TI4100 receivers in the 10-station network. The overall adjustment with UNB's GPS software package DIPOP led to an internal consistency of 10 mm (rms) in baseline length. Simultaneously, part of the deformation network was observed with high precision Electromagnetic Distance Measuring (EDM) equipment. The comparison between GPS and EDM results revealed a significant scale difference of 0.97 ppm and residual baseline length differences of 8 mm (rms). A 116 km baseline was observed on 11 days during the campaign. The analysis of these repeated observations shows considerable contamination of the GPS measurements by signal multipath. The implications of this error in high precision relative positioning is discussed.

INTRODUCTION

Seismically unstable areas in Canada have been traditionally monitored using electro-optical distance (EDM) measurements incorporating atmospheric data measured along the EDM ray-path. The Geodetic Survey of Canada, in an effort to study the feasibility of GPS monitoring of such areas, established the Port Alberni network, on Vancouver Island and observed it simultaneously, with GPS and state-of-the-art EDM instrumentation. The purpose of this paper is to analyse the GPS survey performed on this network. We assess the baseline accuracies achieved with GPS using the "broadcast ephemerides" representation of satellite orbits by exploring the internal consistency of the network, the repeatability of a master baseline and the agreement between the GPS and EDM solutions. Last but not least, we discuss carrier phase multipath as an accuracy limiting factor on GPS results.

THE GPS AND EDM SURVEYS : CAMPAIGNS AND ANALYSIS

The Port Alberni GPS survey was conducted by the Pacific Geoscience Centre of the Geological Survey of Canada and the Geodetic Survey of Canada in August 1986 (Day numbers 230 through 241). The 10 stations surveyed are part of a larger deformation monitoring network in central Vancouver Island. Figure 1 shows the relative locations of the 10 stations of the network. Four Texas Instruments TI4100 dual frequency GPS receivers were operated simultaneously during the survey.

The DIPOP 2.0 software package used for the GPS data reduction was developed by the Geodetic Research Laboratory (GRL) of the University of New Brunswick (UNB). In this package, carrier phase observations are preprocessed to detect and correct cycle slips, to compute satellite coordinates from ephemerides parameters and to create clean double difference files to be used as input for the parametric adjustment. The "cleaned" GPS carrier phase double differences are adjusted in a sequential least squares process. The station coordinates can be constrained to their initial values by assigning them an appropriate covariance matrix.

The primary output is: coordinates of all stations, the associated covariance matrix, and real-valued estimates of carrier phase double difference ambiguities. Detailed documentation of this software can be found in Vanicek et al. (1985), Kleusberg et al. (1987), Santerre et al. (1987).

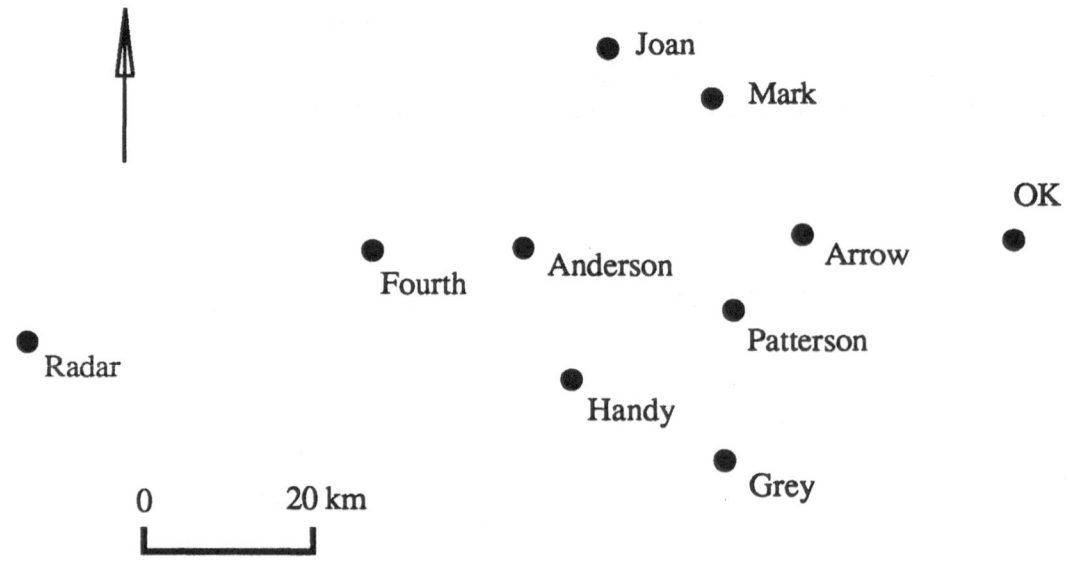

Fig. 1. Port Alberni GPS network

The EDM survey of a part of the Port Alberni network was performed and analysed by the Geodetic Survey of Canada. It is discussed in detail in Beck et al. (1988). Here, we offer a summary of its principal features. Distance measurements were made with a Keuffel & Rangemaster III EDM instrument. The distances surveyed were between 10 and 40 km and the elevations were between 120 and 1800 m. To minimise the effect of refractivity of the air traversed by the light beam, the index of refraction had to be determined along the laser path. For this reason, temperature and humidity were recorded every 20 seconds from probes mounted externally on a helicopter flying along the ray path. The helicopter's course was kept as close as possible to the path of the laser beam by radioing course corrections to the helicopter's pilot. The mark-to-mark distances were combined in a least squares adjustment keeping one azimuth constrained to its initial value. Special care was taken to account for various errors (ibid) resulting to a reduction of the EDM error down to ± (5 mm + 0.2 ppm).

ACCURACY ASSESSMENT

In this section, we assess the accuracy of the GPS network solution by exploring: its internal consistency, the repeatability of day-by-day baseline solutions and the agreement with an external standard of comparable accuracy. We will restrict the discussion to one solution component, namely the baseline length. We do this for reasons of consistency throughout this section, since we will be able to compare only lengths in the third and most crucial part. A discussion on azimuth and height repeatability can be found in Georgiadou (1989).

GPS network internal consistency

To establish the internal consistency of the network we have to adjust each one of selected independent baselines, combine them to an overall network adjustment and analyse the discrepancies. First, a separate adjustment of the L1 and L2 observations served to determine the carrier phase double difference ambiguities for 70% of this set of baselines. The final single baseline result was obtained by combining the L1 and L2 observations to form the L1/L2 linear combination with the purpose of eliminating ionospheric refraction effects. The formal baseline length accuracy, computed from the estimated covariance matrix of the unknowns was 1 and 6 mm for fixed and free ambiguity solutions respectively. The final solution for relative station coordinates was obtained in a simultaneous adjustment of all observations in the network. The a posteriori variance factor obtained in the network adjustment was 1.4 pointing to some inconsistencies between the redundant baseline solutions. This factor is just a gross measure for the compatibility between the individual baseline adjustments. A more representative estimate for the consistency between the single baseline solutions and the overall network adjustment is obtained by analysing all differences in baseline components from the two sets of solutions.

Figure 2 shows length discrepancies between the baseline length from the network adjustment and the individual baseline adjustments as a function of baseline length. All the differences are well below 0.5 ppm or 50 mm with a rms of 10 mm. A more detailed discussion of the Port Alberni GPS survey results has been presented by Georgiadou (1987), Kouba (1988) and Kleusberg et al (1988).

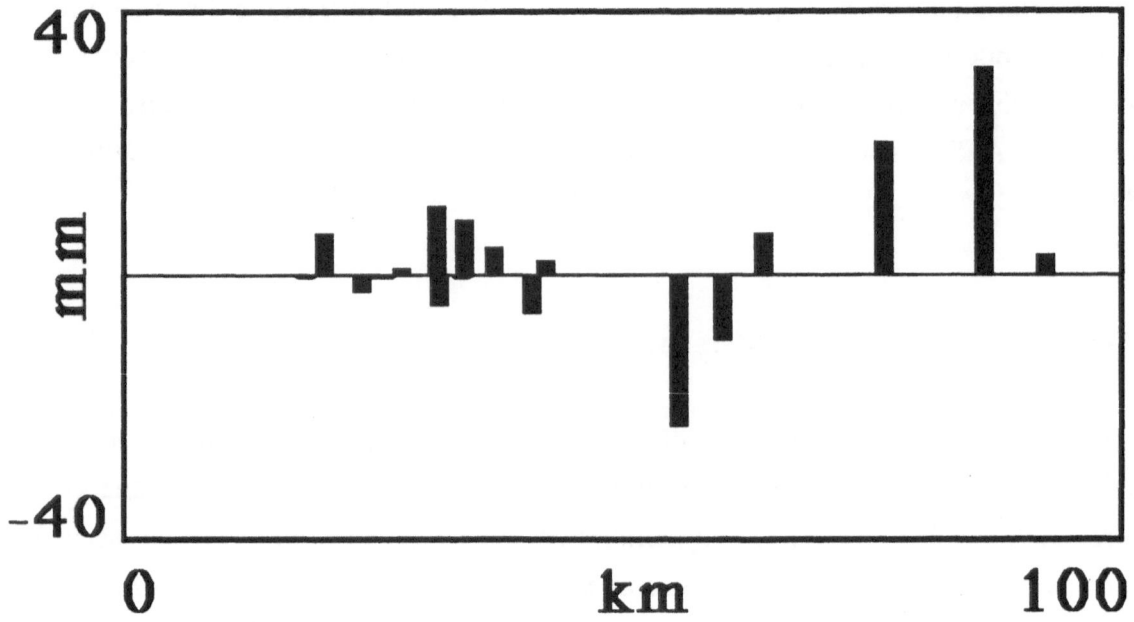

Fig. 2. Baseline versus network solution

Repeatability of the 116 km master baseline

To provide control for the horizontal orientation and scale of the complete network, stations OK and Radar (see Figure 1) were occupied during the entire length of the Port Alberni GPS campaign. The resulting OK-Radar baseline was not explicitly solved for in the previous section. The reason for this was that in selecting independent baselines for each session we opted for the shortest possible connections between stations hoping to resolve as many as possible carrier phase ambiguities. On the other hand, these repeated observations offer an excellent tool for the study of day to day repeatability of baseline length using broadcast ephemerides. Consistent broadcast ephemerides at the 1 ppm level or below will permit the entire GPS baseline processing to be performed in the field.

Least squares adjustments were performed separately for L1 and L2 phase double differences of each OK-Radar session and the carrier phase ambiguity estimates inspected. As anticipated, their integer values could not be unambiguously determined from the real number estimates for this relatively long baseline in the baseline solution mode. Consequently, the observations of this baseline from all eleven sessions were reprocessed simultaneously. In this combined adjustment, the integer values of the ambiguity parameters could be unambiguously determined for six out of eleven sessions. These ambiguities were eliminated from the set of unknowns, allowing a subsequent "fixed ambiguity" solution in both L1 and L2 modes for these six sessions. The final daily adjustments were performed using the L1/L2 ionospheric-free combination.

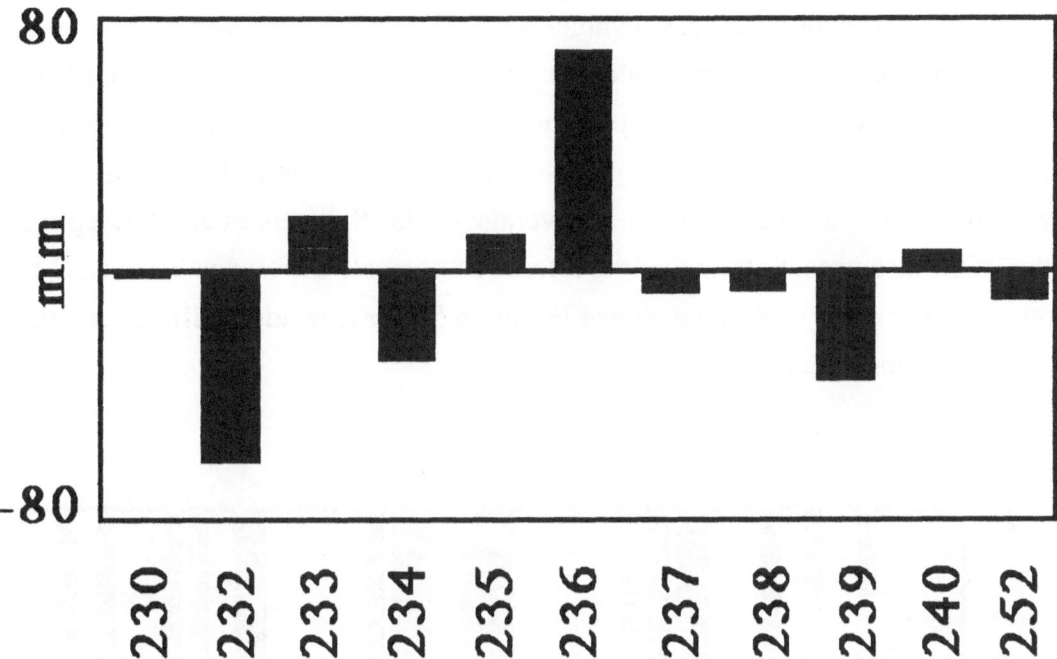

Fig. 3. Baseline length repeatability

The discrepancies in baseline length between the Port Alberni network adjustment and the daily adjustments of the OK-Radar baseline are shown in Figure 3. Differences are larger for Days 232, 234, 236 and 239 where ambiguities could not be resolved. The rms of length differences amounts to 33 mm. Horizontal repeatability is below the 0.1 ppm level if we consider only the six sessions with resolved ambiguities. This is a strong indication that GPS survey results may be considerably better than 1 ppm over a short period (see also Remondi and Hofmann-Wellenhof, 1989).

GPS network external consistency

In the two preceding paragraphs, the consistency of the GPS survey results was assessed by comparing individual baseline results with an overall network adjustment. A more rigorous measure of consistency can be established by comparing GPS results to an external standard of similar accuracy.

In Figure 4, we compare the adjusted values for 22 EDM distances within the Port Alberni network to the mark-to-mark distances computed from the results of the GPS network adjustment. Clearly, a major portion of these differences is systematic in nature. This systematic part of the differences is expressed by a mean scale difference of -0.97 ppm between GPS and EDM distances, i.e. on average the GPS distances are 0.97 ppm smaller than the EDM distances. Subtracting the effect of this scale difference leads us to residual differences with an rms of 8 mm depicted in Figure 5. These residual differences obviously have a quite random nature.

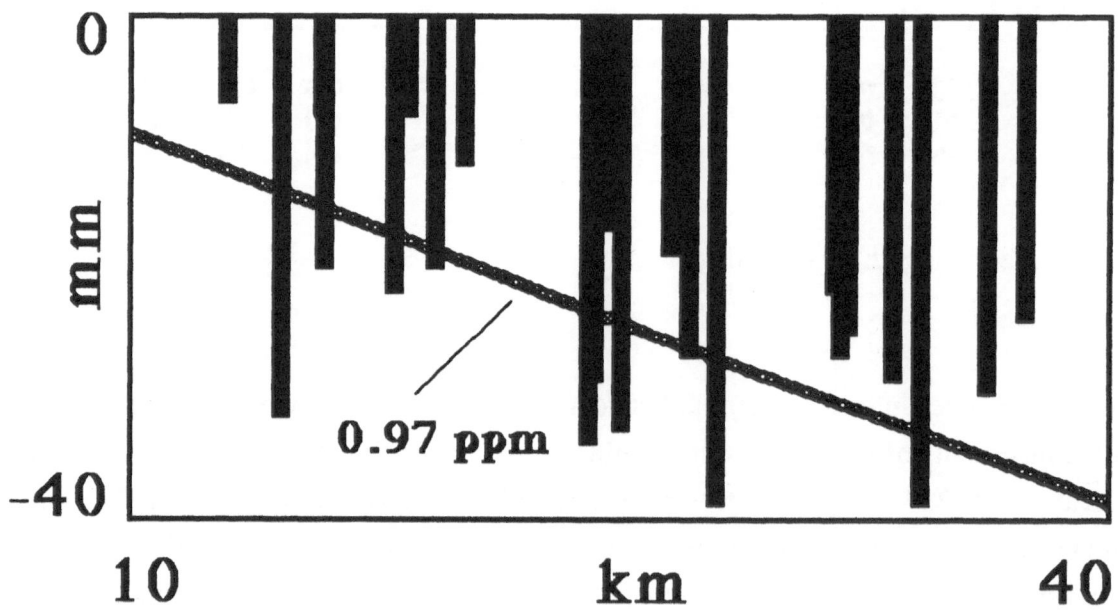

Fig. 4. GPS baseline lengths versus EDM

There are may be several explanations of this scale difference of about 1 ppm. It may be due to inaccuracies in the absolute coordinates of the fixed station OK, or equivalently, to systematic orbital errors throughout the observation period. Another cause may be the different definition and realisation of the scale in the two systems "EDM" and "GPS". "EDM" scale can be calibrated reliably only to a few ppm, and the conversion of kinetic (flown) air temperature to static air temperature can contain systematic errors affecting scale at the level of 0.5 ppm. It is remarkable, however, that the residual differences after scale adjustment between the results of two completely independent systems agree with the estimates of the precision of the GPS network solution.

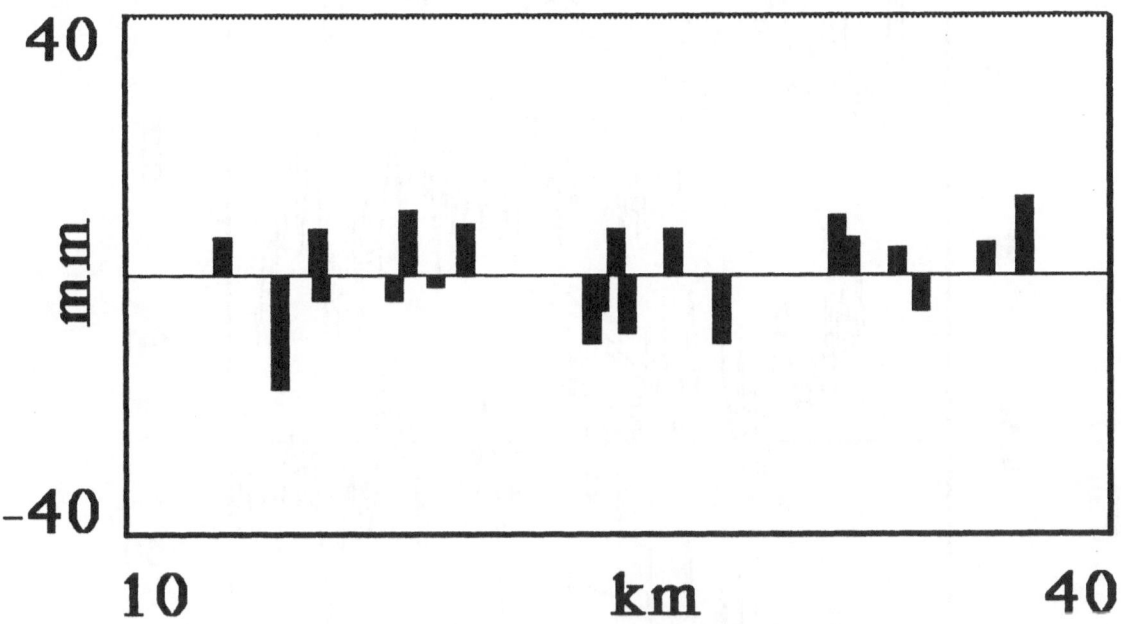

Fig. 5. GPS baseline lengths versus EDM after scale factor removal

CARRIER PHASE MULTIPATH AS AN ACCURACY LIMITING FACTOR

In this section, we have a closer look at carrier signal multipath as an error source affecting the accuracy of relative GPS positioning. Multipath effects in the carrier phase measurements are baseline-length independent and are similar from day to day due to repeated satellite-receiver geometry. In the adjustment of the GPS measurements, parts of these errors will map into the adjustment residuals (Georgiadou and Kleusberg, 1988) and other parts in the estimated baseline components and/or carrier phase ambiguities.

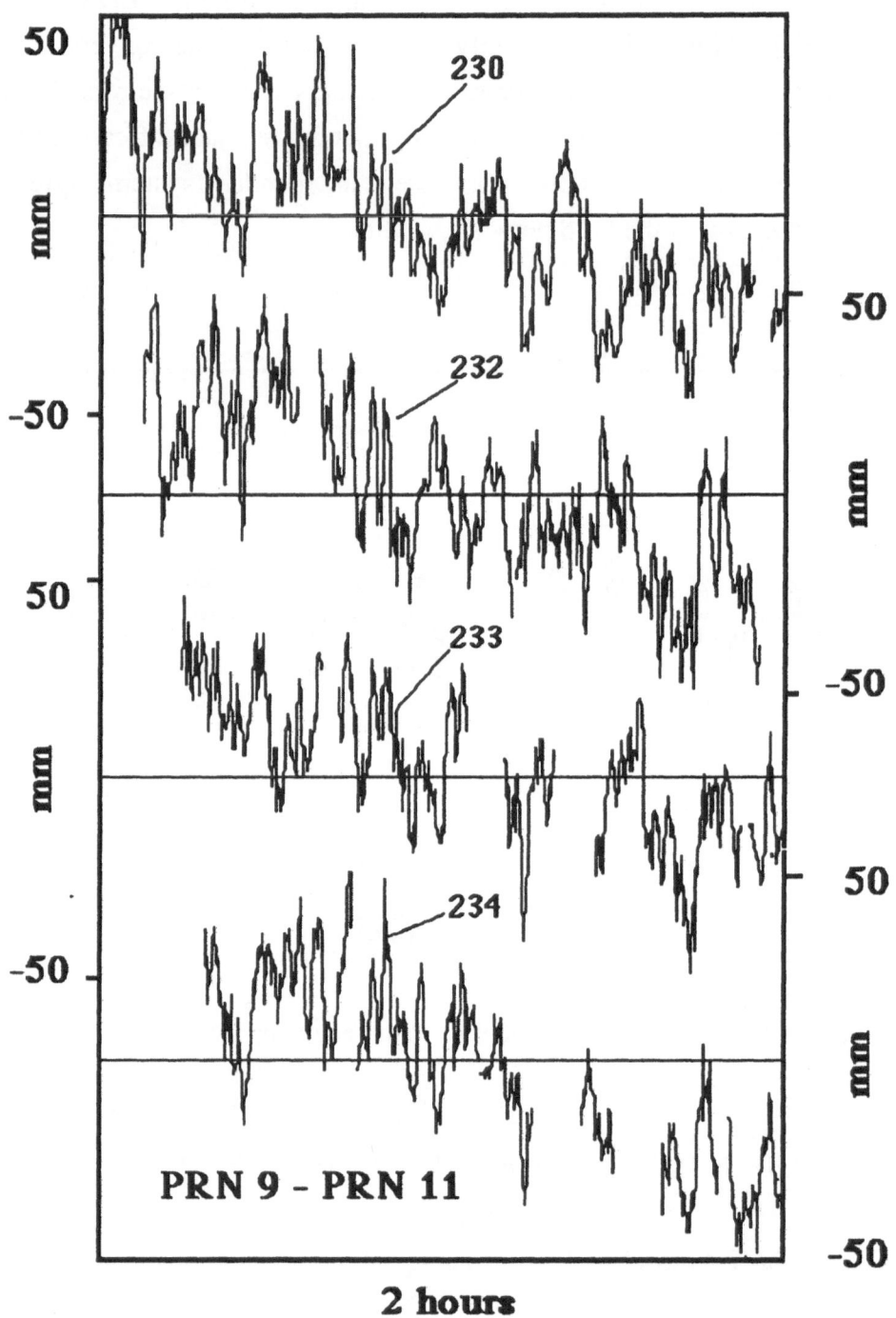

Fig. 6. Adjustment residuals for Days 230, 232, 233 and 234

Inspection of adjustment residuals from Days 230, 232 for the 116 km OK-Radar baseline for the total observation time span of five hours reveals a strong day-to-day correlation between them. This suggests presence of signal multipath leading to a slow non-uniform convergence of the estimated baseline components in the sequential adjustment. Figure 6 shows adjustment residuals, over the first two-hour observation period, for the first four days (230, 232, 233 and 234) appropriately shifted in time for direct visual comparison. If only 2 hours of data had been available for each daily session, an additional bias could have been introduced in the estimated parameters, that would affect the comparison with EDM results.

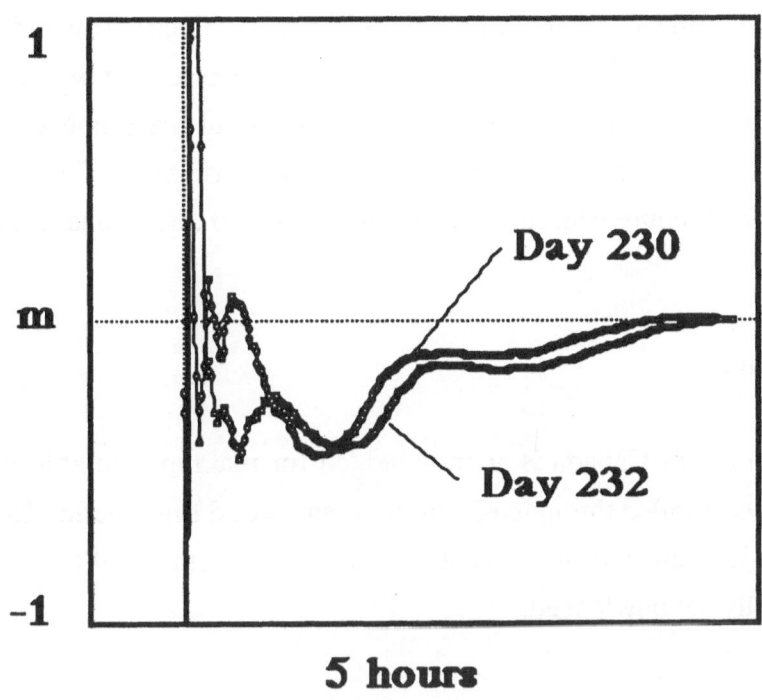

Fig. 7. Baseline length convergence for Days 230 and 232

Figure 7 shows the convergence of baseline length for these two days over a period of 5 hours. The convergence curves exhibit a strikingly similar pattern. Similar convergence curves can be obtained for the remaining days not shown here. Over 3.5 hours of data are necessary to achieve 10 cm accuracy, which is equivalent to ppm. For a more detailed discussion on the effect of carrier phase multipath see Georgiadou and Kleusberg (1988, 1989).

CONCLUSIONS

Relative positioning results obtained from dual frequency GPS observations in a crustal deformation network have been analysed using broadcast orbits for the GPS satellites. Remaining discrepancies of 8 mm rms, after removal of a scale factor between results of the GPS and a concurrent high precision EDM survey, agree with formal accuracy estimates estimates from the GPS network least squares adjustment. The achieved length repeatability is 0.4 ppm (0.1 ppm if ambiguities can be resolved). This indicates that using broadcast orbits may be considerably better than 1 ppm over a short period of several days. Finally, we suggest that observation sessions longer than the longest multipath period should be designed to average out possible multipath interference. Conclusions are based on TI4100 data. Other equipment may be less susceptible to carrier phase multipath.

Acknowledgement

The Geodetic Survey of Canada is acknowledged for making available the Port Alberni data. This work was funded through a Natural Sciences and Engineering Research Council (NSERC) strategic grant entitled "Application of Differential GPS Positioning". This support is gratefully acknowledged.

REFERENCES

Beck N, Duval JR Taylor PT (1988): *GPS versus precise trilateration- Comparison of various processing approaches:* Presented at the May 11-14, 1988 ASCE Specialty Conference "Engineering Applications of GPS Satellite Surveying Technology" held in Nashville Tennessee

Beutler G, Bauersima I, Gurtner W, Rothacher M, Schildknecht T (1988): *Static Positioning with the Global Positioning System (GPS): State of the art.* In Lecture notes in Earth Sciences: GPS Techniques applied to Geodesy and Surveying, pp. 363-380, eds. E. Groten and R. Strauss

Beutler G, Gurtner W (1987): *The influence of atmospheric refraction on the evaluation of GPS observations:* Bericht Nr. 16, Satellitenbeobachtungsstation Zimmerwald.

Dragert H (1987): *The fall (and rise) of central Vancouver Island: 1930 - 1985.* Can.J.Earth Sci. 24, 689-697

Georgiadou Y (1987): *Results of the Port Alberni GPS survey 1986.* Internal Report, Geophysics Division, Geological Survey of Canada, Ottawa, Ontario

Georgiadou Y and Kleusberg A (1988): *On carrier signal multipath effects in relative GPS positioning,* manuscripta geodaetica 13, 172-179

Georgiadou Y and Kleusberg A (1989): *Further studies on carrier phase multipath in relative GPS positioning.* Paper presented at IAG General Assembly, 3-12 August, Edinburgh

Georgiadou Y (1989): *GPS repeatabilty studies using broadcast ephemerides.* Accepted for publication in the Hellenic Journal for Geodetic Sciences "Eratosthenes"

Kleusberg A, Georgiadou Y, van den Heuvel F(1987): *Dual frequency GPS data preprocessing with DIPOP.* Technical Memorandum 21, Department of Surveying Engineering, University of New Brunswick, Fredericton, N.B.

Kleusberg A, Georgiadou Y, Dragert H (1988): *Establishment of Crustal Deformation Networks Using GPS: A Case Study.* CISM Journal AGSGC, Vol. 42, No.4, Winter 1988, pp. 341-351

Kouba J (1988): *Reduction of Port Alberni GPS data.* Presented at the AGU Chapman Conference on GPS Measurements for Geodynamics, Fort Lauderdale, Fla, USA, 19-22 September

Remondi BW and Hofmann-Wellenhof B (1989): *GPS broadcast orbits versus precise orbits: A comparison Study .* Paper presented at IAG General Assembly, 3-12 August, Edinburgh

Santerre R, Craymer MR, Kleusberg A, Langley RB, Parrot D, Queck SH, Vanicek P, Wells DE, and Wilkins F (1987): *Precise relative positioning with DIPOP 2.0.* Paper pres. IUGG XIX General Assembly, IAG Section II Scientific Meeting, Vancouver

Vanicek P, Beutler G, Kleusberg A, Langley RB, Santerre R and Wells DE (1985): *DIPOP: Differential positioning program package for the Global Positioning System.* Technical report No 115, Department of Surveying Engineering, University of New Brunswick, Fredericton, N.B.

Wells DE, Beck N, Delikaraoglou D, Kleusberg A, Krakiwsky EJ, Lachapelle G, Langley RB, Nakiboglou M, Schwarz KP, Tranquilla JM and Vanicek P (1986): *Guide to GPS Positioning.* Canadian GPS Associates, Fredericton, N.B., Canada

EXTENSION OF ACTIVE LOCAL NETWORKS ALONG BOCONO FAULT AND FIRST SATELLITE CONNECTIONS

HEINZ G.HENNEBERG
UNIVERSITY OF ZULIA-ESCUELA DE GEODESIA-FAC.ING.
MARACAIBO - VENEZUELA

ABSTRACT

The geodetic activities to monitor movements along the Caribbean South American Plate Boundary started in 1973 with first local network installations at Mitisus and Mucubaji at heights up to more than 4000 m.At the same time,a tunnel network at Yacambu,crossing Bocono Fault which belongs to the same Plate Boundary,was installed and measured.Repeated measurement campaigns in the following years showed two types of movements:right-lateral strike-slip movements and deformation components toward the fault.Maximum values of displacement vectors showed 32 mm in 5 years at Mucubaji (net of 10 km extension) and 88 mm in 2 years at Yacambu (net of 28 km extension).In total 18 sites were studied and explored for local geodetic networks.Of these,5 are operational and were repeatedly measured.Now,5 other nets are incorporated,monumented and measured. Further,in January-February 88 first satellite observations (GPS) were realized to link the local networks together.The satellite link includes also the connection between the Mucubaji local network and the Geotraverse network of the Andes,which is a German-Venezuelan joint research program.The satellite campaign in Venezuela was a Venezuelan-German venture,and will be repeated.It is also incorporated to the multinational CASA UNO PROJECT coordinated by JPL - Pasadena(USA).

INTRODUCTION

The Bocono-,San Sebastian- and Pilar Fault Systems are part of the most probable Plate Boundary between the Caribbean and South American Plates.Along this fault systems 18 different sites have been explored for geodetic network installations.Of these,10 networks are completed,measured and full operational (fig.1).The Pilar Fault Network,shown in fig.1 is not yet operational.But the next satellite observations with GPS,foreseen for the near future,will incorporate several stations of the Pilar Fault Network.Fig.2 shows the large Mucubaji Network and the outer contours of the small net.Observatorio(Merida) was GPS station in the 1988 campaign.The arrow shows the medium field vector of displacement between 1975 - 1981,representing the displacement of the south side of the Bocono Fault in respect to the north side which was held fixed.Fig.4 shows the principal network at Mitisus,where a hydroelectrical dam is located. The medium field vectors show displacements between 1973 and 1982. At Yacambu (fig.6) the Bocono Fault Zone has two main visible fault traces,as recently discovered: the Bocono Fault itself as known before and one of her branches,the Rio Turbio Fault where the main displacements have taken place (in former publications,this trace was considered as Bocono Fault).The field vector represents the medium displacement from one side of the fault in relation to the other between 1973 and 1975 (after the San Pablo Earthquake early 1975). The network in fig.8 is of recent installation.It belongs to the area of Guarico-Sanare.The error ellipses belong to the first free network adjustment (1989).The network stations are of solid construction as seen in fig.5 and fig.7.Fig 5(A) exhibits the typical station as founded in rock areas with steel bars and observation column.In fig.5(B) is seen the top of the column with the forced centering device.The new network,recently installed as seen in fig.8, has another type of monument.We see this other type in fig.7.This type of structure is considered a very economical form of quick installment,because there is no need of concrete work for the column. Later on,this station could be converted into the same type of station as seen in fig.5,adding the concrete column around the steel pipe.

In 1988(Jan./Febr.) were realized the first satellite connections between the local networks,as proposed in fig.3. Applying GPS measurements,together with the CASA UNO campaign,were observed one station in the Uribante network,2 points in the Andes Geotraverse network (Caja Seca and Barinas and one station of the Mucubaji network (Observatorio-Merida)(fig.9).This was the beginning of the inter connection of all local networks along the plate boundary.For 1990 and 1991 are foreseen the reoccupation of these sites and first observations at stations in Yacambu and Pilar.

HISTORICAL ASPECTS
The first network exploration and installation took place at Mitisus in 1973,together with the amplifying of the research network of the hydroelectrical dam of Santo Domingo at Mitisus.The reobservations in this net gave already evidence of displacements during the first year.Also in 1973 was started the network installation at Mucubaji at heights up to more than 4000 m.This network of about 10 km extension was scaled,applying a Geodimeter 8 Laser and in very difficult accesses the Tellurometer M 1000.The measurements lasted several months due to the very difficult topographycal conditions.The Yacambu network was first installed for the Yacambu Tunnel construction.This tunnel has a length of 24 km.The tunnel crosses the Rio Turbio Branch Fault (part of the Bocono Fault System) at 15 km from the south entrance (fig.6).This case at Yacambu shows the importance to consider that a fault is not always just a single break in the crust,it is shown through this special case that a fault zone can have a very wide range (up to 10 km in Yacambu).The two fault traces unite about 15 - 20 km NE of the tunnel axis.We should speak of fault zones or fault areas in all that cases where the active neotectonic evidences are located offset the commonly known and visible fault trace.

NEW LOCAL NETWORKS
The newest local network installations belong to the area between Bocono - Tostos and Guerico - Sanare.In this range three networks were installed between 1988 and 1989.One of this nets is shown in fig.8 (Guarico-Sanare).The fig.shows too the error ellipses of the first free network adjustment.In this nets were applied the monuments as seen in fig.7.In these new networks no coordenate changes were obtained up to date.The repetition of measurements in these nets are considered not before 1990.

Another of the recent new networks incorporated for the research work along Bocono Fault is the Uribante network.This network belongs to the geodetic control system of the Uribante-Caparo Hydroelectrical ProjectSpecially the two control nets of the Uribante dam,lying south of Bocono Fault and north of the Caparo Branch Fault,were considered as an idel system to monitor relative movements in comparison with the north side of the fault.One of the Uribante stations was incorporated into the GPS campaigns to inter connect the local networks along the plate boundary.

SATELLITE OBSERVATIONS

The International Symposium on Recent Crustal Movements at Maracaibo,Venezuela 1985,gave start of the "Geodynamics Project of the Caribbean".The presentation of three panels and five papers related to this area of high geodynamic and neotectonic activities was the scientific background to begin with an international cooperation (Panel:Ch.Whitten,A.Chrzanowski,S.Bakkelid,H.Drewes,E.Grafarend, E.Groten,H.Henneberg,S.Holdahl,I.Joo,H.Kahle,J.Kakkuri,A.Lazzari, K.Linkwitz,N.Renzetti,B.Schaffrin,R.Vieira,W.Welsch,D.Villalta,A. Vera,L.Ordonez,J.Stock,W.Torge,Ivan Mueller and others).The project started first with a german-venezuelan cooperation.Meetings have taken place in Maracaibo,Merida,Munich,Stuttgart,Hannover (Henneberg, Stock,Hoyer,Drewes,Reigber,Torge,Linkwitz,Kahle).The result of these meetings was the "Venezuelan Andes GPS Network Project".This project was incorporated into the later created CASA UNO Project.CASA UNO was presented in Vancouver 1987 and the field campaign was executed in Jan./Febr.1988.The planning of CASA UNO started with a historical meeting on March 17,1987 at Maracaibo,Venezuela,under the topic: "Horizontal control of the Geodynamic Motion of the Caribbean Plate with respect to the South American Plate by GPS observations", (Venezuela:Benitez,Bravo,Chourio,Henneberg,Ramirez,Hernandez,Hoyer, Stock.Colombia:Ropain,Camargo,Bermudez,Fernandez.USA:Dixon,Kellogg, Ramsey.Germany:Birk,Drewes.Switzerland:Kahle).
(Obs.:The former text in this paragraph is an extract of the report 1989 of the Subcommission -CRCM- for Central and South America presented at the General Meeting of IAG in Edinburgh,Scotland,the 3-12 of August 1989).
Fig.3 shows the general aspect to tie together all local networks along the plate boundary in Venezuela through GPS observations and in fig.9 we observe the 4 stations measured during the CASA UNO campaign in Jan./Febr. 1989 to fullfill this purpose (stations:Uribante, Barinas,Caja Seca,Observatorio-Merida).The stations Caja Seca and Barinas belong to the Geotraverse of the Andes,crossing Bocono Fault (Linkwitz).The fig. shows that 2 stations are south of Bocono Fault and 2 stations located in the north.The coordenates computed in WGS 84 have accuracies between 1 and 2 cm (Drewes).The next satellite campaign foresees the connection to Yacambu and Pilar and the reoccupation of the stations observed in 1988.

ACKNOWLEDGEMENT

Parts of the research work described in this paper were supported by: CADAFE (Venez.Power Organization),Conicit (Venez.Research Counsel) The University of Zulia-Maracaibo.The author thanks the following persons and institutions for their cooperation:C.Schubert,H.Arp,J. Fischer,H.Drewes,J.Stock,Cartografia Nacional,students and personnel of the Geodetic School of Maracaibo University,especially A.Gonzalez,E.Escartin and G.Avila,and all other persons who helped directly and indirectly.On behalf of the satellite campaign I thank all colleagues and persons mentioned in this article,further R.Neilan, S.Rekkedal,K.Stuber and H.Tremel.

REFERENCES

Cluff,L.S.,Hansen,W.R.: 1969. Seismicity and seismic geology of Northwestern Venezuela,Compania Shell de Venezuela,Caracas,2 vol.

Drewes,H.: 1986.Evidence from physical geodesy for geodynamics in the Caribbean area.Tectonophysics,130,77-94

Drewes,H.,et al.: 1988.The Venezuelan Andes GPS Network,IAG,Comm.8, Coordinat.Space Tech.Geod.Geodyn.Bull.,Munich,10,103

Escartin,E.,Avila,G.: 1989.Instalacion de una micro red de precision en el area de la Falla de Bocono correspondiente al sector Gaurico-Sanare.Tesis,Escuela de Geodesia,Fac.Ing.,Univ.del Zulia,Maracaibo.

Fischer,J.: 1981.Diferentes tipos de mediciones de ingenieria para el proyecto de la Represa Yacambu.Tesis,LUZ Fac.Ing.,Esc.Geod., Maracaibo,Venezuela.

Henneberg,H.: 1975.Results of geodetic deformation measurements on large scale structures including some application in geology.Int. Symp.on Defomation Measurements,Krakow,Poland.

Henneberg,H.: 1978.Local precision nets for monitoring movements of faults and large engineering structures.Proc.9th GEOP Conference Dept.Geod.Sciences,The Ohio State University,Columbus,Ohio.

Henneberg,H.: 1981. Geodetic nets installed for monitoring movements along Bocono Fault in Venezuela.First Int.Symp.on Crustal Movements in Africa,Addis Abeba.

Henneberg,H.: 1983.Geodetic control of neotectonics in Venezuela. Tectonophysics,97,1 - 15.

Henneberg,H.: 1986.Neotectonic Geodesy.Tectonophysics,130,95 - 104

Henneberg,H.,Schubert,C.: 1986.Geodetic networks along the Caribbean South American Plate Boundary.Tectonophysics,130,77 - 94.

Kellogg,J.et al.: 1988.Epoch GPS Geodesy in the North Andes-CASA UNO. Journal of South American Earth Sciences

Kellogg,J.et al.: 1989.CASA-Central and South America GPS Geodesy. EOS,June,649.

Linkwitz,K.,Boettinger,W.: Messungen hoher Präzision entlang der Bocono Verwerfung in Venezuela.Int.Symp.Def.Meas.,Krakow,Poland,1975.

Linkwitz,K.: 1986.Geodetic traverse in the Venezuelan Andes,1983 field campaign and first results.Tectonophysics,130.

Neilan,R.et al.: 1988.CASA UNO GPS.Transactions,Am.Geophys.Union,69. Neilan,R.: 1989. Operational Aspects of CASA UNO '88.IEEE Transact. on Instrumentation and Measurement,Vol.38,No.2,April

Renzetti,N.: 1986. A global positioning measurement system for regional Geodesy in the Caribbean.Tectonophysics,130.

Shubert,C.,Henneberg,H.: Geological and geodetic investigations on the movements along the Bocono Fault,Venezuelan Andes.Tectonophysics, V.29,199 - 207, 1975.

Schubert,C.: 1979. El Pilar Fault Zone, Northeastern Venezuela. Tectonophysics,52,447-455.

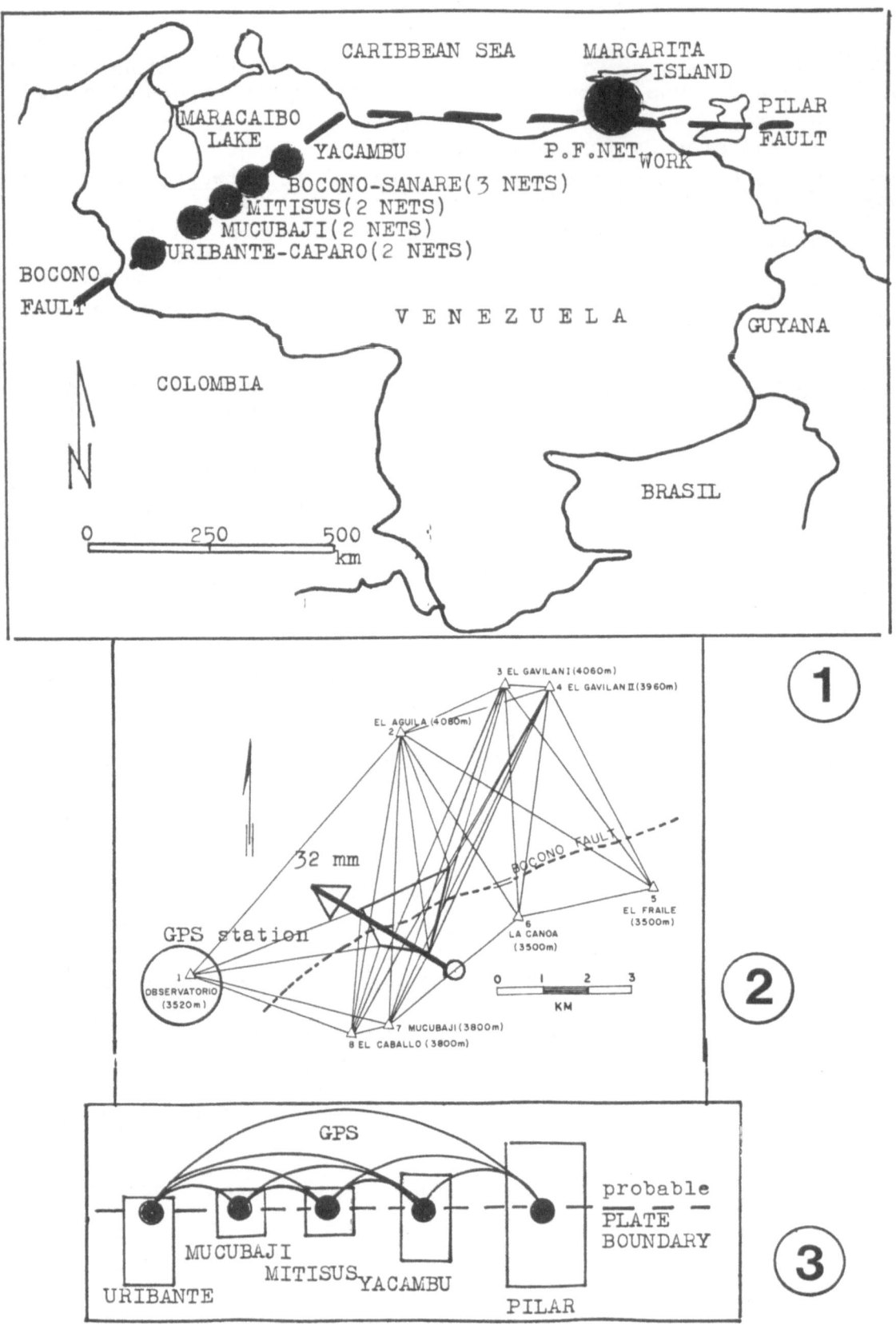

CARIBBEAN SEA

MARGARITA ISLAND

PILAR FAULT

MARACAIBO LAKE

YACAMBU

BOCONO-SANARE(3 NETS)

MITISUS(2 NETS)

MUCUBAJI(2 NETS)

URIBANTE-CAPARO(2 NETS)

P.F.NETWORK

BOCONO FAULT

VENEZUELA

GUYANA

COLOMBIA

N

BRASIL

0   250   500
                    km

①

3 EL GAVILANI (4060m)

4 EL GAVILAN II (3960m)

EL AGUILA (4080m)
2

32 mm

BOCONO FAULT

5
EL FRAILE
(3500m)

GPS station

6
LA CANOA
(3500m)

1
OBSERVATORIO
(3520m)

0   1   2   3
        KM

7 MUCUBAJI (3800m)

8 EL CABALLO (3800m)

②

GPS

probable
PLATE
BOUNDARY

MUCUBAJI

MITISUS

YACAMBU

URIBANTE

PILAR

③

208

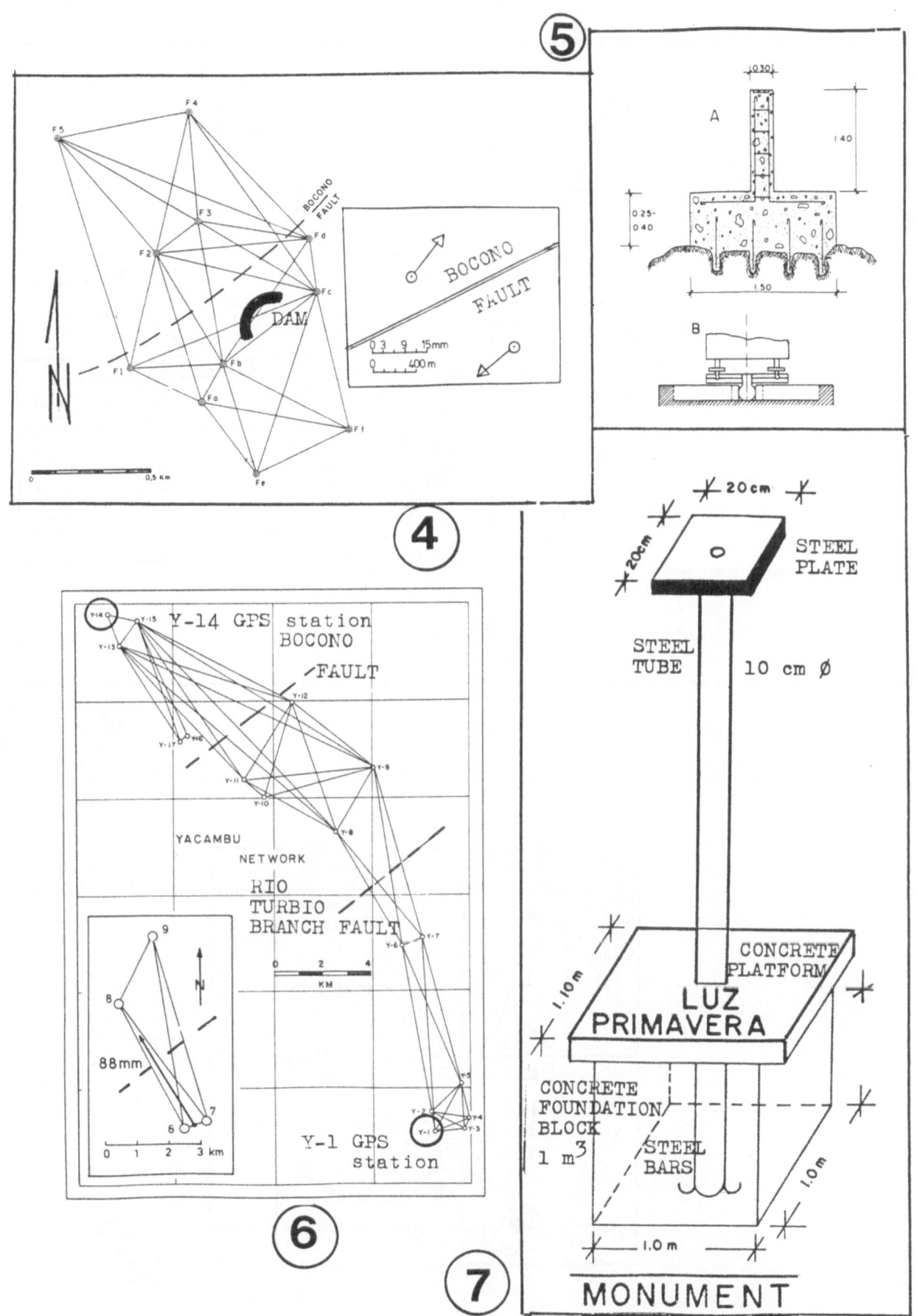

⑤

A

B

④

Y-14 GPS station
BOCONO
FAULT

YACAMBU
NETWORK

RIO
TURBIO
BRANCH FAULT

Y-1 GPS
station

⑥

⑦

20cm

STEEL
PLATE

STEEL
TUBE   10 cm ∅

CONCRETE
PLATFORM

LUZ
PRIMAVERA

CONCRETE
FOUNDATION
BLOCK
1 m³

STEEL
BARS

MONUMENT

209

STATION	MAX mm	MIN mm
PRIMAVERA	3.2	2.7
LA CURVA	2.6	2.1
MATICA	3.1	2.4
JAGUEY	2.5	2.3
SEQUERA	2.5	2.1
LAGUNA	2.9	0.7

8

9

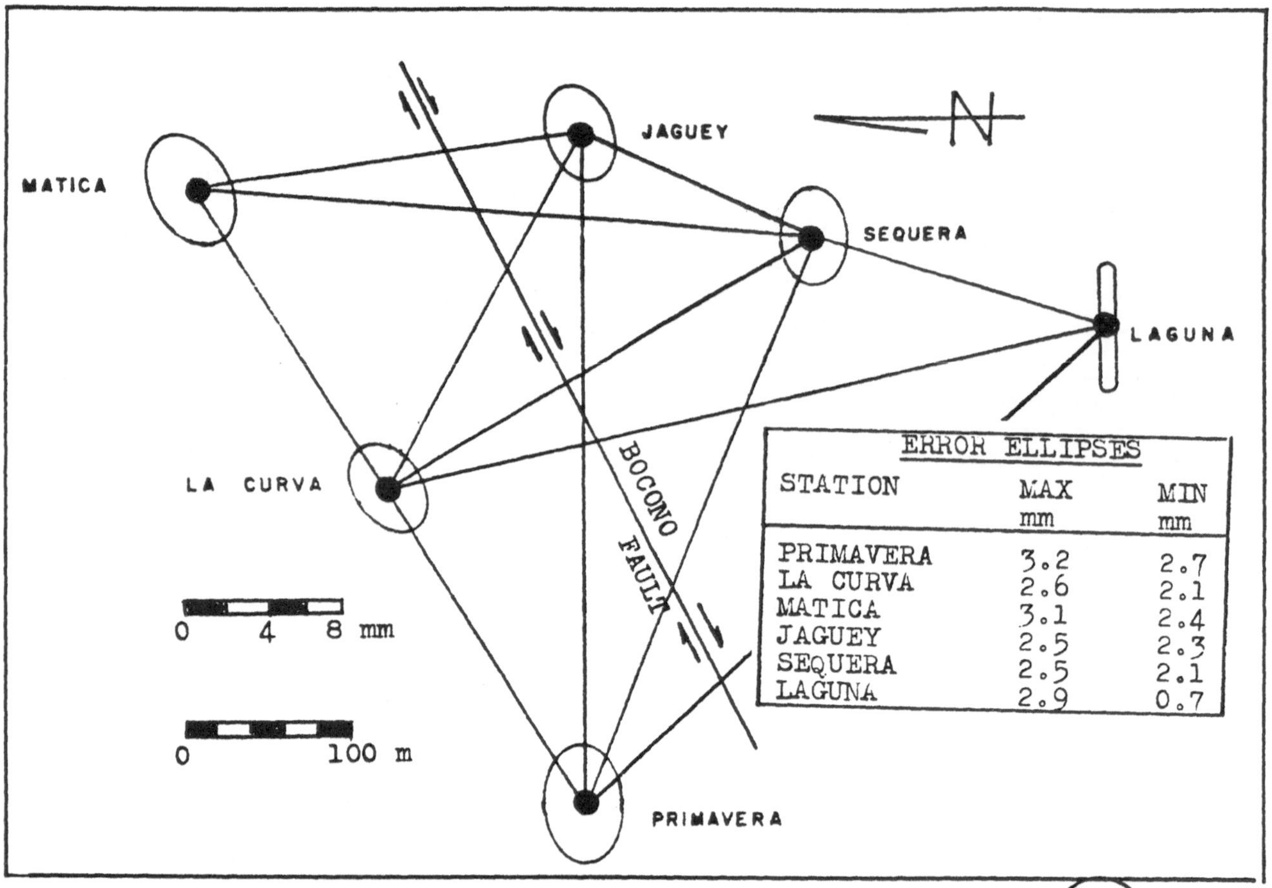

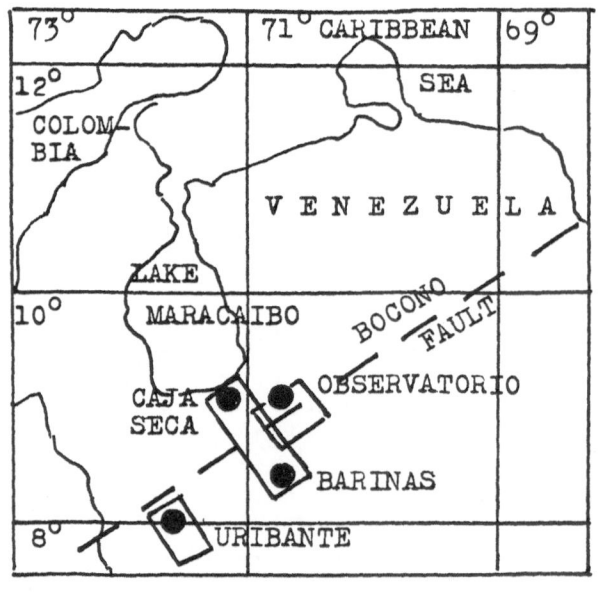

DEFORMATION ANALYSIS OF A LOCAL TERRESTRIAL NETWORK IN ROMANIA WITH RESPECT TO THE VRANCEA EARTHQUAKE OF AUGUST 30, 1986

Günter Schmitt[1]), Trajan Moldoveanu[2]), Valeriu Nica[2]), Reiner Jäger[1])
1) *Geodetic Institute, University of Karlsruhe, Englerstraße 7, D—7500 Karlsruhe*
2) *Institute of Hydroelectrical Studies and Designs (ISPH), Bd. Republicii 29, Bucuresti, Romania*

1 INTRODUCTION

The future Siriu and Surduc hydroelectrical developments under advanced construction are located in the Vrancea area (45,0⁰—46,1⁰ N latitude, 26,0⁰—27,8⁰ E longitude), whose seismicity determines the dominant aspect of the Romanian territory seismicity. The seismicity of the Vrancea area may be compared to the Hindukush montains area (Pamir) and the Bucaramanga region (Columbia). Frequent earthquakes of intermediate depth (60—180 km) occur in this area, mostly within a depth of 130—150 km. The generally high energy of these earthquakes makes them to be recorded in large areas as monokinetic shocks (those of low intensity) and as multiple—phenomena shocks accompanied by aftershocks in case of high seismic events.

The general distribution of the epicenters and the hypocenters indicates a seismogenic area of about 9000 km², if the normal earthquakes are included, and of only 2000 km² for the intermediate earthquakes. The hypocenters show a gradual deepening to the inner arch of the Eastern Carpathians along an arched Benioff surface with an inclination of 60⁰—65⁰ that suggests a paleo (relic) subduction process. That is a subduction of the Carpathian foreland, which presented a linear character before the Sarmatian, when the Eastern Carpathians were rising. But these characteristics gradually disappeared to Quarternary.

Figure 1 presents the location of some strong intermediate Vrancea earthquakes that occured during the last 200 years. The epicentral distances of the strongest earthquakes during this time in relation to the sites of the new dams range between 17 and 50 km.

The study of the March 4, 1977, earthquake rendered evident that the complex fracturing process (multiphenomenon shock consisting of a preshock and three main shocks) propagated on a NE—SW direction up to the neighbourhood of the future development sites zone where the strongest shock occured (M_s=7,2).

The occurence of remarkable activities in the earth's crust was made evident instrumentally during the period following the earthquake of March 4, 1977. In consequence of these observations and of the requirements mentioned above, a high amplifying seismic station and a digital recording system for seismic data were installed in the Surduc—Siriu area, which allow the recording of events of the microearthquake type, and a SMA—1 type accelerograph for recording the large displacements respectively.

Earthquake prediction has come to be one of the central problems of scientific research during the past 10 or 15 years. Evidently it is not only a problem of seismological methods, there are also implied geochemical, geomorphological, geodetic, biological methods a.s.o. Remarkable foregoing anomalies, that may manifest during all prediction stages, are often presented by deformations of the earth's crust, rendered evident by levelling, triangulation or gravimetric measurements, as well as by inclinometers and extensometers.

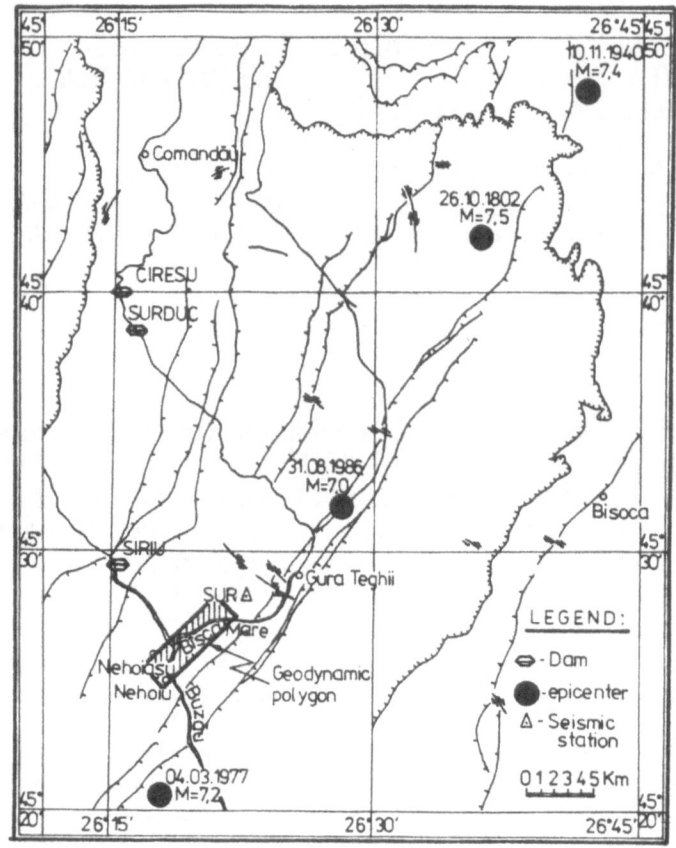

Fig. 1
Location of earthquake
epicenters with $M_s \geq 7.0$

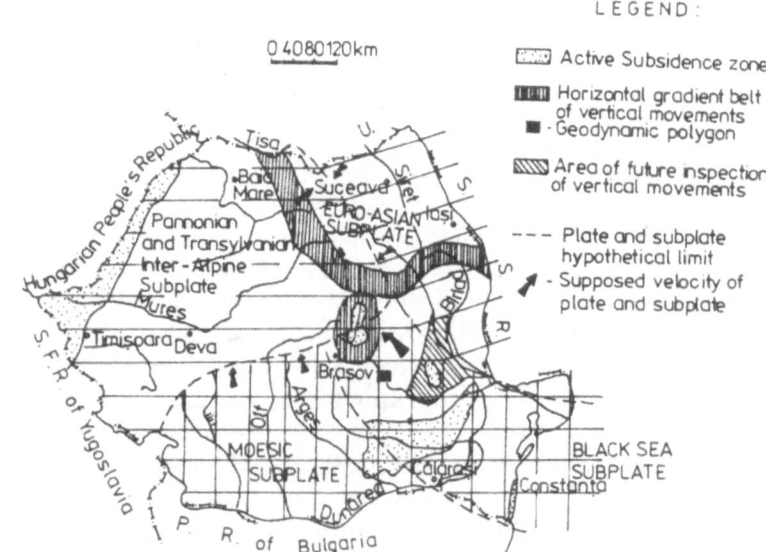

Fig. 2
Tectonics of Romania

As support from geodesy a geodynamic deformation network consisting of 18 pillars was set up in 1985—1986 in Surduc—Siriu area so that the first geodetic measurements could be carried out in August 1986. Figure 1 and 2 present the networks position within the seismic—tectonic framework of Vrancea zone of Romania respectively.

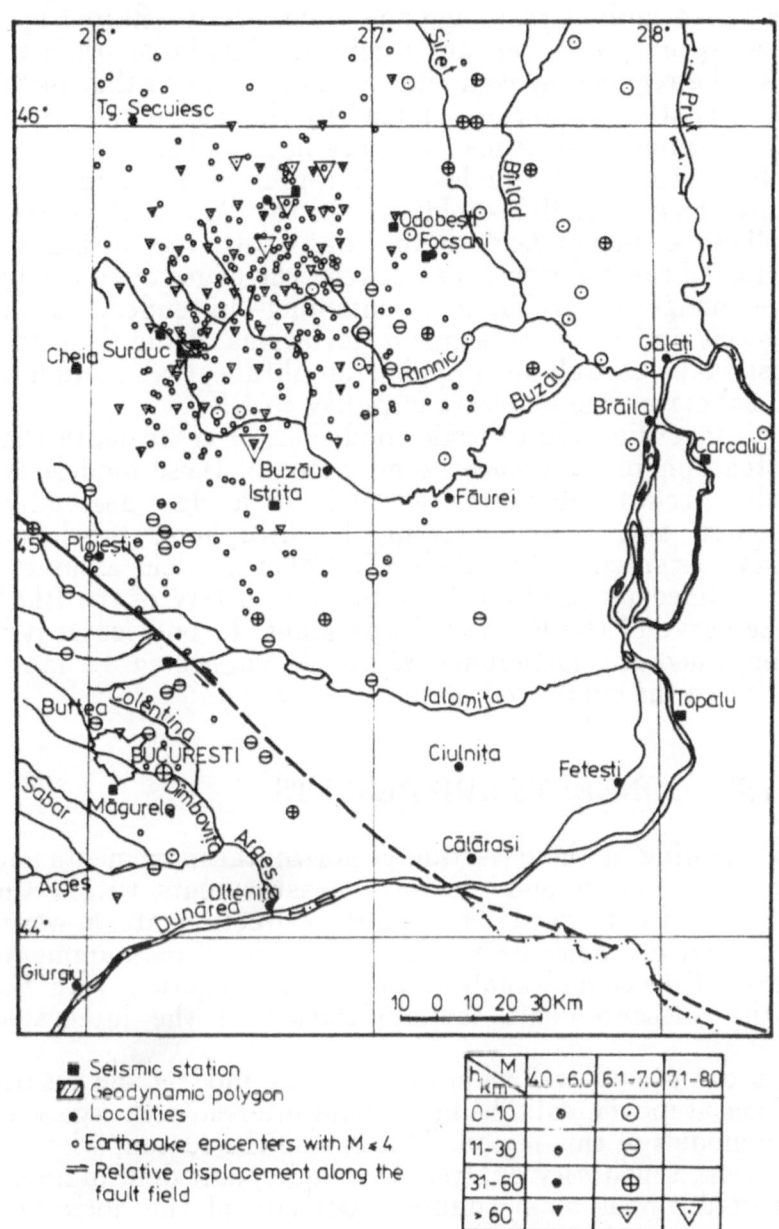

h\M km	4.0 – 6.0	6.1 – 7.0	7.1 – 8.0
0 – 10	⊙	⊙	
11 – 30	•	⊖	
31 – 60	•	⊕	
> 60	▼	▽	▽

■ Seismic station
▨ Geodynamic polygon
● Localities
○ Earthquake epicenters with M ≤ 4
⇌ Relative displacement along the fault field

Fig. 3
Location of the geo—dynamic network and distribution of earthquake epicenters

2 SEISMIC—TECTONIC CONDITIONS

The Surduc—Siriu geodynamic network is located near the frontal part of the Black Sea subplate. This subplate subduces in the Carpathian curving zone the Intra—

Alpine subplate with its Pannonian and Transsylvanian Segments and is limited to NE by the Euro—Asian and to SW by the Moesic subplate (see fig. 2). Figure 3 shows that the network located in the Buzau valley is practically set up in the Pleistocene seismic zone of the Vrancea earthquakes.

From the point of view of the regional tectonic conditions the network is located on the Paleogene flysh of the median sheet of Tarcau, overthrust on the marginal flysh during Miocene, that is 12—14 million years ago and transversely affected by a series of fallings—out of modest spread, with rupturing deformations larger than the main overlapping movements. The regional geologic information suggests that in the southern part of the Eastern Carpathians curving (crossed by the Buzau valley) the medium flysh thrust over the marginal flysh tends to reduce in amplitude, the super overlapping plane tendency breaking up into modest overlapping disposed in relay. At the same time the structures formed of flake—folds within the Tarcau sheet tend to deepen axially to SW. These elements lead to the idea that even within the profound tectonic characteristics of the collision zone between the edges of the above mentioned subplates direction changes occur. From the summing—up tendency of the jumps on the traverse falling—out and from the sigmoid deviations of the flysh fold axes it results that the post collision sub—overlapping amplitude (subduction of alpine type) of the Black Sea subplate edge increases form SW to NE.

More models were drawn up regarding the tectonic conditions of great depth that concern the lithosphere — astenosphere for vrancea zone. Some of these models are based on the idea of a paleo (relic) subduction exerted by a slab descending vertically into the astenosphere and detached form the litosphere, previously entailed in a subduction plane oriented either NW—SE. Other models assume a subduction still active on an inclined plane of 60⁰ by subthrust to NW of the Black Sea subplate in relation to the curve of the Eastern Carpathians. In both cases it is estimated today that the intermediate earthquake source is connected to faults (usually inverse) from the inside of the subducted plate.

3 DEFORMATION ANALYSIS, CONCEPTS AND RESULTS

The analysis concept is mainly aiming at the detection of horizontal movements and the strain filed respectively by direction and distance measurements within the network. The zenith—angle measurments used for distance—reduction are however appropriate for performing a vertical deformation analysis. These measurements would principally allow a three dimensional analysis model and together with the additionally performed gravity measurements the application of the integrated model.

In the following we restrain our considerations however to the models and results due to two—dimensional horizontal movements, being of main interest with regard to existing geological microplate models of this region. This is also the reason, why the network conception (optimization, sensitivity analysis for single point deformations, fig. 4, (a)) is essentially directed versus an optimum sensitivity of the horizontal scene.

The deformation analysis has been carried out with respect to the results of direction and distance measurements in the following four epochs:

Epoch 1: finished on 29.8.1986 — just before the earthquake of 30.8.86,
Epoch 2: starting on 30.8.1986 — just in the morning after the earthquake,
Epoch 3: performed on august 1987 and
Epoch 4: performed on august 1988 .

Because the chain structure of the network — being imposed by the local topography of the area — it is unavoidable to have a rather low sensitivity concerning the detectable amount of deformations directed versus the networks 'weak(est)–form' (JÄGER, 1989), fig. 4, (b). This fact does not principally attack the quality of the

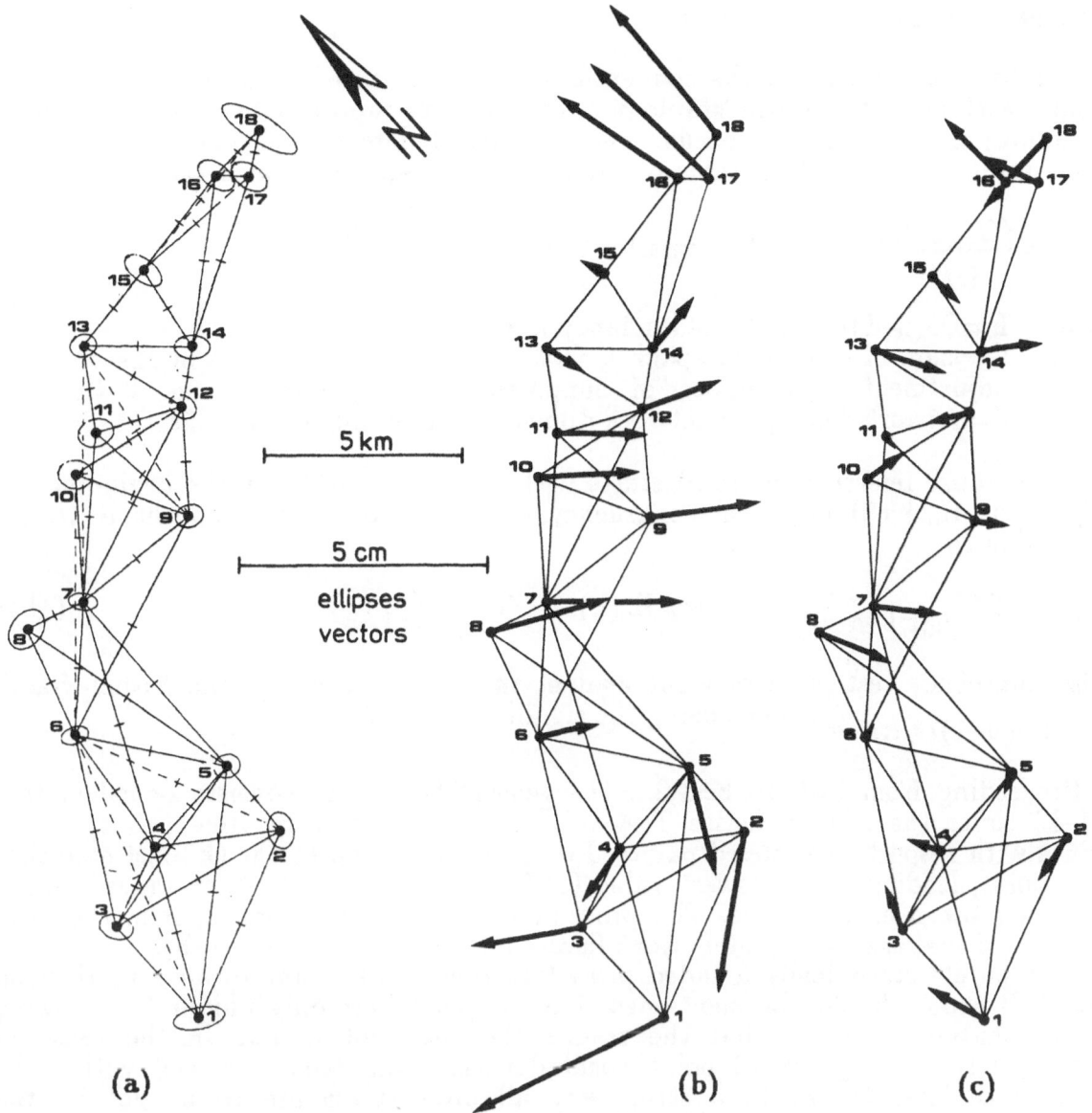

(a) (b) (c)

Fig.4 The networks sensitivity (power ß=80%)
    (a)   Sensitivity for twodimensional single point movements (α=5%)
    (b)   Shape and amount of sensitivity default directed to the 'weak–form' of the deformation network (α=5%)
    (c)   Typical deflections between two epochs by the view of a Helmert transformation

deformation analysis or even that of the network concept, because deformations pointed out in fig 3.1 (b) are unlikely to occur with respect to the geological models.
    Vice versa however, the shapes according to fig. 4, (b) turn out to express themselves as those 'pseudo–deformations' fig. 4, (c) being received by applying a

Helmert–transformation between the coordinates of the epochs. In order to find the behaviour with regard to the searched 'true deformations' the coordinate differences require to be filtered.

3.1 Absolute deformation analysis

Let us first take a look at the conventional analysis procedure of dececting 'moved points' with resepct to those 'stable points' which are supposed to have remained fix (absolute) in all regarded epochs. The concept of detecting in the k–th step the desired number of $s_k = M-k$ stable points consists of sequential calculation of $T(s_k)$:

$$T(s_k) = \frac{(\Omega_{c,k} - \Omega)/f_k}{\Omega/r} \qquad (\text{for small r: } T(s_k) = \frac{(\Omega_{c,k} - \Omega)/f_k}{\sigma_0^2}), \qquad (1a)$$

with $f_k = D \cdot (n-1) \cdot s_k - d$; r_i=redundancy in epoch i; $r = \Sigma r_i$;
Ω_i=sum of squares in epoch i; $\Omega = \Sigma \Omega_i$; Ω_c=sum of squares of the common adjustment; s_k=number of supposed stable points in the k–th step, d=network defect, D=network dimension, n=number of epochs

The procedure referring to (1a,b) starts with k=0 and s_k=M. At the beginning the hypothesis $H_o(k=0)$ of global congruency of all M points in n epochs is tested. According to

$$H_o(k) \rightarrow T \leq F_{f_k, r; 1-\alpha} \qquad (\text{or } H_o(k) \rightarrow T \leq F_{f_k, r=\infty; 1-\alpha}) \qquad (1b)$$

this congruency test is carried out against the critical value of the Fisher–fractil $F_{f_k = D \cdot (n-1) \cdot M, r; 1-\alpha}$, $(1-\alpha)$ being the significancy level.

Proceeding from k=0 to k=1,2,.. the general k–th step consists of calculating (M–k) times the test–statistics $T(s_k = M-k)$, by taking every time (M–k) stable points with respect to j moved points. From the k–th step to the (k+1)st step that very point, leading to the lowest value for $T(s_k)$ is regarded to have moved and is furtheron taken as nonstable. The procedure is sucessfully finished, if in the k–th step one of the regarded constellations finally fulfills the unequality (1b).

This purely statistically founded procedure takes no account to an a piori given model. It should further be mentioned that the results are only unique by following the sequential procedure, but the results they are not unique in the sense of alternatively existing (M–k) point constellations being stable as well within the significance level $(1-\alpha)$. In a strict way all alternatives are to be got by the calculation of all M!/(M–k)!/k! stable point cases of s_k=(M–k) points (e.g. for M=49 and k=6 this would mean an 'lottery–game' consisting of a impracticable calculation of a number of 13.983.816 $T(s_k=6)$–tests of adjustments!

The figure 5 shows the interesting result that only point number 8 has moved immediately after the earthquacke, the result is global unique. As this movement is additionally connected with a simultaneous and corresponding change in height and gravity, the movement itself could be regarded as a simple 'sliding downwards' although it also suits with the final results of the strain field.

The final analysis between epoch 1 and epoch 4, three years after the earthquake, leads within procedure (1a,b) after 161 steps to the stable points 2,5,9,15, the result is shown in fig. 5. It does not harmonize with the geological model and in the sense of the above mentioned global nonuniqueness of the procedure (1a,b) it gives a wrong or at least a non satisfactory impression of the real deformation behaviour.

It is worth to be mentioned as a general statement resulting from the authors experience that single point deformation analysis is not a model suited for relative crustal movements and in case of noise within the 'stable points' and should only be applied to the monitoring of deformations of objects such as dams, buildings etc. To get a deeper insight to the final deformation models, we now start very simply, by taking a look at the coordinate differences referring to the adjusted coordinates of epoch 1 as a unique reference datum.

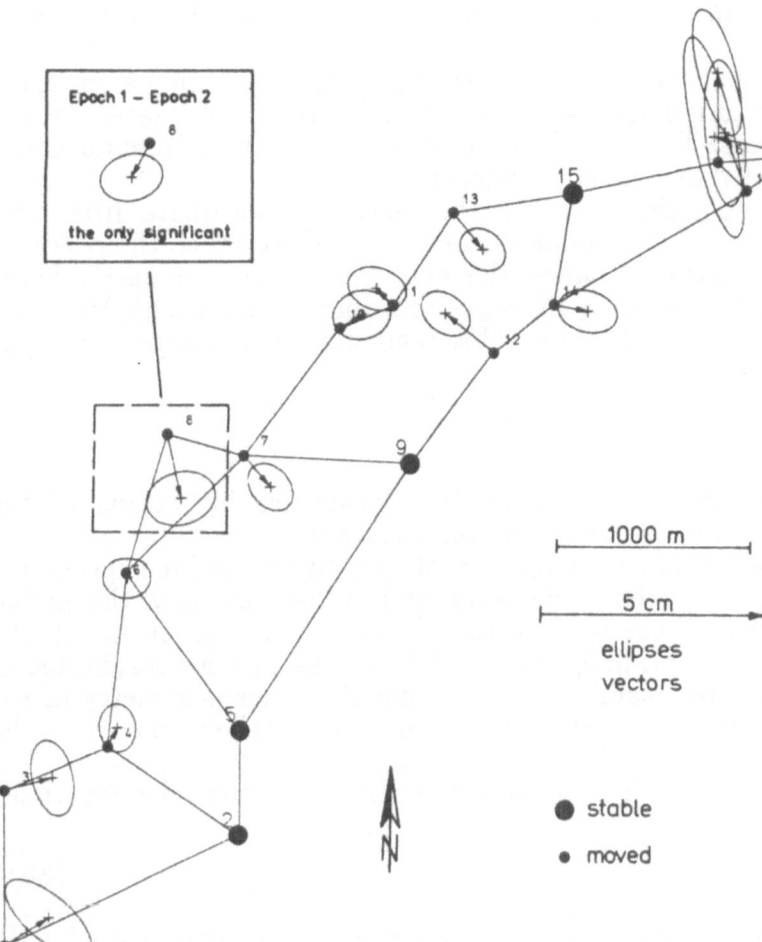

Fig. 5:
Results from single point deformation analysis between epochs 1 and epoch 4. The small picture shows the only significant movement from epoch 1 (before...) to epoch 2 (..immediately after the earthquake)

3.2 Relative deformation analysis

3.2.1 Filtering the weak forms

It is known from the theory of spectral network analysis (JÄGER, 1988) that the most likely deflection or the deflection of maximum size between the 'true shape' and the 'realized shape' of a geodetic network is to appear in the shape of the so called 'weak form' of the eigenvector x_{max} belonging to the first nonzero eigenvalue λ of the normalequation matrix. The inverse value $1/\lambda$ gives the expectable variance or the expectable size of the deflection x_{max} and its pointwise components, x_{max} being standardized as $\|x_{max}\|=1$. The ratios between $1/\lambda$ and the subordinate inverse eigenvalues characterize the likelihood of appearance, what means the dominance of the 'main weak form' x_{max} over the subordinate weak forms (JÄGER, 1989). A geodetic deformation network is least sensitive for those deformations v between two epochs k and l, which are directed versus the eigenvector d_{max} (2) of the largest eigenvalue of the covariance—matrix $C_d = C_{x_k} + C_{x_l}$ of the coordinate differences each in the common pseudo—inverse $(+)$ datum, see fig. 4, (a). Usually d_{max} coincides with the networks weak—form $x_{max_{k,l}}$

$$d_{max}(C_d) = d_{max}(C_{x_k} + C_{x_l}) \cong x_{max}(C_{x_{k,l}}) \tag{2}$$

with respect to identical shape and similar design, especially of course for a univariate epoch design.

Regarded vice versa the eigenvector d_{max} (2) gives the very shape of those residuals of a Helmert transformation between the adjusted coordinates of the epochs k and l, which are most likely and expectable to appear as 'pseudo deformations' even when the regarded epochs are congruent.

Our deformation field $r = x_4 - x_1$ (fig. 4, (c)) as the vector of coordinate differences between epochs 1 and 4 — due to the common datum being identical to the residuals of a mutual Helmert transformation — shows the expectable trend of partly being superimposed by pseudodeformations d_{max} or x_{max} according to fig. 4, (b). Even the absolute deformation result (fig. 5, (b)) shows this tendency. This suggests to split off a part $c \cdot d_{max}$, c being unknown,

$$v = r - c \cdot d_{max} \tag{3a}$$

from the given deformation field r to get more information by means of the remaining filtered — and considered as 'true' — deformations v.

The filtering off of the long waved shape d_{max} (with regard to the inner geometry d_{max} contains the least part of relative geometry and influences only the global network geometry (JÄGER, 1988)) changes in the desired way the outfit of the relative deformation field r, while adding $\pm d_{max}$ partly to the epoch coordinates x_i implies least strain. From the filter result v — preserving the inner geeometry of r — we therefore expect to recognize our relative deformation analysis model to be investigated in the following.

By performing a least squares filtering $v'v = min$ the remaining 'true deformations' v are resulting as

$$v = (I - 1/\lambda \cdot d_{max}' \cdot C_d^+) \cdot r \tag{3b}$$

The filtered v (fig. 6) shows better than r a classifiable deformation field with respect to the implied geological scene as the second decisive component for the choice of the deformation models.

From the filtered scene alone we recognize: rigid block movements shown by point—groups obviously carrying out a common translation and rotation (e.g. 1–3–4–5–2, 13–14–15 or 16–18–17), local fault—zones (e.g. between 13–14 and 12–11) and zones of high strain (the area 9–10–11–13–15–14–12 of the fault scence in another interpretation).

It is important in this context to point out that the filtering has been used only as a tool for getting a graphical overview concerning a suitable model for relative deformations. The filtering assures a model — built up with the impression of the areas of 'inner deformations' (fig. 6) — which is free from global effects like this of a 'bended banana' (fig. 4 (b),(c)), in the case of our chain—shaped network. The processing and statistical testing of the model parameters has been carried out however with the originally unfiltered r. This is regarded next with respect to the strain—parameters.

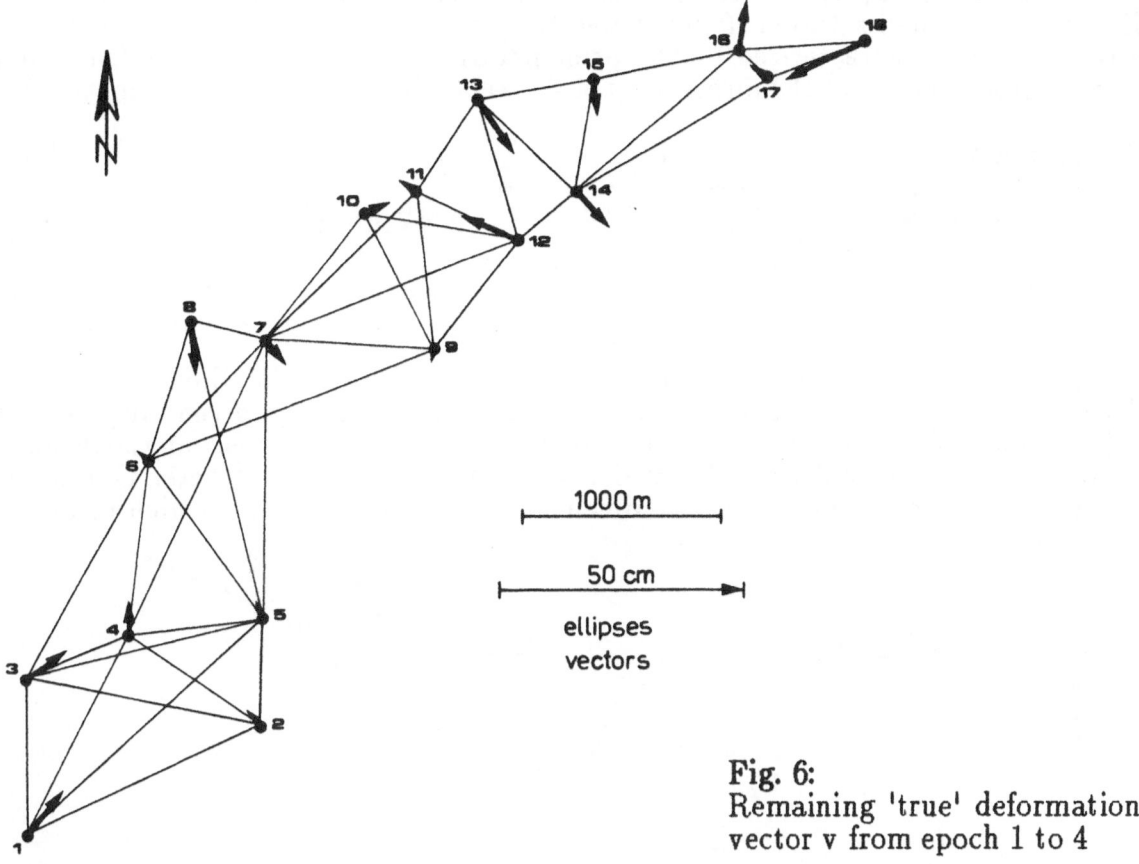

Fig. 6:
Remaining 'true' deformation
vector v from epoch 1 to 4

3.2.2 Strain analysis

By means of the deformation field (fig. 6) several statistical significant deformation models (e.g. fault zones and block movements) were investigated and must be followed with respect to a further maintaining of their trends and to future epochs simultaneously. Their preliminary results can not be described here because of the limited extension of this paper. Therefore we concentrate our considerations to the results of the strain–analysis.

Supposing r (3a) being the discretisation of the two dimensional locus–function of displacements in (x,y)

$$r = r(x,y) = \begin{bmatrix} u(x,y) \\ v(x,y) \end{bmatrix} \qquad (4a)$$

– u(x,y) and v(x,y) being the displacements of a point P(x,y) directed to the x–and y axis – the strain parameters at P(x,y) are given by the partial derivatives $u_x(x,y)$, $u_y(x,y)$, $v_x(x,y)$ and $v_y(x,y)$ of r(x,y) (ZIENKIEWICZ, 1984).

The strain–field is characterized by the so called 'deformation matrix' $V = R \cdot E$ giving furtheron the link to the stress field $\sigma(x,y)$. V(x,y) is the product of a rotation matrix R(x,y) and the <u>symmetric strain tensor E(x,y)</u>. With $E = R' \cdot V$ we have

$$E = \begin{pmatrix} \epsilon_{xx} & \epsilon_{xy} \\ \epsilon_{yx} & \epsilon_{yy} \end{pmatrix} = R' \cdot V = \begin{pmatrix} \cos\omega & -\sin\omega \\ \sin\omega & \cos\omega \end{pmatrix} \cdot \begin{pmatrix} u_x & u_y \\ v_x & v_y \end{pmatrix} \underset{(*)}{=} \begin{pmatrix} u_x & 1/2 \cdot (u_y + v_x) \\ 1/2 \cdot (u_y + v_x) & v_y \end{pmatrix}, \quad (4b)$$

219

where the angle $\omega(x,y)$ in R (4b) is determined by its property to symmetrize $E(x,y)$. For a small rotation ((*):$\tan\omega=\omega$) the wellknown relation (4b)/(*) holds. The <u>strain ellipse</u> is characterized by the principal axis — the eigenvalues A (maximum) and B (minimium) and the direction φ of A, respectively — of the <u>strain matrix</u> S:

$$S(x,y) = E(x,y) + I \qquad , \text{ and} \qquad (4c)$$

$$A,B = 1 + \left(\frac{\epsilon_{xx}+\epsilon_{yy}}{2}\right) \pm \frac{1}{2}\cdot\sqrt{(\epsilon_{xx}-\epsilon_{yy})^2+4\cdot\epsilon_{xy}{}^2}\ , \quad \tan(2\varphi)=\frac{2\cdot\epsilon_{xy}}{\epsilon_{xx}-\epsilon_{yy}} \quad \text{and for} \quad (4c.1)$$

$$(*)\ A,B = 1+\left(\frac{u_x+v_y}{2}\right) \pm \frac{1}{2}\cdot\sqrt{(u_x-v_y)^2+(u_y+v_x)^2}\ , \quad \tan(2\varphi)=\frac{\sin 2\varphi}{\cos 2\varphi}=\frac{u_y+v_x}{u_x-v_y}\ . \quad (4c.2)$$

The function r(x,y), whose partial derivates deliver the strain ellipses (4c) is to be derived as a function interpolating somehow the discretely given displacements $u_i(x,y)$ and $v_i(v,y)$ of r (3a) at each point P_i. There are various methods for the local or global interpolation of r which are not to be discussed here. Our method applied here is to employ linear twodimensional polynoms in (y,x), which means a

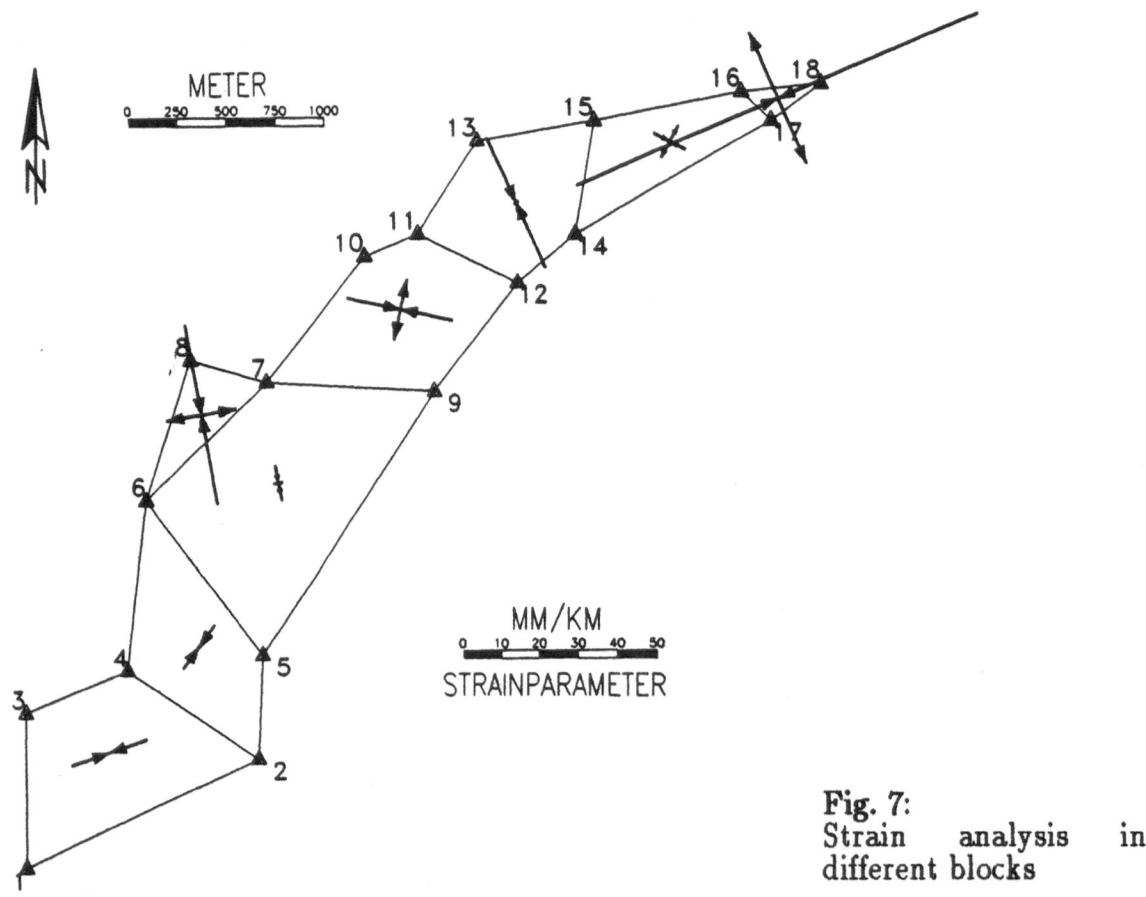

Fig. 7:
Strain analysis in different blocks

consistent and continuous interpolation of displacements in connected triangels. For this case the derivatives (4b) and the axes A,B and φ (4c), respectively can be derived by the parameters of an overdetermined or consistent affine transformation

$$(a):\ x_k=\begin{bmatrix}a & b\\ c & d\end{bmatrix}\cdot x_l+\begin{bmatrix}t_x\\ t_y\end{bmatrix}, \qquad (b):\ r_{k,l}= x_k-x_l = \left(\begin{bmatrix}a & b\\ c & d\end{bmatrix}-\begin{bmatrix}1 & 0\\ 0 & 1\end{bmatrix}\right)\cdot\begin{bmatrix}x\\ y\end{bmatrix}_{(l)}+\begin{bmatrix}t_x\\ t_y\end{bmatrix} \quad (5a,b)$$

(5a) between the corresponding point groups x_k and x_l from epoch k to epoch l or by the modified affine transformation (5b) which is referring difference vector $r_{k,l}$. By ignoring the index (l), the derivatives

$$u_x=a-1., \quad u_y=b, \quad v_x=c \quad \text{and} \quad v_y=d-1 \tag{5c}.$$

are directly visible from (5b). In case of the — mostly differential character of the affine transformation (5a) — the strain ellipse can be calculated according to (4c.2) with a,b,c,d (5a) otherwise a finite ω has to be splitted of from the affine transformation matrix according to (4b). The axes A,B and φ of the strain ellipse are identical with those of the Tissot—indicatrix of the affine transformation. The significance of strain is examined by applying the law of erorr propagation to A,B (4–c.1, 4–c.2), with respect to the coefficients (5c) and their covariance matrix, and then (A—1.0) and (B—1.) are tested against zero.

The above fig. 7 shows the significant strain ellipses due to a block division built up from v in fig. 6.

The last example is given by the global strain ellipse for the whole network region (see also fig.2 and 3) calculated from an overdetermined system (5a).

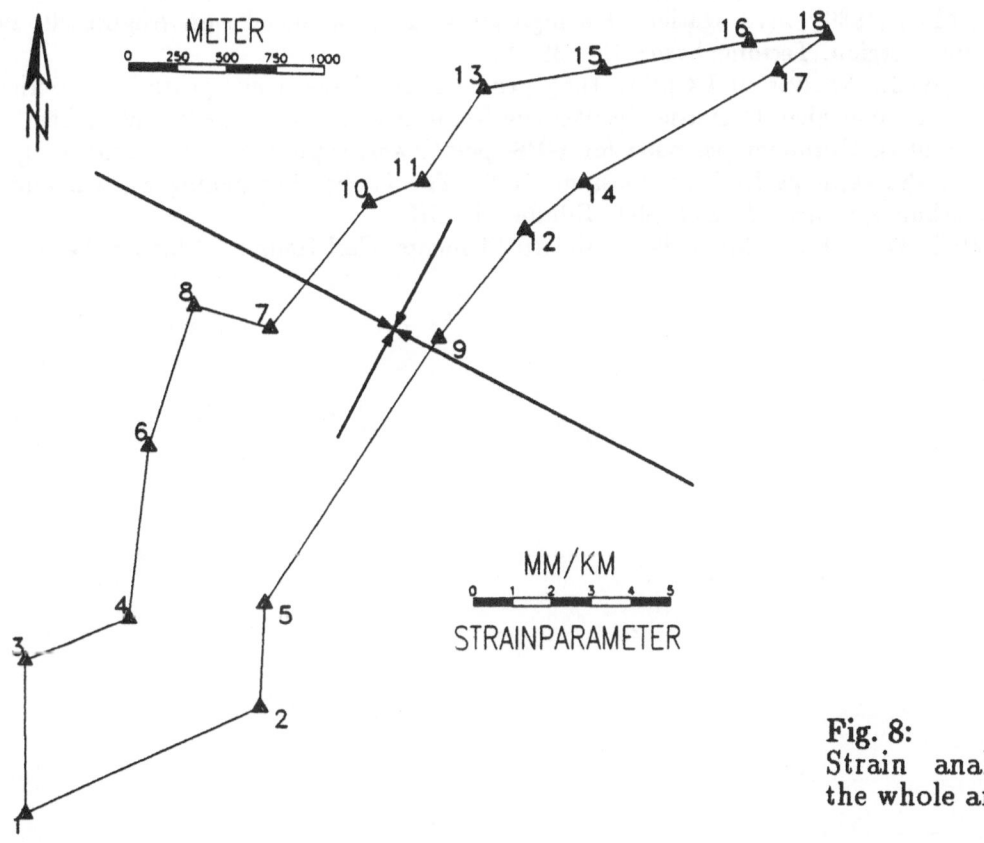

Fig. 8:
Strain analysis for the whole area

This more general information shows a good accordance with the tectonic model of the subduction of the Black Sea subplate under the Inter—Alpine subplate, even if the stress rate seems to be very high.

CONCLUSIONS

The late results of the strain analysis in the Surduc–Siriu geodynamic network over two years after the 1986 event coincide in a high degree with the general tectonic model of this area and with the sourceplane solution of the 1986 earthquake, which is characterized by a strike angle of 233^0, a dipping of 65^0 and a slip angle of 92^0 (Oncescu, 1989). It is in the moment an open question if the stress recovered in the network is due to a continuous motion in the Vrancea area or if it was stimulated by the earthquake, what would imply a stress reduction within the next time. In order to answer this question, the network will be remeasured in the future in time intervals of two or three years.

Acknowledgement The geodetic project is embedded in the bilateral cooperation between Romania and the FRG concerning scientific research and technological development, project–no. 054.2, 'Recent Crustal Movements'.

REFERENCES

ONCESCU, M.C. (1989): Investigation of a high stress drop earthquake on August 30, 1986 in the Vrancea region. Tectonophysics 163, 35–43.

JÄGER, R. (1988): Analyse und Optimierung geodätischer Netze nach spektralen Krierien und mechanische Analogien. Deutsche Geodätische Kommission. Report C–342, München.

JÄGER, R. (1989): Optimum positions for GPS–points and supporting fix–points in geodetic networks. Proceeedings IAG–Symposium S102 "The Global Positioning System and other radio tracking systems", 7.–8.08.1989, Edinburgh, UK.

ZIENKIEWICZ, O.C. (1984): Methode der finiten Elemente. Carl Hanser, München/Wien.

MICROCLIMATE STUDIES IN EXTREME CONDITIONS AND THEIR APPLICATIONS TO THE MONITORING OF RECENT CRUSTAL MOVEMENTS

P. Vyskočil

International Centre on Recent Crustal
Movements, Zdiby, Prague, Czechoslovakia

A. Tealeb and K.O. Sakr

National Research Institute of Astronomy
and Geophysics, Helwan, Cairo, Egypt

ABSTRACT

One of the outer conditions which greatly affect the accuracy of the terresterial geodetic measurements is the air temperature gradients above the earth's surface. These gradients affect the coefficient of refraction which causes for instance curvature of the sight beam between the levelling instrument and levelling rods. Therefore, the study of the geodetic refraction in the extreme conditions and calculations of the related corrections plays especially an important role for the performance of precise geodetic measurements aimed at recent crustal movement studies.

   Geodetic measurements, to monitor recent crustal movements, were carried out in the area southwest of Aswan lake since 1984. The study of the microclimate in the area was begun laterly in 1987. Temperature gradients were measured on six levels above the earth's surface at three different site localities. Refraction corrections were calculated for several levels and different azimuth of the sun.

INTRODUCTION

In order to determine the coefficient of refraction necessary for the corrections of the levelling measurements which were performed at Aswan region since 1984, the study of the problem of geodetic refraction in levelling for the microclimate conditions of Aswan desert was begun in 1987. At the northwestern part of Aswan lake (fig.1), local levelling lines were established in the desert area crossing parts of seismoactive faults. The regional levelling line was established along the main asphaltic roads in the area.

For the comparison with Aswan conditions, refractions were also studied at Helwan region (fig.1), 30 km to the south of Cairo. The effect of clouds, rains and pollutions are studied. Comparisons were carried out between the refraction results of Aswan and Helwan with the results given before for the refraction in Central Europe.

SITE SELECTIONS AND TEMPERATURE MEASUREMENTS

Three sites were selected, for temperature measurements, of the air close to the earth's surface, in the area to the northwest of Aswan lake. These sites are: the Aswan desert, the Aswan asphaltic roads in the desert area and the Aswan desert close to the Aswan lake. One site only was selected, near to Helwan Observatory, at Helwan region.

At all sites, the temperature of the air close to the earth's surface were measured at six levels: 0.01, 0.5, 1.0, 1.5, 3.0, 4.0 meters above the surface (fig.2). Wooden tower and portable unit of electronic measuring instruments type THERM 2230-1 connected to six sensitive PT 100 temperature probes and channel junction selector (THERM 2601-30) were used (Instruction Mannual). The air temperatures at these six levels were measured simultaneously each 15 minutes during the day-time (one hour before sunrise - one hour after sunset), using the Egyptian mean time. The measurements were continued for selected days each month to cover the whole year.

TEMPERATURE GRADIENTS

For each site of measurements, six temperature gradients G_1, G_2, G_3, G_4, G_5 and G_6 were approximated using the equation given by Vyskočil 1966,1982 in the form:

$$\frac{dt}{dh} = a + b \cos Z + c \cos^2 Z + e \cos^3 Z \tag{1}$$

where $\frac{dt}{dh}$ is the measured temperature gradient. a,b,c and e are constants and Z is the zenith angle of the Sun for the site of measurements.

The zenith angles of the Sun (Z), for the sites of the temperature measurements, were calculated from their geographic latitudes and the mean local time of measurements using a group of Astronomical formulae (Vyskočil and Ramboucik 1987, Sakr 1989).

The diagrams representing the temperature gradients G2, G3 and G4 with the zenith angle of the Sun (Z) for the sites of the Aswan desert, Aswan asphaltic roads and Aswan desert close to the lake are given in figures 3,4 and 5 respectively. According to these figures, the area inside

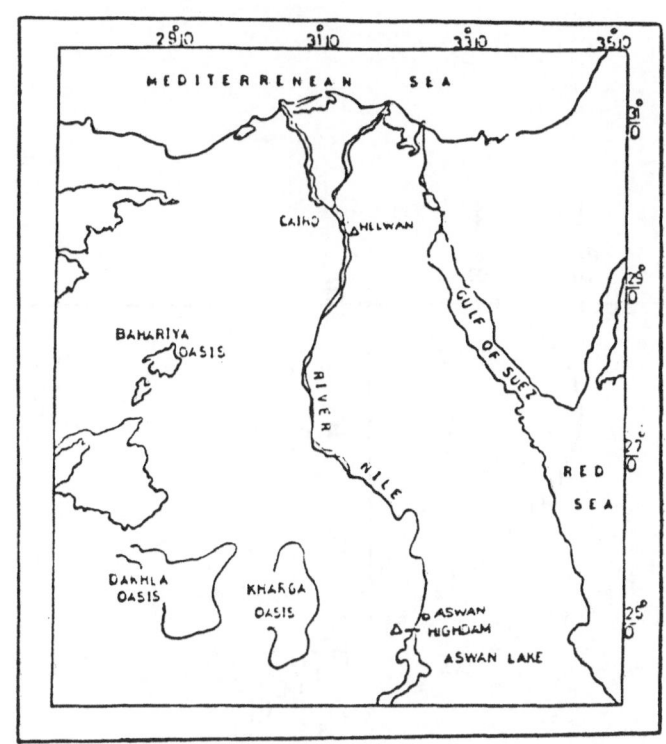

FIG. 1. Sites of refraction studies in Egypt.

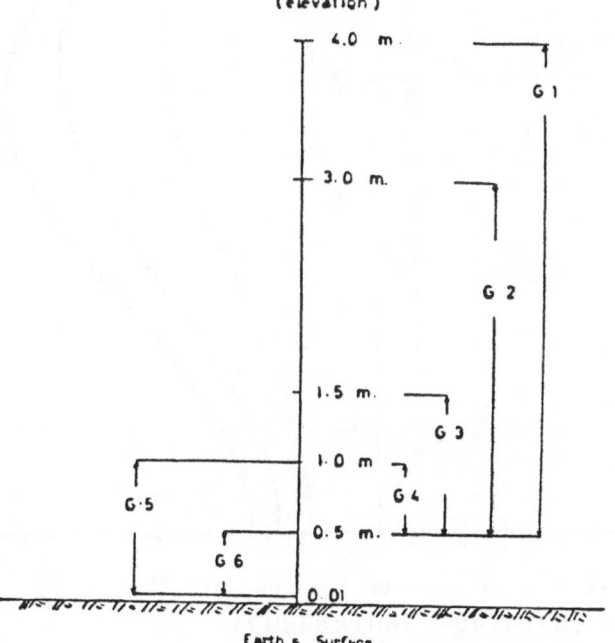

FIG. 2. Elevations of temperature probes and diagram for temperature gradients.

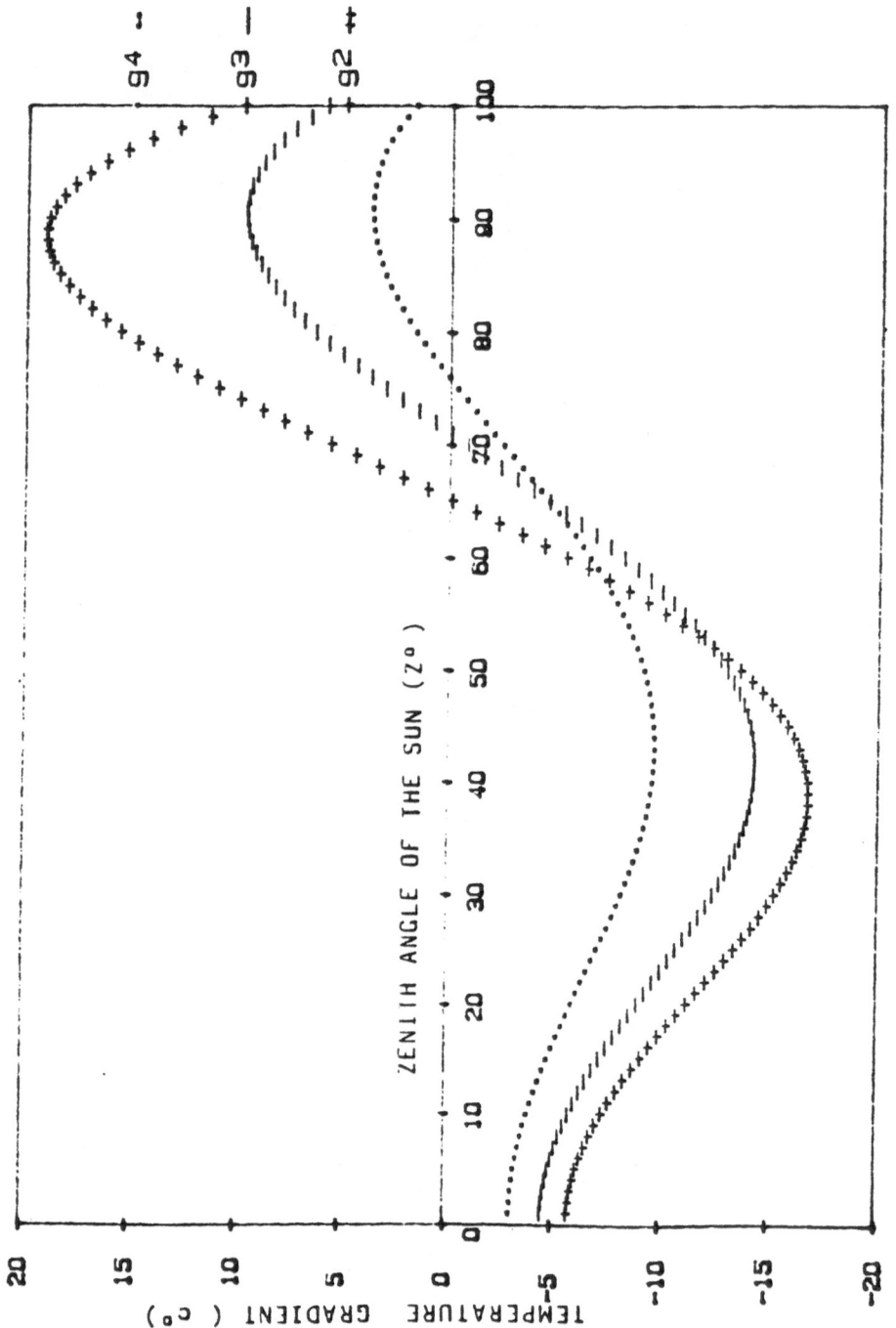

FIG. 3. Temperature gradients G_2, G_3 and G_4 for Aswan desert.

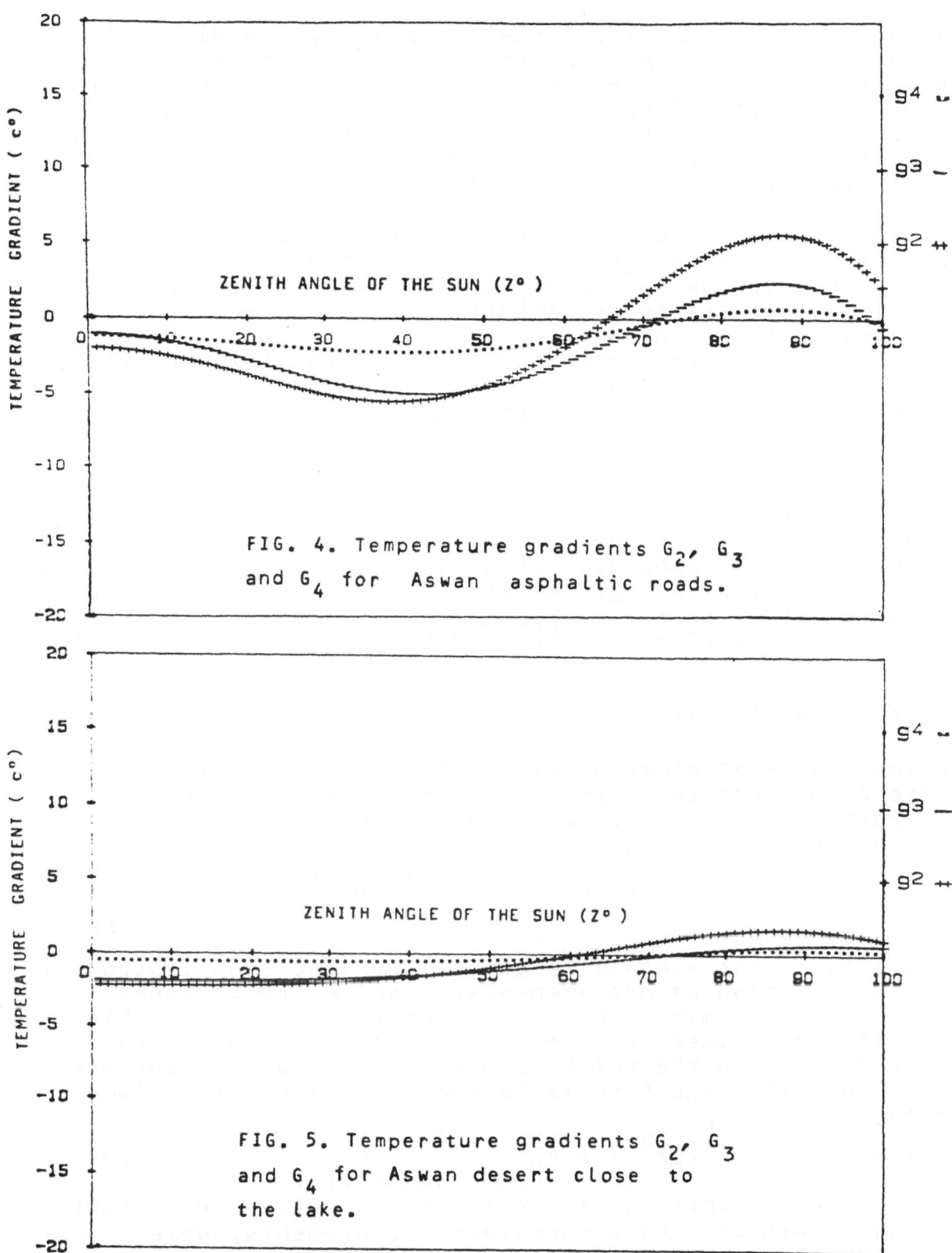

FIG. 4. Temperature gradients G_2, G_3 and G_4 for Aswan asphaltic roads.

FIG. 5. Temperature gradients G_2, G_3 and G_4 for Aswan desert close to the lake.

Aswan desert has mostly higher temperature gradients (fig.3) than the areas of the asphaltic roads (fig.4) and Aswan desert close to the lake (fig.5) . Here it can be concluded that the effect of refraction are greater in the desert area of Aswan than the area close to the asphaltic roads and Aswan lake. For the desert area close to Aswan lake lower temperature gradients (fig.5) of few changes during day-time were recorded. This, of course, are due to the evaporation of water from the surface of the lake and the circulations of homogeneously distributed humidt air in the area close to the lake and its vicinity. The temperature gradients for the Aswan asphaltic roads in the desert area (fig.4)are lower than those for the Aswan desert (fig.3) and higher than the same gradients for the Aswan desert close to the lake (fig.5) .

At Helwan region, the temperature measurements were carried out in different climate conditions. Some measure-ments were performed in completely sunny days and others were carried out on cloudy and rainy days. Pollutions from industry are affected greatly the whole region. In figures 6 and 7, the temperature gradients G2, G3 and G4 for the sunny and the cloudy days respectively are represented against the zenith angle of the Sun. According to these figures, the temperature gradients for the cloudy days (fig.7) are smaller than those for the sunny days (fig.6) .

REFRACTION COEFFICIENT

Using the temperature gradients G2, G3 and G4, for each site of measurements, coefficients related to the refraction error, refraction coefficient (C), were calculated for various elevations from the earth's surface. For the calculations of the refraction coefficient, the following equation (Vyskočil 1966, 1982, 1989) was used :

$$C = c + 3e\bar{h}_0 \qquad (2)$$

where $\bar{h}_0$ is the average height of the levelling instrument in the section of measurements. c and e are costants.

For the determination of the constants c and e, cubic equation described the propagation of the temperature (t) with respect to the height (h) above the earth's surface was used. This equation is in the form (Vyskočil 1966, 1982) :

$$t = a + b.h + c.h^2 + e.h^3 \qquad (3)$$

where a, b, c and e are unknown constants.

The coefficients of refraction (C) , which were calcula-ted for each site of temperature measurements, were tabulated[*]with respect to the zenith angle of the Sun (Z).

----

* According to the limited space given for publishing, the full text including the tables should be published later.

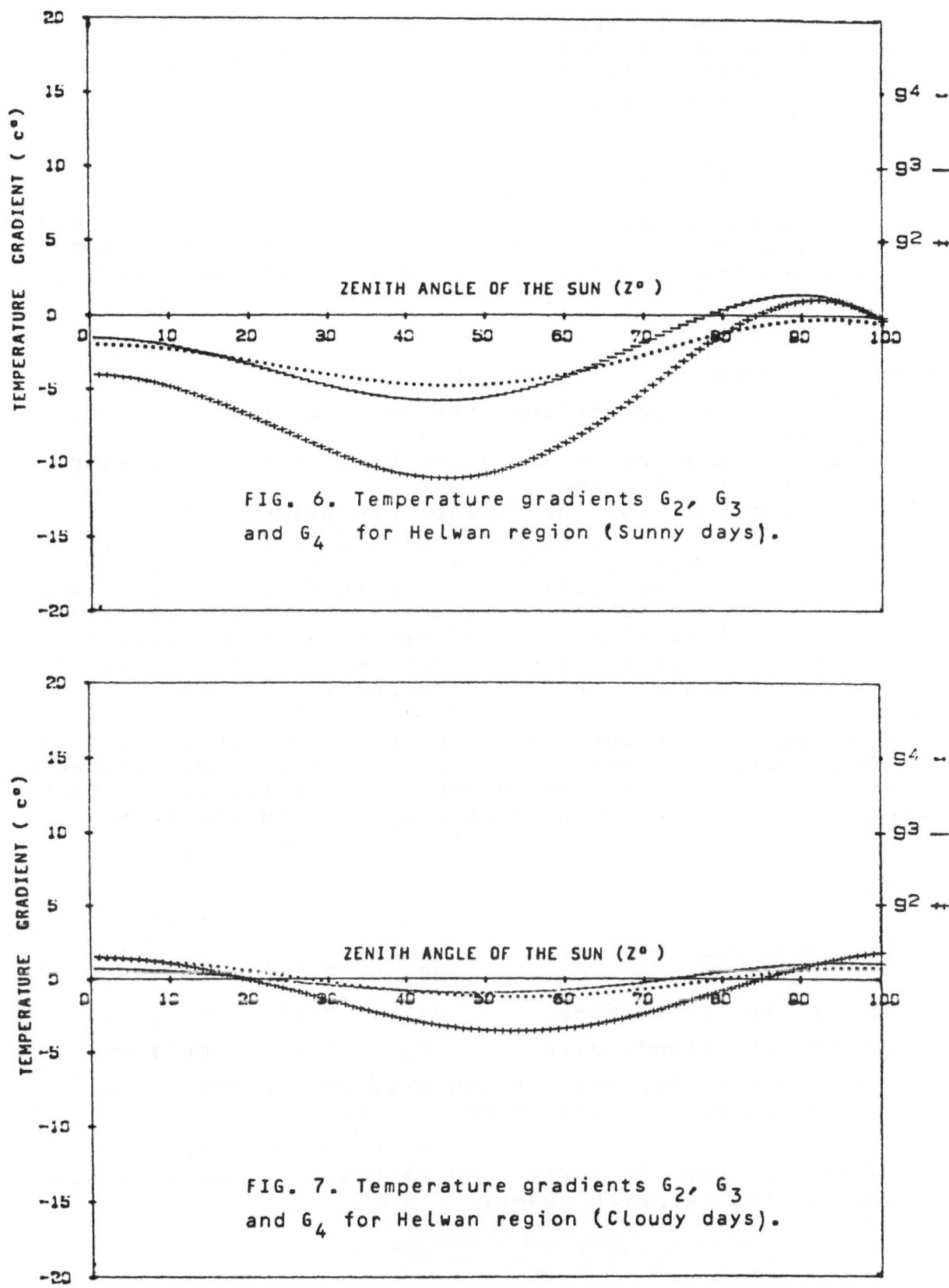

FIG. 6. Temperature gradients G_2, G_3 and G_4 for Helwan region (Sunny days).

FIG. 7. Temperature gradients G_2, G_3 and G_4 for Helwan region (Cloudy days).

These tables can be applied to correct the levelling measurements carried out at these sites and also for weather conditions similar to the conditions of the initial measurements.

DIFFERENTIAL REFRACTION

The tables calculated for the refraction coefficient (C) can be also used to estimate the average correction (dR_c) in each section by applying the formula for the differential refraction (Vyskočil 1966, 1989), in the form :

$$dR_c = \frac{1}{2} K.\bar{S}^2. \quad - C \quad . \Delta h \tag{4}$$

where $K = 8\times10^{-7}$,

$\bar{S}$ the average distance between the instrument and rods,

Δh the difference in height between two levelling benchmarks, and

C constant for the average elevation of the levelling instrument, taken from the tables.

Using an average value for the time of the levelling measurements at one section, mean height of the levelling instrument about 1.5 m , distance between the instrument and rods equal to 25 m and difference in height between two levelling benchmarks equal to 1 m, the mean value of dR_c were calculated for the different sites from the formula (4). The mean value of the coefficient of refraction (C) for the zenith angle of the Sun (Z) between 30-80 degrees was calculated for each site of measurements from the tables mentioned before. The results of the mean value of refraction correction (dR_c) are in the form :

For Aswan desert $\qquad dR_c = 13\times10^{-4}$ mm/1 m.

For asphaltic roads in Aswan desert $\qquad dR_c = 3\times10^{-4}$ mm/1 m.

For Aswan desert close to the lake $\qquad dR_c = 2\times10^{-4}$ mm/1 m.

For Helwan Sunny days $\qquad dR_c = 6\times10^{-4}$ mm/1 m.

For Helwan Cloudy days $\qquad dR_c = 1\times10^{-4}$ mm/1 m.

The corresponding value calculated before for Central Europe given by Vyskočil 1966 in the form :

$$dR_c = 3\times10^{-5} \quad \text{mm/1 m,}$$

is very low than the correction values which were calculated for the Egyptian desert.

COCLUSIONS

To correct the levelling measurements, which were carried out at seismoactive areas in Aswan desert, south Egypt, aimed to monitor the crustal deformations, the study of the refraction problem for the microclimate conditions of Aswan region was of great necessity. Three sites were selected at Aswan region for the temperature measurements and one site was selected at Helwan region in order to compare the results. Because, there no temperature measurements were systematically carried out before in the desert conditions of Egypt and North Africa, this work is considered as the first study of refraction in Africa.

The development of temperature and temperature gradients for the desert area, asphaltic roads and desert close to Aswan lake are significantly different. The temperature gradients for the desert conditions are more higher than those for other conditions, i.e. asphaltic roads and close to the lake.

The refraction coefficient were calculated for each site with respect to different elevations and the zenith angle of the Sun. The average corrections for the levelling measurements were also calculated for the different sites of measurements. Comparisons between these corrections and the correction for Central Europe was carried out. The average corrections for the different sites in Egypt are very high than the correction for Central Europe.

Similar studies should be extended in future to cover other areas in Egypt, such as the Nile Delta area and the desert areas close to the Mediterranean and Red sea coasts, taking into considerations the various surfaces of the ground.

REFERENCES

Instruction Mannual of THERM instruments. (1986). THERM 2230-1, THERM 2601-3 and temperature probe PT 100. Holzberg, West Germany.

Sakr, K.O. (1989). Microclimate studies at Aswan region and their applications to monitoring recent crustal movements. M.Sc. Thesis, Qena Assuit University, Egypt.

Vyskočil, P. (1966). Study of the possibilities of deminishing of microclimate effect at the levelling measurements. Desertation work in Czeck language, Technical University, Prague. not published .

Vyskočil, P. (1982). Refraction in levelling. Sbornik Výzkumných praci VUGTK, Svazek 14, 63-87.

Vyskočil, P. (1989). Procedures for monitoring recent crustal movements. Part 2, IAG, CRCM, ICRCM, Prague, 62 p.

THE DISTRIBUTION OF LENGTH AND DIRECTION OF TWO-DIMENSIONAL RANDOM VECTORS

W. Caspary, W. Haen, V. Platz
Universität der Bundeswehr München
Werner–Heisenberg–Weg 39, D–8014 Neubiberg

INTRODUCTION

Geodetic deformation analysis is usually based on conventional stochastic assumptions and Least–Squares (LS)– estimation. In this case direction and length of the received residual position error vectors (RPEV) are random variables with known distribution. Therefore these quantities can be used to establish a statistical test of the model assumptions. Because of the sensitivity of the LS–estimation to local violations of the assumptions, estimation methods, more robust in this respect should be considered, as for example Maximum–Likelihood–type estimators (M–estimators). Since these M–estimators are non–linear, a straightforward derivation of the distribution of direction and length of RPEVs is not possible.

In this paper the distributions of directions (Offset–Normal (ON) –distribution) and lengths (χ–distribution) of two–dimensional RPEVs under LS–estimation are given. In a simulation study the influence of local deformations on these distributions and on those of neighbouring points in a network type model is investigated. The results are compared to the corresponding outcome of M–estimation.

DISTRIBUTION OF THE LENGTH OF RPEVs

In the two–dimensional case the length $\bar{d}$ of a RPEV is computed from x– and y–residuals v_x and v_y at a point:

$$\bar{d} = \sqrt{v_x^2 + v_y^2} \quad \text{with:} \quad \bar{v} = \begin{bmatrix} v_x \\ v_y \end{bmatrix} \sim N(0; \Sigma_{\bar{v}})$$

and
$$E(\bar{v}) = 0 \; ; \qquad \Sigma_{\bar{v}} = \begin{bmatrix} \sigma_{v_x}^2 & \sigma_{v_x v_y} \\ \sigma_{v_y v_x} & \sigma_{v_y}^2 \end{bmatrix} \qquad (2.1)$$

A standardized vector v can be derived:

$$v = A \, \bar{v} \sim N(0; I) \qquad (2.2)$$

by multiplication with a matrix A satisfying:

$$A \, \Sigma_{\bar{v}} \, A^T = I; \quad I : \text{unit matrix.} \tag{2.3}$$

Then $\quad\quad d \quad = \, ||v|| \tag{2.4}$

is the square root of the sum of squares of two independent standard–normal random variables. Consequently d is χ–distributed with two degrees of freedom (d.f.) (χ_2–distribution). The density function f_{χ_n} of the central χ–distribution with n d.f. is given by eq. 2.5 (cf. Prohorov, Rozanov 1969):

$$f_{\chi_n}(d) \quad = \quad \begin{cases} \dfrac{d^{n-1}}{2^{\frac{n-2}{2}} \, \Gamma(\frac{n}{2})} \cdot e^{\frac{-d}{2}} & ; \, d > 0 \\[2mm] 0 & ; \, d \leq 0 \end{cases} \tag{2.5}$$

This density function can be derived from the χ_n^2 density function with n d.f. via:

$$f_{\chi_n}(d) \quad = \quad 2d \cdot f_{\chi_n^2}(d^2) \tag{2.6}$$

In Fig. 2.1 the density functions of χ_2– and χ_2^2–distribution with 2 d.f. are shown.

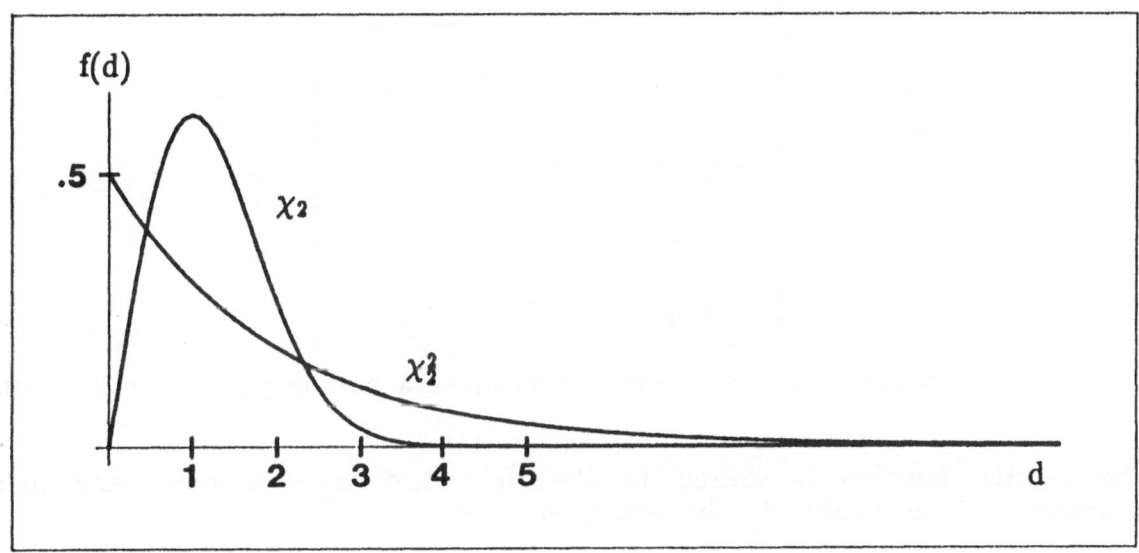

Fig.2.1. $\quad \chi_2$ – and χ_2^2 – density function

If the residuals v_x and v_y have non–zero expectation, d follows a noncentral χ_2–distribution. The effect of increasing noncentrality on f_{χ_n} is depicted in Fig. 2.2.

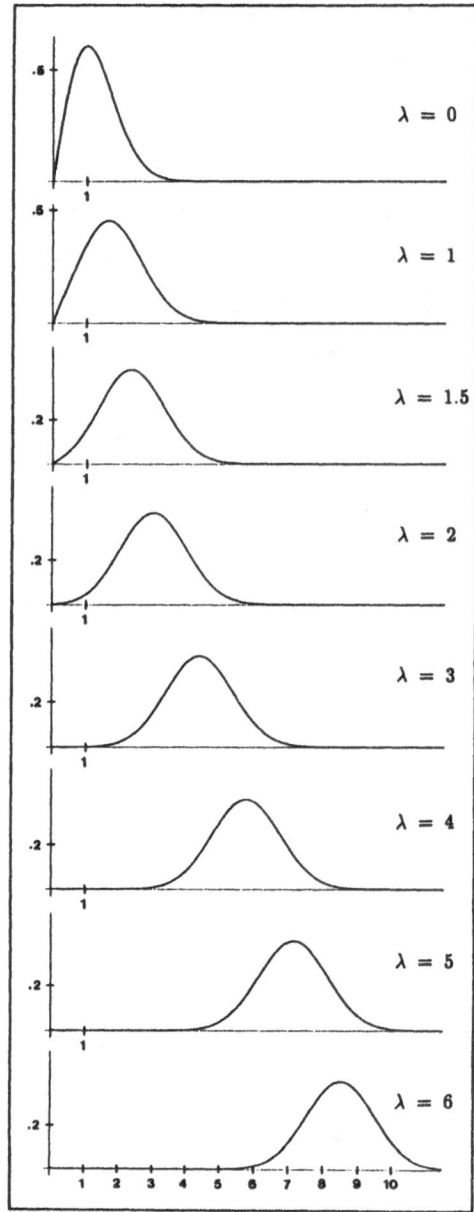

Fig. 2.2. Effect of a noncentrality parameter λ on the χ_2–density function

The density function is shifted to the right and takes a more and more symmetrical form, similar to the normal–density.

DISTRIBUTION OF THE DIRECTION Θ OF RPEVs

The direction of a two–dimensional random vector is a circular stochastic variable. It takes values between 0 and 2π. In the special case of normally distributed components v_x and v_y the direction Θ is offset–normal (ON)–distributed:

$$\Theta = \arctan\left[\frac{v_y}{v_x}\right] \sim \text{ON} \quad \text{if} \quad v = \begin{bmatrix} v_y \\ v_x \end{bmatrix} \sim \text{N}\left(\begin{bmatrix} \xi \\ \eta \end{bmatrix};\Sigma_v\right); \quad \begin{bmatrix} \xi \\ \eta \end{bmatrix} = \text{E}(v) \quad (3.1)$$

234

with density function: f_{ON}:

$$f_{ON}(\Theta; \xi, \eta, \sigma_{v_x}, \sigma_{v_y}, \rho) =$$

$$C(\Theta)^{-1}\left[\varphi_2(\xi, \eta; 0, \Sigma_{\bar{v}}) + a \cdot D(\Theta) \cdot \Phi_1\big(D(\Theta)\big) \cdot \varphi_1\Big(a\big(C(\Theta)\big)^{-\frac{1}{2}}\big(\xi \cdot \sin\Theta - \eta \cdot \cos\Theta\big)\Big)\right] \quad (3.2)$$

with
$$C(\Theta) = a^2 \cdot \left(\sigma_{v_y}^2 \cdot \cos^2\Theta - \rho \cdot \sigma_{v_x} \cdot \sigma_{v_y} \cdot \sin 2\Theta + \sigma_{v_x}^2 \cdot \sin^2\Theta\right)$$

$$a = \left(\sigma_{v_x}\sigma_{v_y}\sqrt{1-\rho^2}\right)^{-1}$$

$$D(\Theta) = a^2\, C(\Theta)^{-\frac{1}{2}}\left(\xi \cdot \sigma_{v_y}(\sigma_{v_y} \cdot \cos\Theta - \rho \cdot \sigma_{v_x} \cdot \sin\Theta) + \right.$$
$$\left. \eta \cdot \sigma_{v_x}(\sigma_{v_x} \cdot \sin\Theta - \rho \cdot \sigma_{v_y} \cdot \cos\Theta)\right)$$

and

ξ, η : expectation of v_x, v_y

$\sigma_{v_x}, \sigma_{v_y}$: standard deviation of v_x, v_y

ρ : correlation of v_x and v_y

$\Sigma_{\bar{v}}$: covariance matrix of $\bar{v}$

$\varphi_1(\cdot)$: density function of the N(0;1)–distribution

$\Phi_1(\cdot)$: distributon function of the N(0;1)–distribution

$\varphi_2(\cdot, \cdot, 0, \Sigma_{\bar{v}})$: density function of the two–dimensional normal–distribution with expectation 0

The distribution of Θ depends on the expectation ξ and η of the components v_x and v_y, their variances and their correlation. In Fig. 3.1–3.3 the influence of these parameters is shown. A circular representation is used, so the density curves are closed but the encompassed area is not unity in this case.

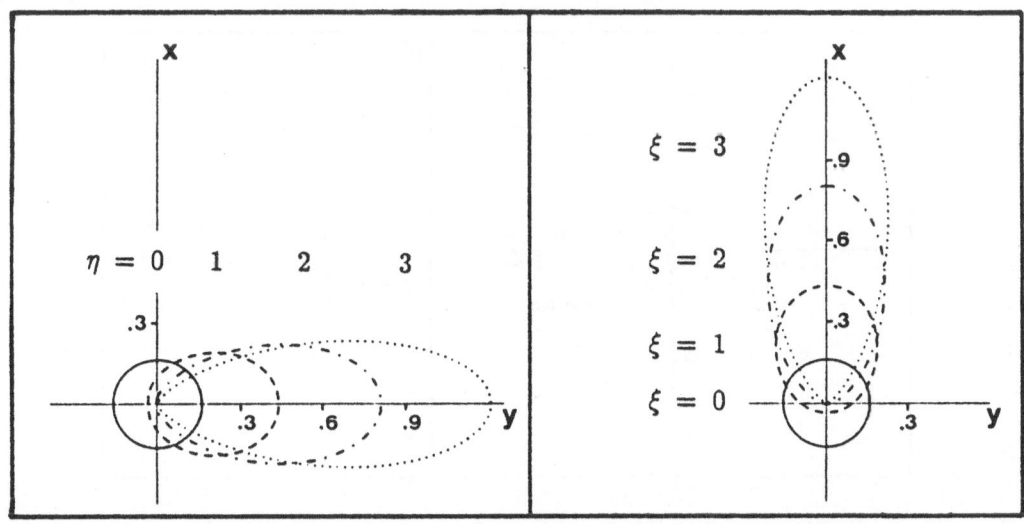

Fig 3.1.   ON–density function: variation of the expected values ξ and η

The simplest special case of the ON–distribution ($\xi=\eta=\rho=0$, $\sigma_{v_x}=\sigma_{v_y}=1$) is the uniform distribution: the circular representation becomes a circle with radius $1/2\pi$. In Fig. 3.1 the expectation of one component is gradually increased whereas the expectation of the second one is kept zero, and $\rho=0$, $\sigma_{v_x}=\sigma_{v_y}=1$. The circle is deformed correspondingly, following the direction of the increased expectation.

Another variation is shown in Fig. 3.2. The ratio of variances is changed by increasing σ_{v_x} or σ_{v_y} respectively while the second standard deviation is kept fixed ($=1$) and $\xi=\eta=\rho=0$:

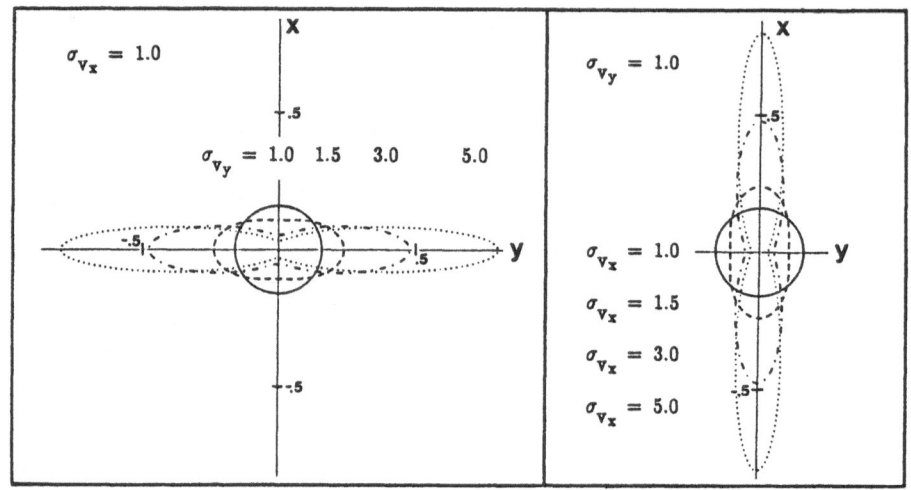

Fig. 3.2. ON–density function: variation of the variances

This results in bimodal distributions with modes in the direction of the larger variance component.

Variation of correlation as well yields bimodality. The modes now are points on a diagonal straight line (cf. Fig. 3.3a).

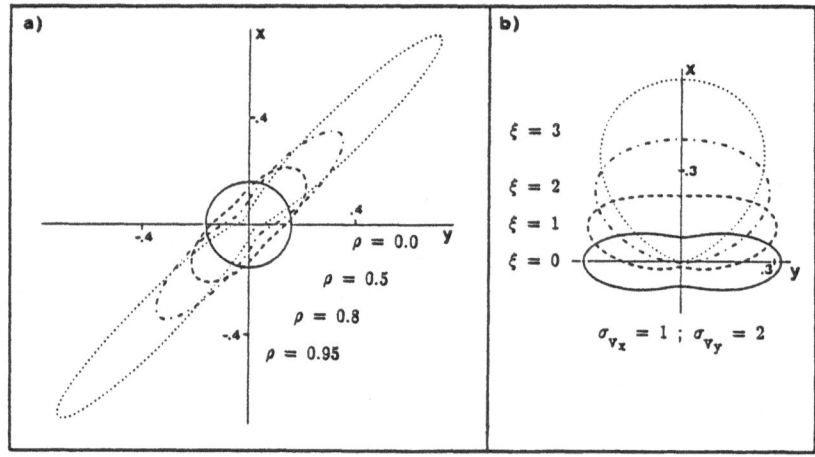

Fig. 3.3. ON–density function: a) variation of correlation, b) variation of variance ratio, starting from non–uniformity

So far one parameter has been varied while all others were fixed at their standard values. Fig. 3.3b shows the effect of increasing ξ, when $\sigma_{v_x}=1$ and $\sigma_{v_y}=2$. The initially bimodal distribution becomes unimodal with a maximum on the x-axis which however is wider in comparison to that of Fig. 3.1 ($\sigma_{v_x}=\sigma_{v_y}=1$).

In practice all parameters of the ON–distributions will have values which deviate from the standard quantities. Therefore, the corresponding ON–distributions must be computed for every special case. In general it can be stated that small variations of the parameters have only a minor effect on the density function.

A SIMULATED EXAMPLE

A study has been carried out where the theoretical distribution of the polar components of two–dimensional RPEVs have been compared with the corresponding empirical distributions. Two sets of RPEVs have been obtained from simulated deformation analyses using LS–estimation and robust M–estimators (Huber– and Hampel–estimators) respectively. The objective functions of the latter have been modified according to Caspary, Chen, König (1983) in order to warrant invariance of the estimated parameters with respect to rotations of the chosen reference system.

Two samples of "observations" (raw deformations) have been generated, the first sample conforming to the mathematical model, the second sample containing an outlier which can be looked upon as a single point movement or a deviation from the assumed stochastic model of normality. All simulations are based on the network of Fig. 4.1 and a sample size of 1000.

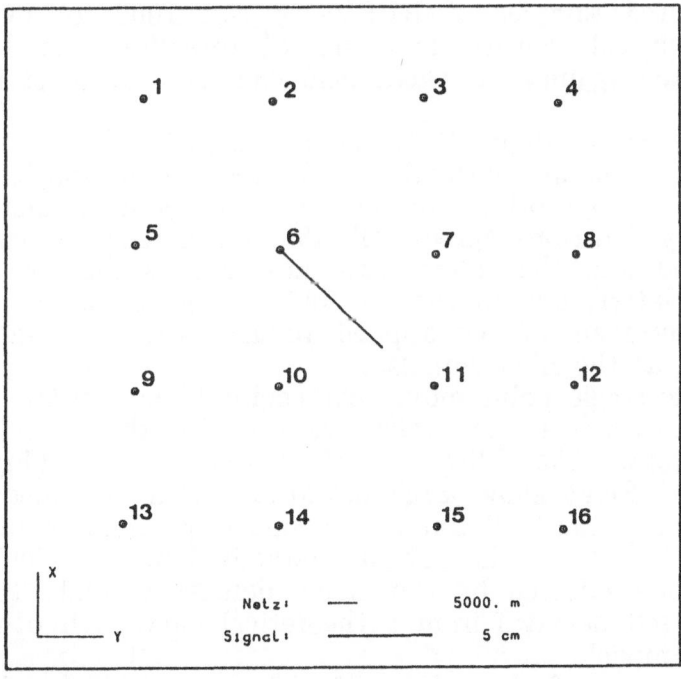

Fig. 4.1. Simulated network with deformation of point 6

The vector Δ of raw deformations (input data of deformation analyses) has been computed as the difference of two independent sets x_1 and x_2 of standard normal variables:

$$x_1 \sim N(0;I) \; ; \quad x_2 \sim N(0;I) \tag{4.1}$$

$$\Delta \quad = x_2 - x_1 \sim N(0;\Sigma=2I) \tag{4.2}$$

The deformation model is the planar rigid body model with two translations, one rotation and a scale parameter to be estimated. With these parameters as components of of the vector t and with the design matrix H the LS–estimator takes the form:

$$\hat{t} \quad = (H^T P \; H)^{-1} H^T P \; \Delta \tag{4.3}$$

where $P=\Sigma^{-1}$ is the weight matrix.
The robust estimates have been computed via iteratively reweighted LS–estimation:

$$\hat{t}^{(i)} = (H^T W^{(i-1)} P \; H)^{-1} H^T W^{(i-1)} P \; \Delta \tag{4.4}$$

with pseudoweights $w_j^{(i)}$:

$$W^{(i)} = \text{diag } w_j^{(i)}; \quad w_j^{(i)} = \frac{d\rho(d_j^{(i-1)})}{dd_j^{(i-1)}} / d_j^{(i-1)} \tag{4.5}$$

$$j \quad = 1,...,2n \; ; \quad \text{n: number of points}$$

For more details on this estimation method see Caspary, Borutta (1987). The tuning constants of the Huber–estimator has been fixed at 1.5 and those of the Hampel–estimator at 1.5, 3.0, 6.0.
From the computed samples of RPEVs of size 1000 for every point of the network the empirical density functions of direction and length have been derived and plotted against the theoretical density. The most interesting results are:

1st case: The assumptions of the model are satisfied.
If the raw deformations are normally distributed and no outliers are present the rigid body model is appropriate. In this case the empirical distributions and the theoretical density functions agree well. Small differences between the outcome of the LS–method and the robust estimates are caused by a certain loss of efficiency of the latter, due to the selected tuning constants. The results show that the polar residuals of the applied robust estimation methods follow the same distribution as the LS–residuals.

2nd case: One single point movement (point 6) is present.
In this situation which is not taken care of by the model the results are completely different. The LS–empirical density curves (Fig. 4.2a) of the directions of the RPEVs show large deviations from the theoretical curves for all points. Obviously the well known LS–smearing effect is present and makes the results rather useless. The robust methods however yield estimates (Fig. 4.2b) which are not effected by the model deficiency. Only the density at the deformed point itself deviates from a theoretical curve. All others are the same as in the correct model.
The good performance of the robust M–estimators can also be shown for the distribution of the lengths. As an example a plot of the empirical distribution

of the standardized lengths of the RPEVs, which were obtained from a Hampel estimation, and the corresponding χ–density functions are shown in Fig. 4.3.

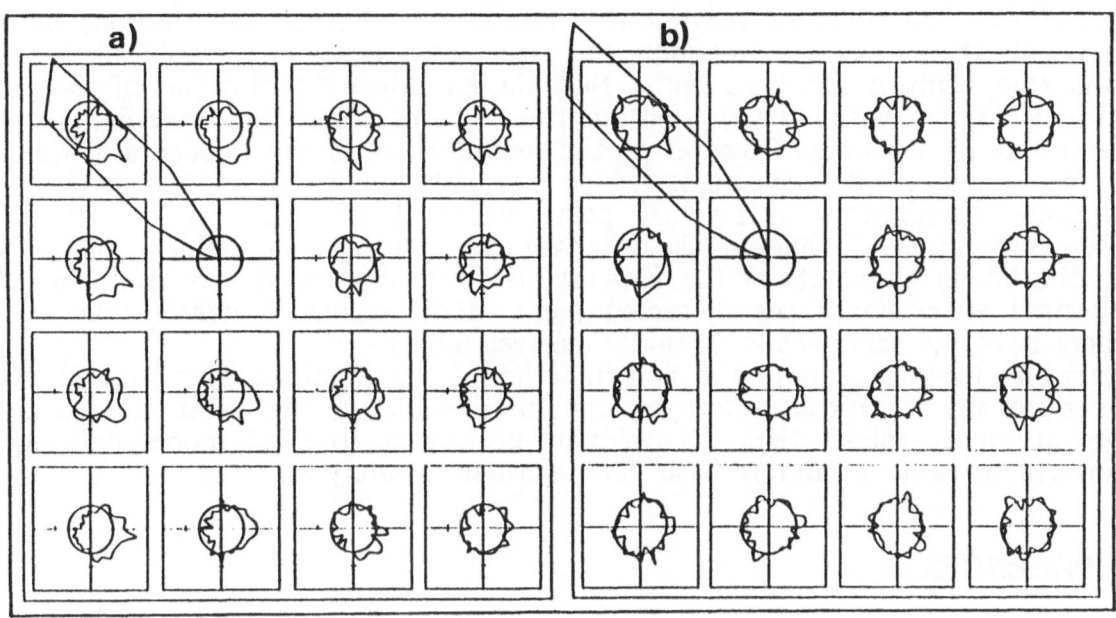

Fig 4.2. Theoretical and empirical density functions of RPEV directions
a) resulting from LS–estimation
b) resulting from robust Huber–estimation

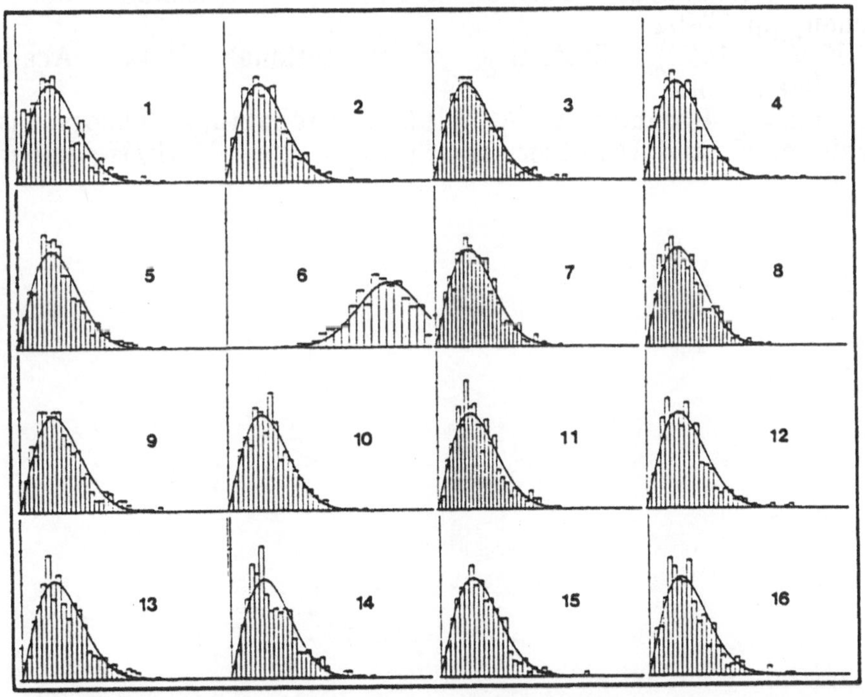

Fig. 4.3 Empirical and theoretical density functions of RPEV lengths (standardized), resulting from robust Hampel–estimation

CONCLUSIONS

It has been pointed out that the direction and length of the RPEV have ON- and χ–distribution respectively if the orthogonal error components are N(0;1)- distributed. The effect of varying the parameters of these distributions on the shape of their density functions has been demonstrated graphically. In a simulation study it has been found that the empirical distributions of the polar residuals agree with theory, no matter whether LS- or modified robust M-estimators are used. However if the assumptions of the model are violated e.g. by single point movements which are not considered in the model, the empirical distributions of residual polar errors of LS- and robust estimation differ considerably. While in the LS case the deformations influence the error distribution in all points of the network, the robust methods are free from this undesired effect. Only the deformed points show deformed density curves, all others have the same shape as in the correct model.

According to these results the non–linearity of the modified M–estimators does not have any significant effect on the distribution of the polar residuals. As they are more robust than LS–residuals in respect to local model deficiencies they can serve as a suitable basis for deformation analyses.

REFERENCES

Caspary, W., Borutta H. (1987). Robust Estimation in Deformation Models, Survey Review, Vol. 29, No. 223, pp. 29–45

Caspary, W. Chen, Y.Q., König, R. (1983). Kongruenzuntersuchungen in Deformationsnetzen durch Minimierung der Summe der Restklaffungsbeträge, in: Welsch W.(ed.): Deformationsanalysen '83, Schriftenreihe des wissenschaftlichen Studienganges Vermessungswesen, Hochschule der Bundeswehr München, pp 77–94

Mardia ,K.V. (1972). Statistics of Directional Data, Academic Press, London/New York

Prohorov, Yu.V., Rozanov Yu.A. (1969). Probability Theory, translated by Krickeberg, K. and Urmitzer, H., Springer, New York/Heidelberg/Berlin

CRUSTAL DEFORMATION MEASUREMENTS ON VANCOUVER ISLAND, BRITISH COLUMBIA: 1976 TO 1988 [1]

H. Dragert
Geological Survey of Canada
Pacific Geoscience Centre, 9860 West Saanich Rd
Sidney, B.C.,   V8L 4B2   Canada

M. Lisowski
U.S. Geological Survey
345 Middlefield Road, Menlo Park, CA 94025   USA

Abstract. Over the past 12 years, repeated precise geodetic surveys along with longterm tidal monitoring have been used to measure crustal strain on Vancouver Island, a highly seismic area on Canada's west coast. Precise levelling and gravity surveys as well as mean-sea-level trends have provided the basic data for the estimation of vertical deformation rates, whereas laser trilateration and GPS surveys have provided the geodetic control data used to estimate horizontal strain rates. The overall regional pattern that appears to be emerging is one of accumulating strain. Vertical uplift rates vary significantly across the coastal areas, indicating crustal tilting at rates of up to 0.1 microradians per year. Average horizontal shear strain is accumulating at rates of about 0.2 ppm/yr, with relative shortening in a northeast direction, parallel to relative plate motion. This surface deformation is consistent with a model of a locked subducting Juan de Fuca Plate underlying southern Vancouver Island. However, in the central Vancouver Island area, temporal and spatial changes in strain rates have been measured which do not conform to a simple, two-dimensional elastic/visco-elastic tectonic model. This area overlies the landward extension of the Nootka Fault Zone, which is the transform boundary between the Juan de Fuca and Explorer Plates. It is also the area where 2 large (M>7) earthquakes have occurred this century. These complicating factors may be generating transient strain signals that tend to be focused in the crust overlying the Nootka Fault.

INTRODUCTION

Tectonic Setting

Vancouver Island, located on the southwestern coast of British Columbia, overlies the northern part of the Cascadia Subduction Zone (Fig.1). The young (<10My) subducting oceanic crust is comprised of two microplates in this region. To the south, the Juan de Fuca Plate converges with the North American Plate margin at a rate of about 4.5 cm/yr in a northeast direction. To the north, the Explorer Plate converges at a rate of about 2 cm/yr in a direction that likely varies as a function of position because of the close proximity of its pole of rotation. The boundary between these microplates is the Nootka Transform Fault zone and its extension underlies central Vancouver Island. Most recent tectonic models propose that the Explorer Plate, because of its relative youth and small size, is no longer subducting into the upper mantle and may be underplating the North American margin (Riddihough, 1984).

---

[1] *Geological Survey of Canada Contribution 29389*

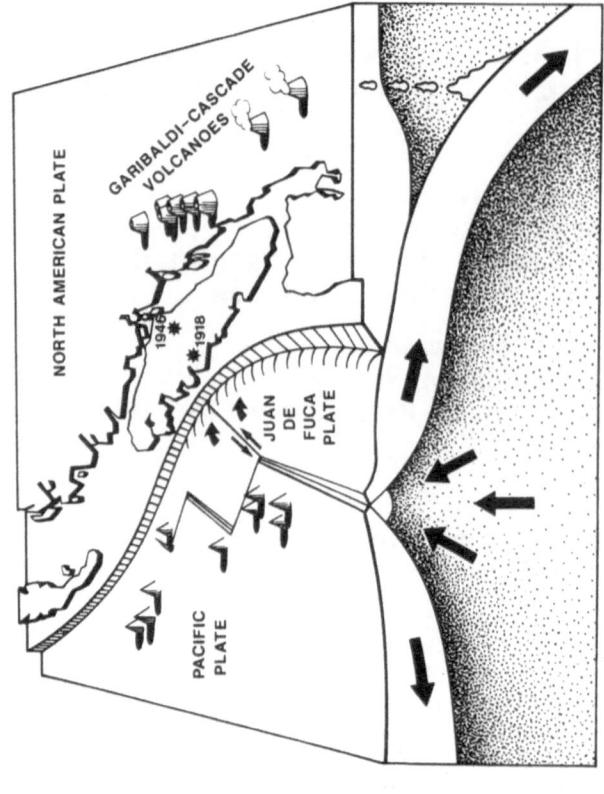

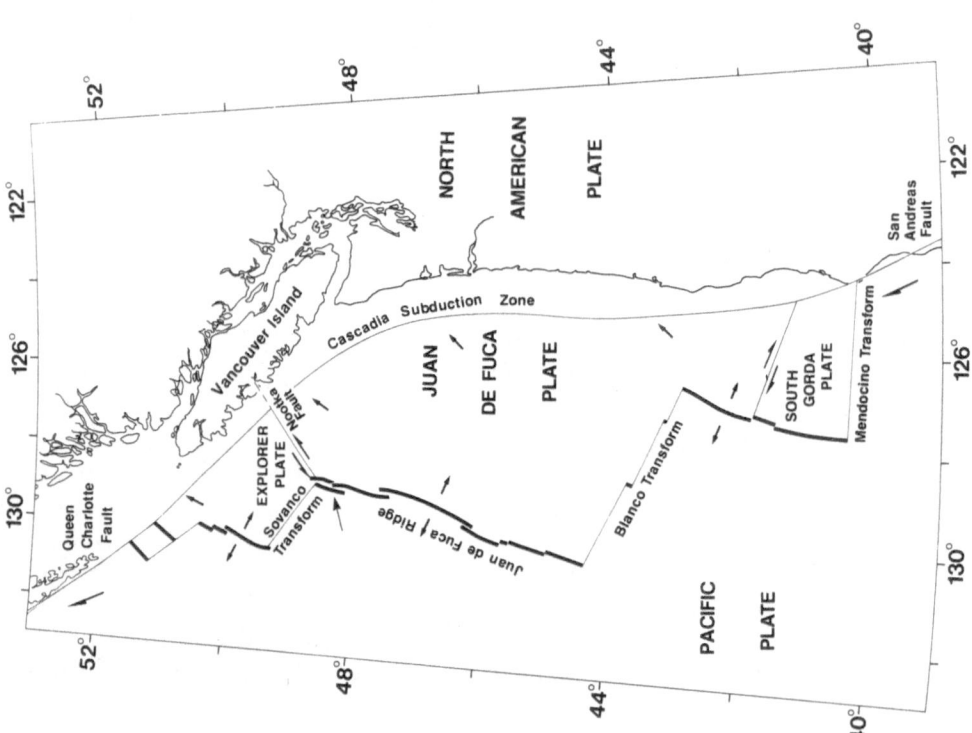

Fig. 1. Plan view and cartoon cross-section of the tectonic elements in the region of the Cascadia Subduction Zone. Stars mark epicentres of the two large (M>7) crustal earthquakes which have occurred on Vancouver Island during this century (adapted from Riddihough, 1978).

Seismicity

The low-level seismicity of this region is extensive as shown by the distribution of earthquakes observed for the period 1980 to 1985 (Fig.2, Rogers and Horner, 1989). Two principal aspects of the nature of the seismic activity are as follows:

1. Central Vancouver Island has been the site of two major (M>7) crustal earthquakes during this century (see Fig. 4). The most recent was a magnitude 7.3 event in 1946 (Rogers and Hasegawa, 1978).

2. Although the Cascadia Subduction Zone is similar to other young subduction zones around the world which have experienced large thrust earthquakes, no "megathrust" events have occurred here over historic times (i.e. the past 200 yrs.); Indeed, present-day monitoring of seismicity shows earthquakes occur in the overlying crust and the down-going plate (Fig. 3), but there is a distinct absence of even low-level thrust events on the subduction interface (Rogers, 1988).

Program Objectives

A program to monitor crustal deformation on Canada's west coast was initiated in 1976. This program consists of repeated precise gravity and levelling surveys, augmented by mean-sea-level records, to monitor vertical deformation, and repeated "special order" laser-ranging and GPS surveys to monitor horizontal strain (Fig. 4). The primary geographic focus for these measurements to date has been central Vancouver Island, in the area of the 1946 earthquake. The main objectives of this program can be summarized as follows:

1. From the observed surface crustal strain, provide constraints for or suggest alternate models of local plate geometry and plate dynamics.

2. From the observed regional deformation patterns, address the question of the potential of a megathrust earthquake in this region.

3. From more detailed, temporal strain data in the central Vancouver Island region, assess the likelihood of another large (M=7) crustal earthquake.

METHODS

Vertical Deformation

Precise gravity and levelling surveys are used as complementary tools to monitor relative elevation changes and possible changes in bulk densities of rock. In the Campbell R./Gold R. region of central Vancouver Island, gravity surveys were carried out on a semi-annual basis between 1977 and 1984, with additional single surveys in 1985 and 1988. These surveys have utilized anywhere from 2 to 4 LaCoste & Romberg Model-D gravimeters simultaneously, and through rigorous field procedures and network schemes of repeated measurements, relative gravity values with a precision of 2 microgals have been determined as a function of time (Dragert *et al.*, 1981).

Repeated precise levelling traverses have also focused on the central Vancouver Island area: i.e. from Parksville to Campbell R. (Line 145), and from Campbell R. to Gold R. (Line 1C76). Most frequent surveys have been carried out on Line 145 which was levelled in 1930, 1946 (after the earthquake), 1977, 1984, and 1988. Post-1978 surveys used Zeiss Ni002 levels and conformed to "special-order" survey specifications, resulting in an estimated standard deviation of 0.7 mm $\sqrt{K}$ for the elevation difference between two benchmarks. (K is the distance between benchmarks in km).

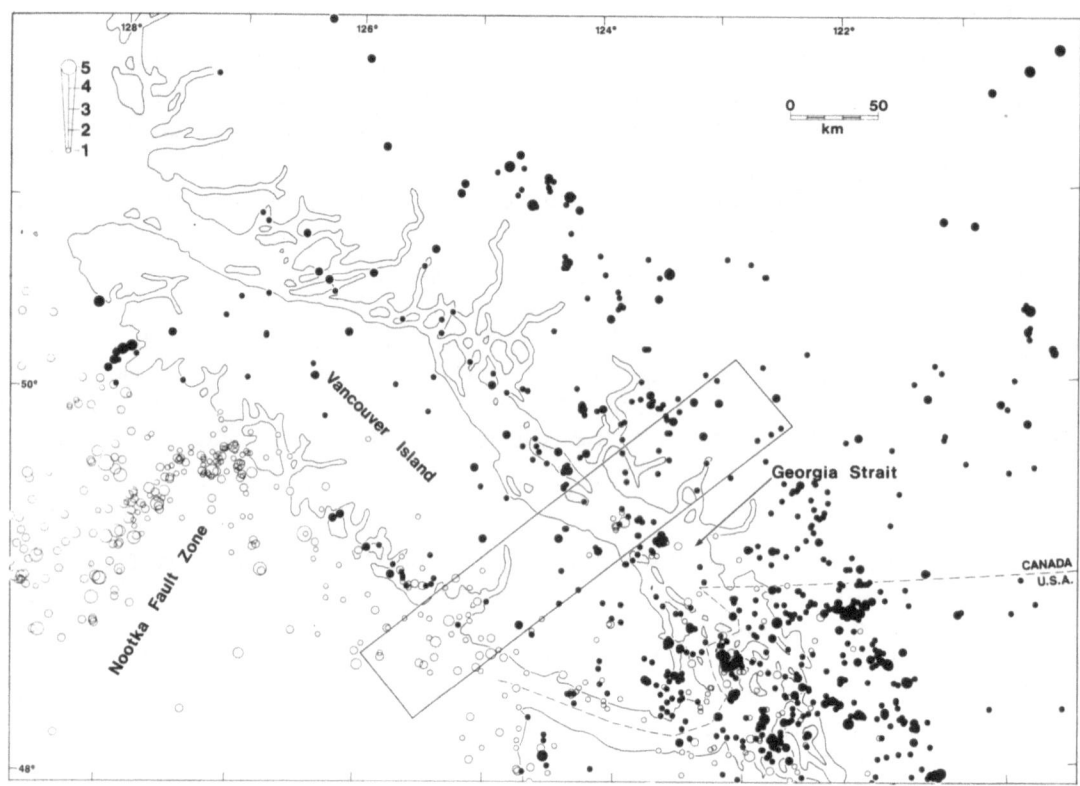

Fig. 2. Distribution of seismicity in southwestern British Columbia for the period 1980 to 1985. Solid circles indicate epicentres of earthquakes occurring in the overlying crust (< 30 km depth). Open circles show location of earthquakes occurring within the descending oceanic plate. Rectangle shows location of the vertical cross-section of Fig. 3 (from Rogers and Horner, 1989).

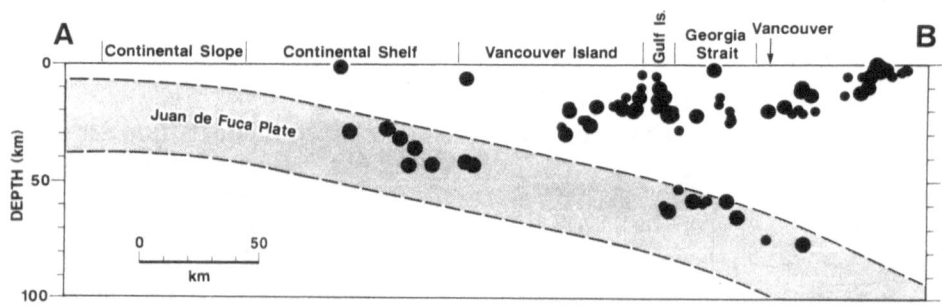

Fig. 3. Vertical distribution of seismicity in a section cutting across southern Vancouver Island. No thrust earthquakes have been recorded near the subduction interface (from Dragert and Rogers, 1988).

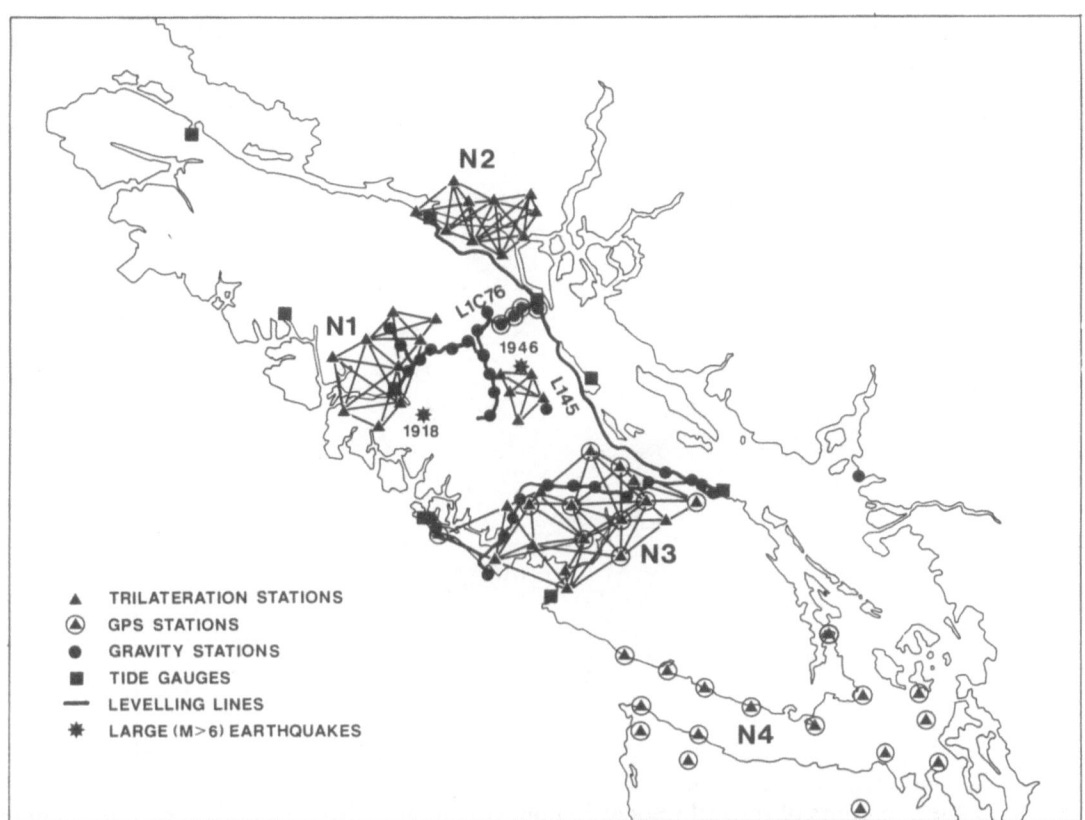

Fig. 4. Map showing the locations of repeated geodetic and gravity control surveys on Vancouver Island up to 1988. Circled dots indicate 4 eastern gravity stations with temporal gravity variations shown in Fig. 8. Tide gauges operated by the Canadian Hydrographic Survey are also indicated. The stars show the epicentres for the 2 large (M>7) earthquakes that occurred in central Vancouver Island.

Tide gauges, which are incorporated along levelling traverses, are used to provide an independent check of relevelling data as well as to establish vertical deformation rates at coastal locations with respect to mean-sea-level (MSL). The oldest tide gauges in this region have been in operation for about 80 years. Differencing tide gauge records with a reference station or a regional average of several gauge records allows the resolution of relative uplift rates of the order of 1 mm/yr if 10 yrs. of continuous data are available.

Horizontal Deformation

Ground-based electronic distance measurement (EDM) techniques, in particular laser ranging, have been used on Vancouver Island since 1982 in the establishment of 3 horizontal strain networks (see Fig. 4):

N1. the Gold River Network, surveyed in 1982, 1985, and 1988;

N2. the Johnstone Strait Network, surveyed in 1984 and 1988;

N3. the Port Alberni Network, surveyed in 1986.

The first two networks were able to utilize an adequate number of original triangulation control points (dating back as far as 1918) in order to allow estimates of net horizontal shear strain over the past few decades. The measurement technique involves a Rangemaster III which is set up over the control monument on one

245

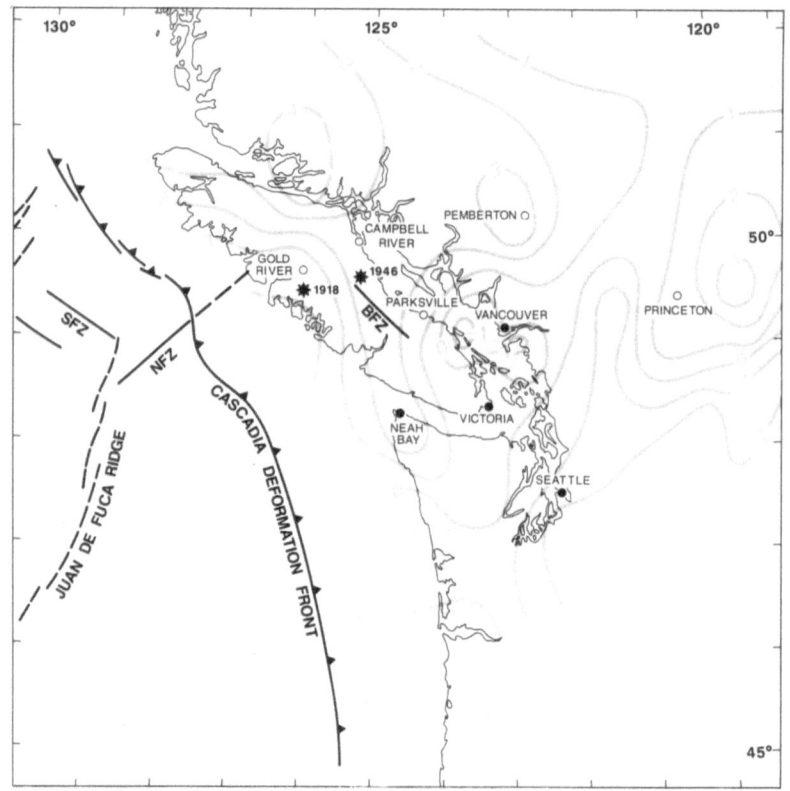

Fig. 5. Smoothed contours of vertical velocity in mm/yr computed from a combined adjustment of tide gauge records spanning the past several decades and of repeated levelling traverses. Average velocities were assumed to be constant over the period of analysis (after Holdahl *et al.*, 1989).

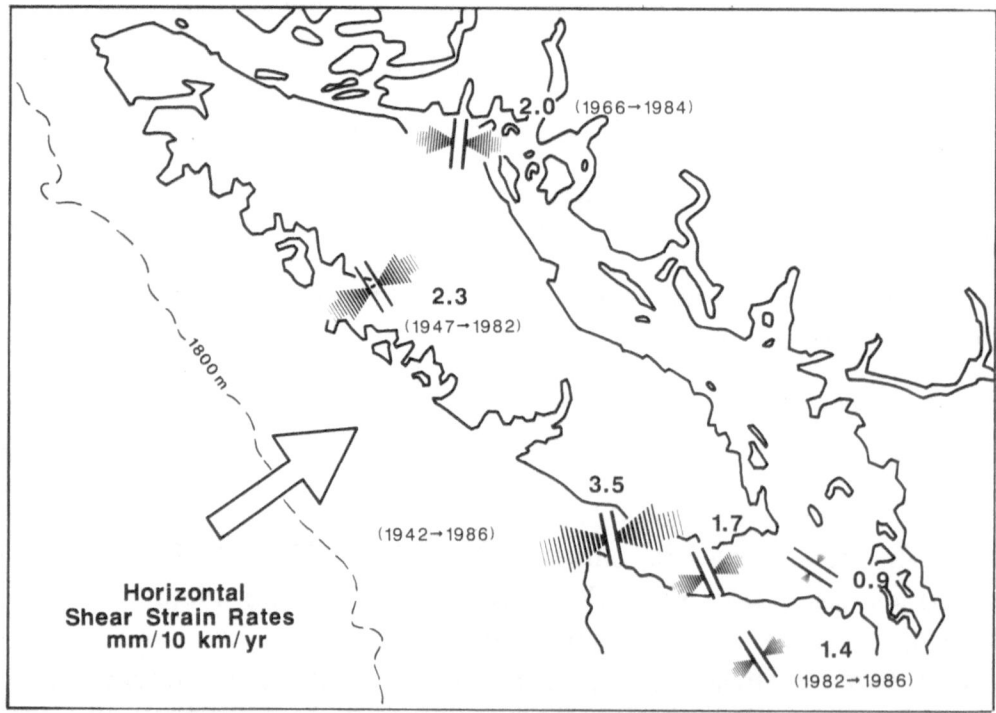

Fig. 6. Annual net horizontal shear strain rates estimated from horizontal control surveys repeated in the indicated years. Strain rates are in parts per ten million per year. The large arrow shows the direction of the convergence vector between the Juan de Fuca Plate and the North American Plate.

mountaintop, and a reflector target, consisting of up to 34 corner cubes, which is set over a second control point. During the half-hour of continuous ranging measurements, a helicopter flies along the line-of-sight to monitor temperature and humidity. Within these networks, this EDM technique has attained precisions of 0.2 to 0.4 ppm for baselines ranging from 10 to 40 km.

GPS surveys for the express purpose of monitoring crustal strain were initiated on Vancouver Island in 1986. 16 original triangulation points, some dating back to the turn of the century, were recovered to form the Juan de Fuca Strain Network (N4). 10 control points of the Port Alberni EDM Network were also occupied with GPS receivers to facilitate accurate comparisons between ground-based and satellite-based survey techniques. In 1987, a 9-station subset of the Juan de Fuca Network was surveyed, while in 1988, 8 sites of the Gold River/Johnstone Strait EDM Networks were surveyed, again to allow more comprehensive comparisons with precise laser measurements. All GPS surveys used anywhere from 4 to 7 TI4100 GPS receivers (tracking 4 satellites for 4.5 hrs. with 15 sec. sampling for each daily session) with at least double occupations of each site. Data analysis has been carried out with various software packages (DIPOP, GPS22, BERNESE) with resulting baseline precisions of 0.3 to 0.5 ppm using broadcast ephemerides. Precisions for the 1987 Juan de Fuca survey were improved by a factor of 5 through the use of precise orbits.

RESULTS AND DISCUSSION

Regional Strain Patterns

Analysis of both vertical and horizontal control data reveals significant regional strain rates when averaged over the past few decades. Combining mean-sea-level trends and all available relevelling data within this region, and assuming constant vertical velocities over the period spanned by the observations (about 80 yrs.), derived uplift rates (Fig. 5) vary from -2mm/yr near Seattle, to 2.7 mm/yr at Neah Bay at the northwest tip of the Olympic Peninsula, to over 3.0 mm/yr in the area northwest of Campbell River (Holdahl *et al.*, 1987). Determination of a constant velocity in this latter region is complicated by about 80 mm of coseismic subsidence associated with the 1946 earthquake, and temporal variations in the uplift rate over the past decade (Dragert, 1988).

Results of the EDM and GPS surveys enable estimates of horizontal shear strain rates to be made by comparison with original triangulation data (Fig. 6). In general, all areas surveyed to date show a shortening of lines roughly parallel to the plate motion vector at rates comparable to those measured in highly seismic regions around the world. Nominal horizontal distances of 10 km striking in a northeasterly direction are shortening about 2 mm each year with respect to orthogonally directed distances held fixed.

Qualitatively, these deformation patterns are consistent with features of the current plate convergence model:

1. The rapid uplift of the region north of central Vancouver Island is consistent with the overriding of the young (<6My), buoyant Explorer Plate;

2. The ridge of uplift extending from the Neah Bay area north across Vancouver Island to Campbell River and the relative crustal shortening parallel to the plate convergence vector are consistent with a pattern expected from a locked subduction zone underlying this coastal region.

In conjunction with the lack of thrust earthquakes on the subduction interface, this latter observation implies a definite potential exists for a future megathrust earthquake (Dragert and Rogers, 1988).

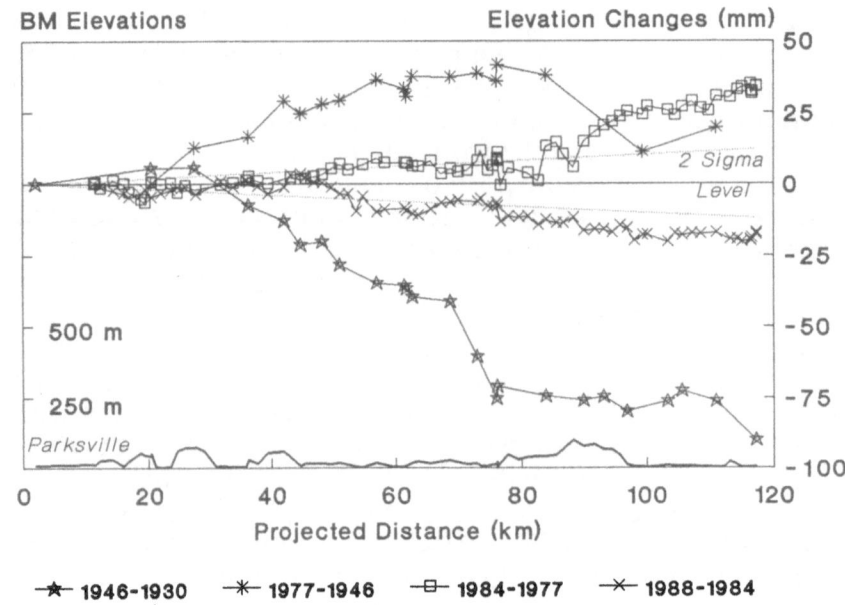

Fig. 7. Elevation changes measured along Line 145 (Parksville to Campbell River) for the indicated epochs with the assumption that the southern terminus (Parksville) has remained fixed. Each plotted line represents the difference between consecutive surveys.

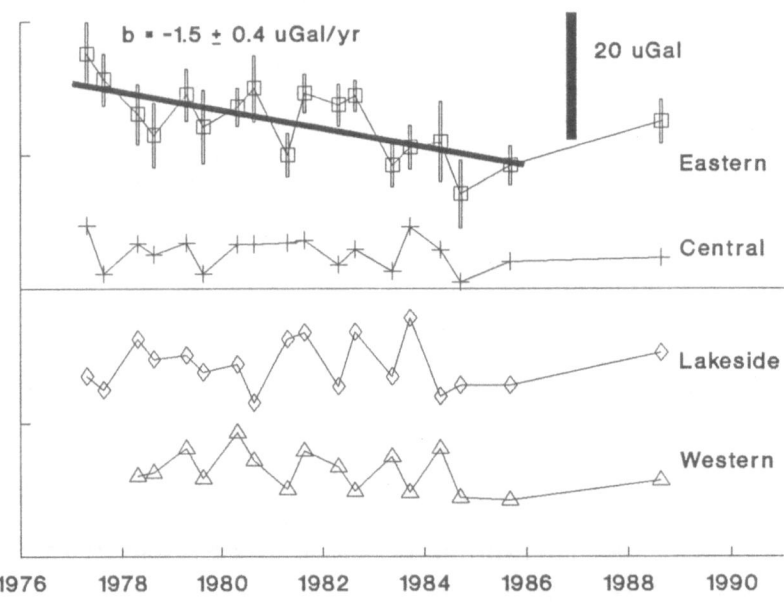

Fig. 8. Regional gravity variations observed in the Campbell River/Gold River precise gravity network. Plots show average gravity trends for different geographic portions of the network. Only the eastern stations show a significant linear trend from 1977 to 1985 and an offset from this trend in 1988. Error levels are similar throughout the network, but error bars (2 sigma) have been plotted only for the eastern averages.

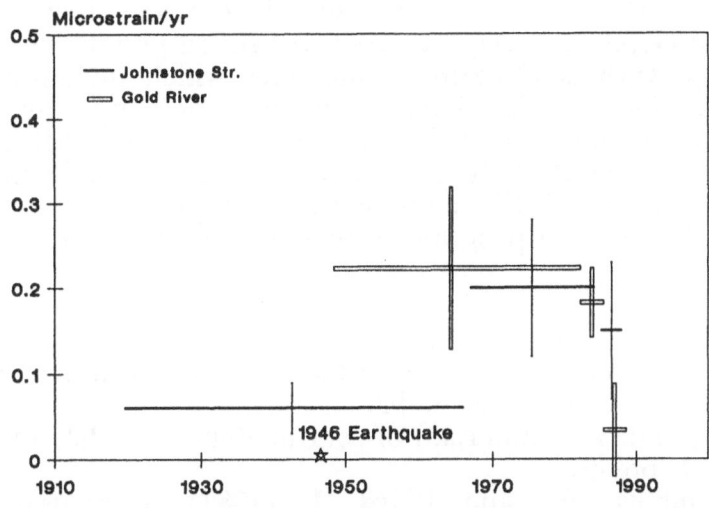

Fig. 9. Net horizontal shear strain rates for the Gold River and Johnstone Strait regions between 1918 and 1988. Error bars are 2 sigma estimates. The star marks the time of the 1946 earthquake.

Temporal Strain Changes

The greater number of repeated gravity, levelling, and trilateration surveys in the central Vancouver Island region has allowed more detailed resolution of temporal changes in local crustal strain. The changes in elevation of benchmarks along Line 145 in the Campbell River region relative to a fixed southern terminus show significant variations in net uplift rates between successive levelling surveys (Fig. 7). A pronounced subsidence of about 80 mm in 1946, likely coseismic, is followed by a gradual uplift of about 1 mm/yr to 1977, an accelerated uplift of close to 4 mm/yr between 1977 and 1984, and subsequently a relative subsidence of about 4mm/yr between 1984 and 1988. Gravity changes observed at 4 sites in the Campbell River area between 1977 and 1988 are consistent with these levelling results, showing a similar change in the sense of deformation over the last three years (Fig. 8).

Horizontal shear strain rates also show variations with time for the central Vancouver Island region (Fig. 9). In particular, strain analysis of EDM results for the Gold River Network show a cessation of strain accumulation over the past three years, while the Johnstone Strait region shows almost no accumulation of strain over the period spanning the 1946 earthquake, a level of 0.2 ppm per year from 1966 to 1984, and a (marginal) decrease in shear strain accumulation over the last 4 years.

The exact cause of these most recent changes in both vertical and horizontal deformation is unknown but differential plate motions across the underlying Nootka Fault is suspected. Similar changes in the sign of crustal movement have been documented as aseismic phenomena in some areas such as Palmdale, California (cf. Castle, 1978), and as precursors to large crustal earthquakes in other locations such as Niigata, Japan (cf. Mogi, 1985). For earthquakes similar in magnitude to those that have occurred in central Vancouver Island, the time between the initiation of the change in the sense of motion and the actual occurrence of the earthquake is estimated at several years (Rikitake, 1976).

Acknowledgement. All vertical and horizontal control surveys in Canada were carried out by the *Geodetic Survey of Canada*. Precise gravity surveys were carried out by contract to various Canadian companies which included *SIAL*, *Usher & Associates*, and *McElhanney Geosurveys*. Canadian tide gauges are operated by the *Dept. of Fisheries & Oceans*. The analysis of crustal vertical velocities was done by *S. Holdahl, NGS*. The DIPOP analysis of GPS data was carried out at the *Geodetic Laboratory, UNB*, and the GPS22 analysis was done by *J. Kouba, GSC*. Horizontal strain was calculated using a program provided by *H. Bibby, DSIR*, New Zealand.

REFERENCES

Castle, R.O. (1978). Leveling surveys and the southern California uplift, *U.S. Geol. Surv. Earthquake Inf. Bull.* **10**, 88-92.

Dragert, H. (1987). The fall (and rise) of central Vancouver Island: 1930-1985, *Can. J. Earth Sci.* **24**, 689-697.

Dragert, H., Lambert, A., and Liard, J. (1981). Repeated precise gravity measurements on Vancouver Island, British Columbia, *J. Geophys. Res.* **86**, 6097-6106.

Dragert, H. and Rogers, G.C. (1988). Could a megathrust earthquake strike southwestern British Columbia?, *GEOS* **17 No. 3**, 5-8.

Holdahl, S.R., Faucher, F., and Dragert, H. (1989). Contemporary vertical crustal motion in the Pacific Northwest, in *Slow Deformation and Transmission of Stress*, S. Cohen (ed.), AGU Monograph, in press.

Mogi, K. (1985). *Earthquake Prediction*, Academic Press.

Riddihough, R.P. (1978). The Juan de Fuca Plate, *EOS* **59 No. 9**, 836-842.

Riddihough, R.P. (1984). Recent movements of the Juan de Fuca plate system, *J. Geophys. Res.* **89**, 6980-6994.

Rikitake, T. (1976). *Earthquake Prediction*, Elsevier Scientific Publishing Co.

Rogers, G.C. (1988). An assessment of the megathrust potential of the Cascadia subduction zone, *Can. J. Earth Sci.* **25**, 844-852.

Rogers, G.C. and Hasegawa, H.S. (1978). A second look at the British Columbia earthquake of June 23, 1946, *Bull. Seism. Soc. America* **68**, 653-675.

Rogers, G.C. and Horner, R.B. (1989). An overview of western Canadian seismicity, in *DNAG Volume GSMV-1: Neotectonics of North America*, in press.

PRELIMINARY DYNAMIC MODEL FOR KALABSHA AREA
AT ASWAN FROM GEODETIC MEASUREMENTS

P. Vyskočil
International Centre on Recent Crustal
Movements, Zdiby, Prague, Czechoslovakia

R.M. Kebeasy, A. Tealeb and S.M. Mahmoud
National Research Institute of Astronomy
and Geophysics, Helwan, Cairo, Egypt

ABSTRACT

A preliminary dynamic model for Kalabsha area was established from the analysis of five repeated horizontal geodetic
measurements performed by the Kalabsha network. The
Kalabsha fault is of a right-lateral strike slip motion on
an east-west plane. This is in a good agreement with those
obtained from the fault plane solutions. For the different
epochs of geodetic measurements, the magnitudes of
movements are variable. These indicates that the Kalabsha
area has got strained from one epoch to another, with
different forces.
   Moreover, on the basis of the determined stress and
strain fields, for the different epochs, within the Kalabsha
area, the area was divided into zones of tensions and
compressions. Zone of high tensions was prevailed to the
southwest of the network.

INTRODUCTION

On November 14, 1981 a moderate earthquake with a local
magnitude of 5.6 was occured in Kalabsha area (fig.1),70 km
southwest of Aswan City (Kebeasy et al. 1982). The
Kalabsha area lies on a large western embayment of Aswan
lake. Since then, the seismic activity are continued to
occur, with different magnitudes, in that area. These earthquakes were estimated to have occured near the epicentre of
the main earthquake of 1981 which was located along the
Kalabsha fault, mainly near the wide area of the lake
(Kebeasy et al. 1987).

In Kalabsha area a sequence of sedimentary rock units of a maximum thickness of about 500 meters, consists mainly from the Nubian Sandstone formation are prevailed (Issawi, 1978). These Nubian formation overlies unconformably the Pre-Cambrian basement rocks. The Kalabsha area is characterized by the presence of a main right-slip fault system of an E-W trend and a total length of about 300 km (the Kalabsha fault). The Kalabsha fault was identified as the source of the 14 November 1981 earthquake (Kebeasy et al. 1982, Kebeasy et al. 1987). The Kalabsha fault is crossed by two system of faults trending in a N-S direction (fig.1).

The predominant focal mechanism of the seismic activity at Kalabsha is a right-lateral strike slip faulting on an east-west fault plane (Kebeasy et al. 1987). The activity is believed to be triggered by the lake (Kebeasy et al. 1982), because there is no significant earthquakes were reported to have take place in the area of the lake throughout history and the area was regarded before as aseismic area.

In order to monitor the crustal deformations at Kalabsha area and to investigate their possible association with the earthquake activity and water loading in the lake, a local geodetic network of 16 geodetic points were established in November, 1983 (Vyskočil and Tealeb 1985, Mahmoud 1988) around an active part of the Kalabsha fault (fig.1 and 2). The initial geodetic measurements were carried out in December, 1984. The repeated geodetic measurements were performed twice a year (Vyskočil et al. 1987, 1988).

The adjustment of five repeated horizontal geodetic measurements (distances and angles) were performed in 1988. The displacement vectors of the Kalabsha network, for each epoch of measurements, were calculated from the coordinate changes. On the basis of the displacement vectors a preliminary dynamic model for Kalabsha area was established and the stress and strain fields were estimated.

GEODETIC RESULTS

The displacement vectors for the epochs 1986, 1987 I, 1987 II and 1988 I are represented in figures 3,4,5 and 6 respectively. Considering the confidence limit, most of these displacement vectors can be attributed mainly as the movement occured within the Kalabsha area in the different epochs of measurements. The initial information for the original displacements at the points of the network in the figures 3,4,5 and 6 were used for the determination of the displacements in grid of 1x1 km. The interpolated displacements of horizontal movements for the epochs 1986, 1987 I, 1987 II and 1988 I are shwon in figures 7,8,9 and 10 respectively. In figure 8 (epoch 1987 I) an increase of rates of displacements was recorded to the west and southwest of the Kalabsha network. These rates are decreased

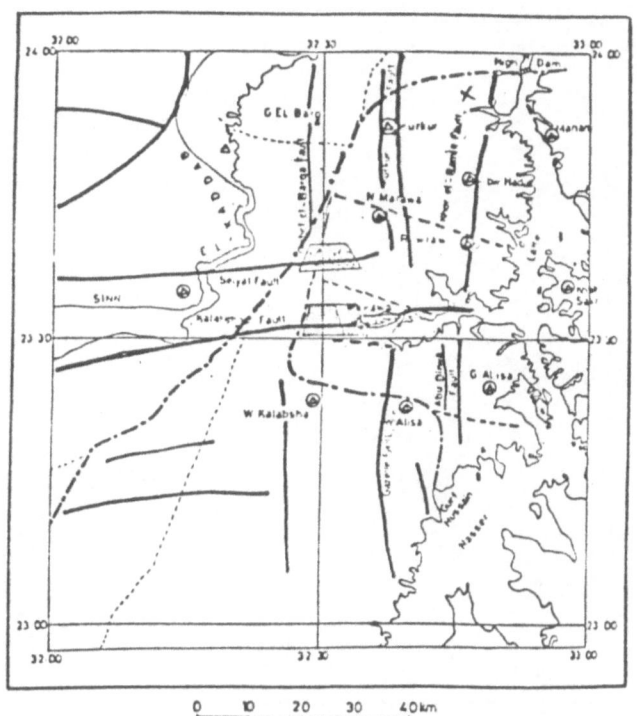

- - - Levelling Line
-.-.-Aswan-Wadi Halfa Road
── Faults
⋮⋮⋮ Seiyal network

▲ Regional geodetic point
⊕ Regional geodetic point &
   Seimic station .
▨ Kalabsha network

FIG. 1. Local and regional geodetic networks in the northwestern part of Aswan lake.

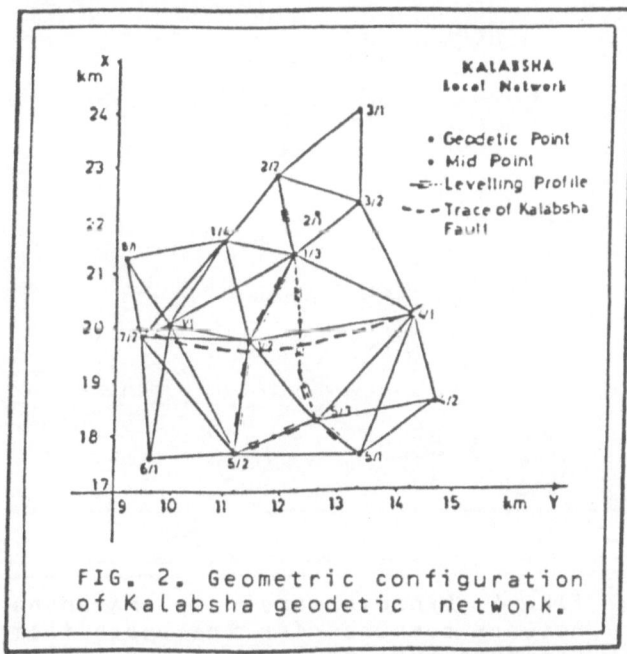

FIG. 2. Geometric configuration of Kalabsha geodetic network.

253

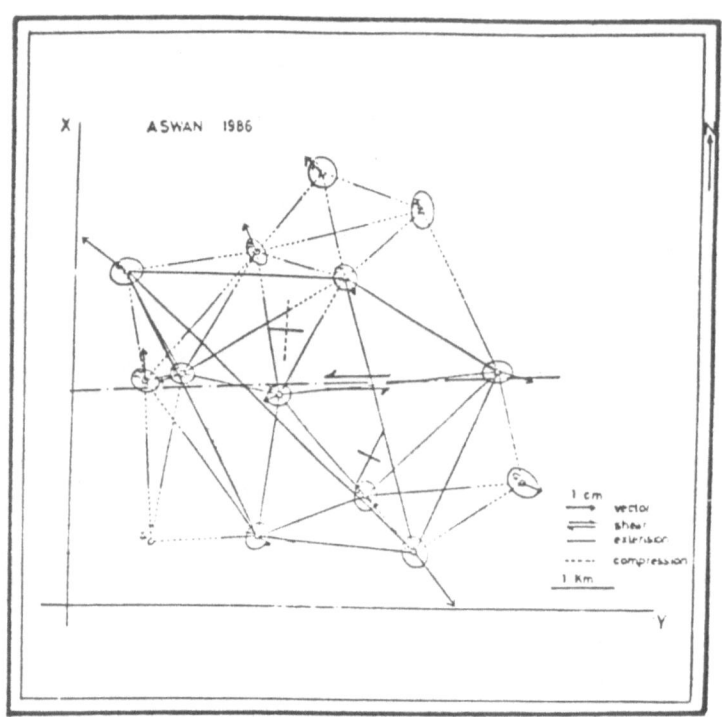

FIG. 3. Dynamic model for Kalabsha
network computed for the epoch from
December 1984 to February 1986.

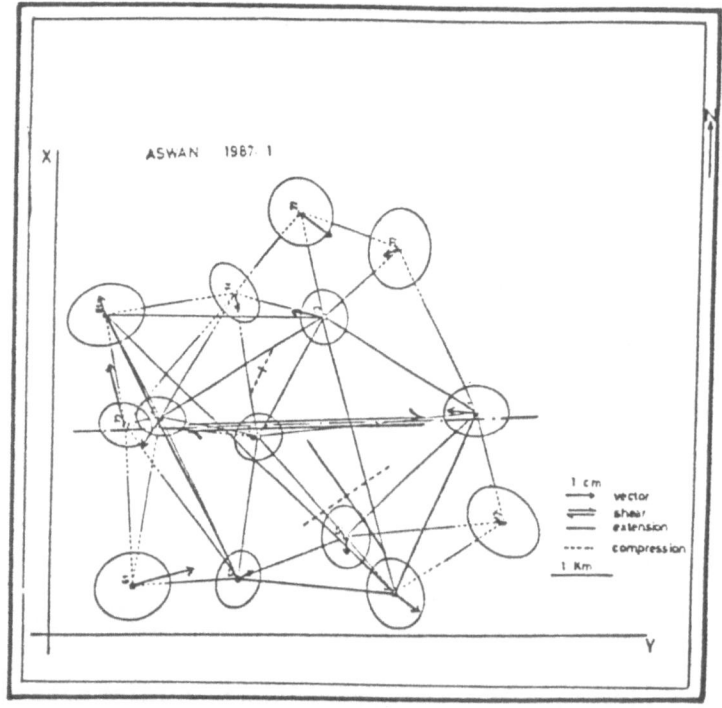

FIG. 4. Dynamic model for Kalabsha
network computed for the epoch from
December 1984 to January 1987.

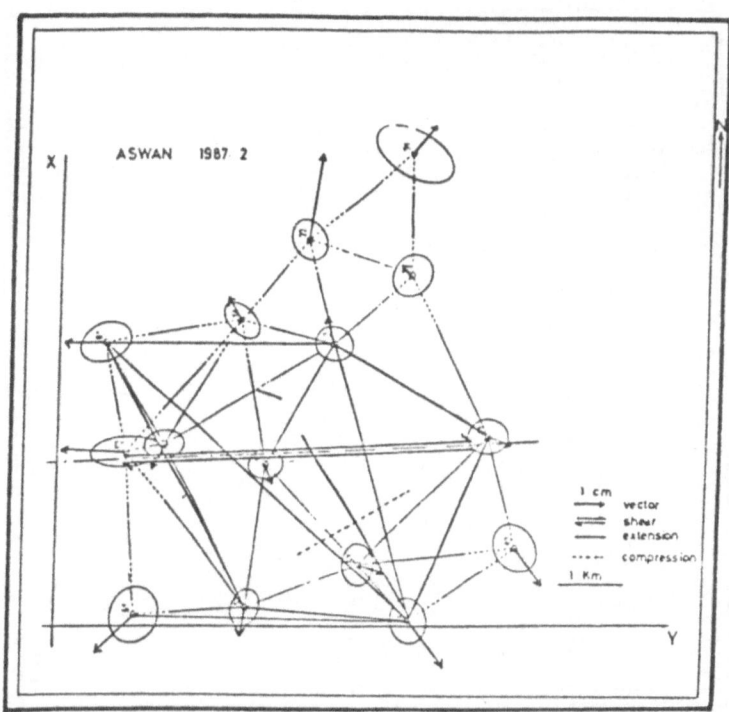

FIG. 5. Dynamic model for Kalabsha
network computed for the epoch from
December 1984 to September 1987.

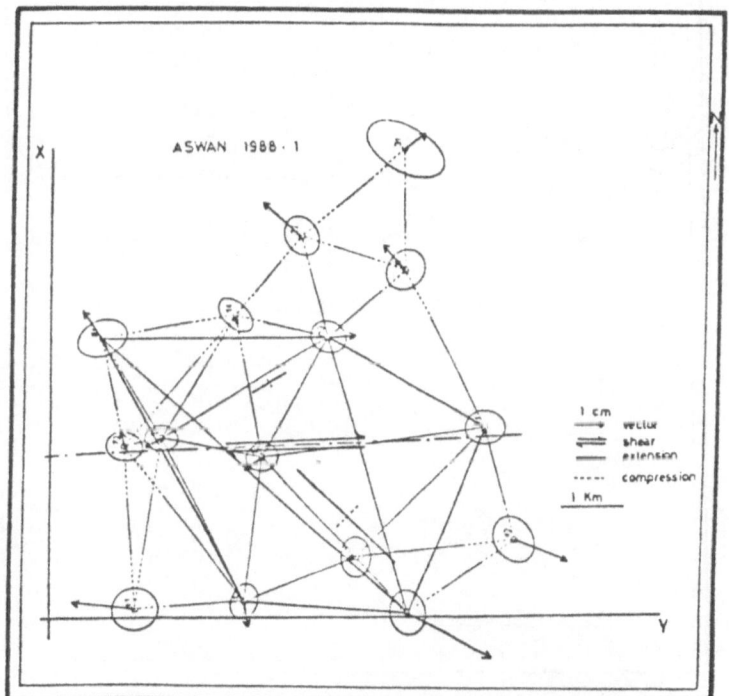

FIG. 6. Dynamic model for Kalabsha
network computed for the epoch from
December 1984 to January 1988.

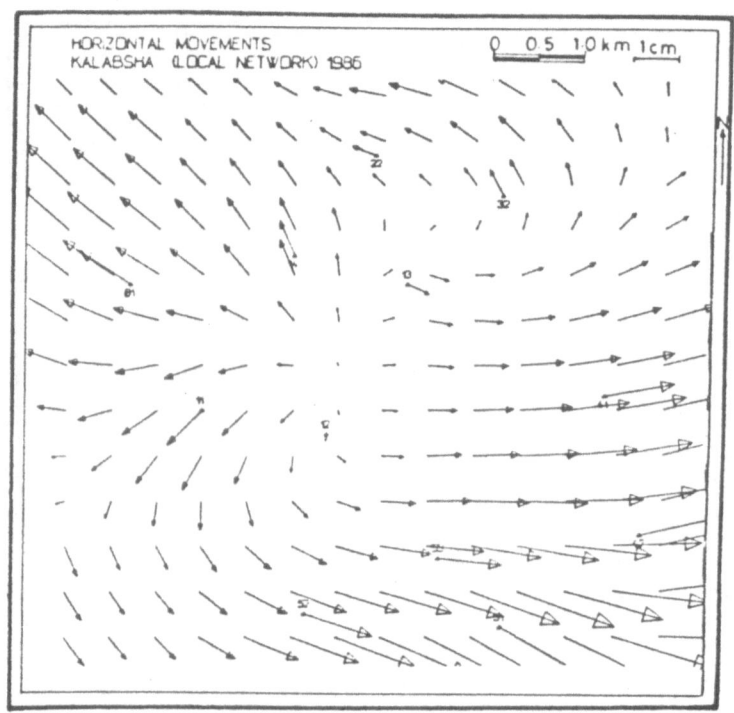

FIG. 7. Horizontal movements of the network for the epoch from December 1984 to February 1986.

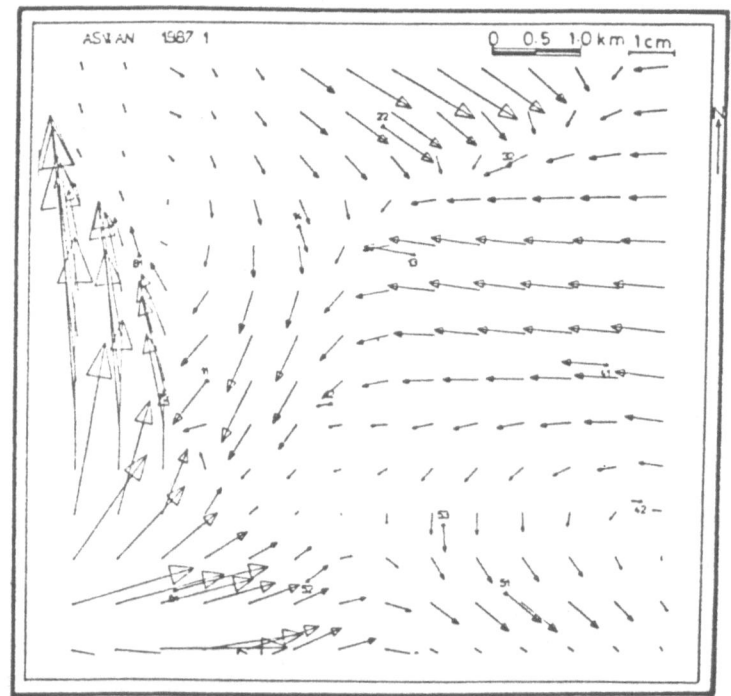

FIG. 8. Horizontal movements of the network for the epoch from December 1984 to January 1987.

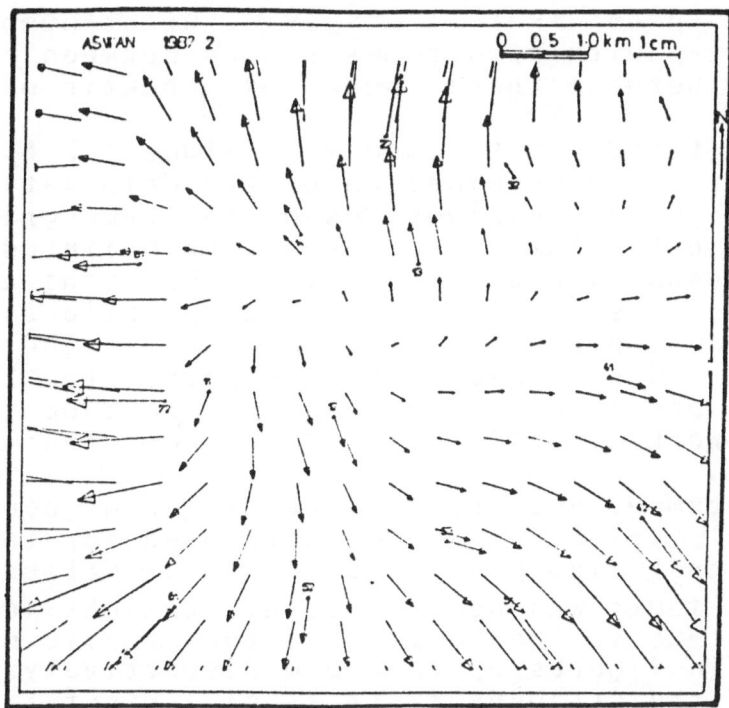

FIG. 9. Horizontal movements of the
network for the epoch from December
1984 to September 1987.

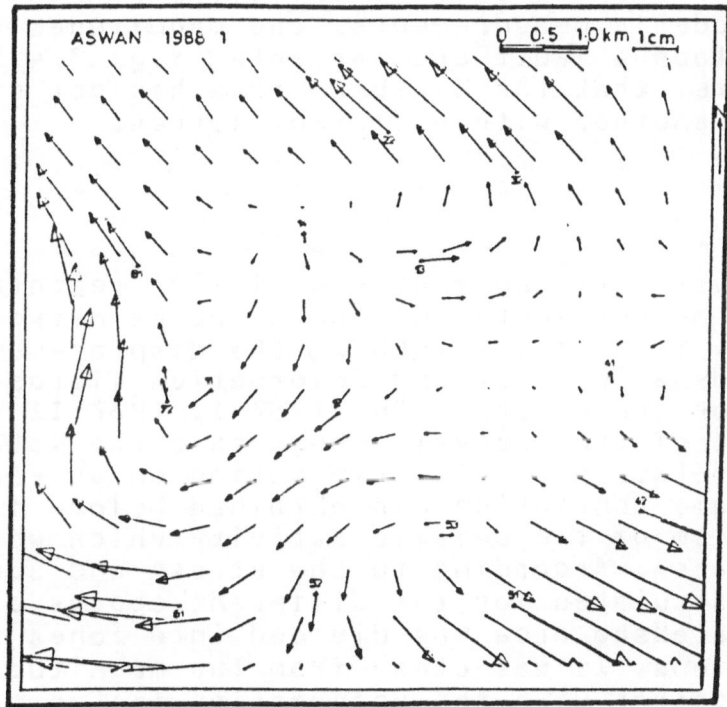

FIG.10. Horizontal movements of the
network for the epoch from December
1984 to January 1988.

again in the epoch 1987 II (fig.9). Great number of earthquake activity, with magnitudes ranges between 2.1 and 3.1,
were occured between these epochs of geodetic measurements
(June, 1987).

The interpolated displacements in figures 7,8,9 and 10
were used for the determination of the deformation fields
(extentions and compressions) using the same grid. On the
basis of the determined stress and strain fields, for the
different epochs of measurements, the Kalabsha area was
divided into zones of tensions and compressions as
represented in figures 11,12,13 and 14, for the epochs
1986, 1987 I, 1987 II and 1988 I respectively. Zone of high
tensions was prevailed to the west and southwest of the
Kalabsha network (fig.12) before the June, 1987 earthquake
activity.

The displacement vectors for the different epochs of
measurements (fig. 3,4,5 and 6) were used for the establishment of a preliminary dynamic model for Kalabsha area.
The Kalabsha fault was considered as straight-forward fault
trending east-west. The results of the dynamic model were
represented in figures 3,4,5 and 6 respectively. Refering to
these model, the Kalabsha fault is of a right-lateral strike
slip motion on an east-west plane. These results are
correlated with the results of the focal mechanism which
were obtained from the seismic activity. For the different
epochs of geodetic measurements, the magnitudes of movement
along the Kalabsha fault are variable (fig. 3,4,5 and 6).
These indicates that the Kalabska area has got strained from
one epoch to another with different forces.

CONCLUSIONS

From the adjustments and analysis of five repeated horizontal geodetic measurements for the Kalabsha network (1984,
1986, 1987 I, 1987 II and 1988 I) the displacement vectors,
Preliminary dynamic model and deformation fields were
calculated for the epochs 1986, 1987 I, 1987 II and 1988 I.
The results of the analysis shows that the Kalabsha fault
is of a right-lateral strik slip motion on an east-west
plane. The same conclusion was obtained before from the
focal mechanism of the seismic activity which were located
at Kalabsha area. According to the stress and strain fields
which were calculated for the different epochs of measurements, the Kalabsha area was divided into zones of tensions
and compressions. It was clear from the magnitudes of
movements together with the deformation fields that the
Kalabsha area has got strained from one epoch to another,
with different forces.

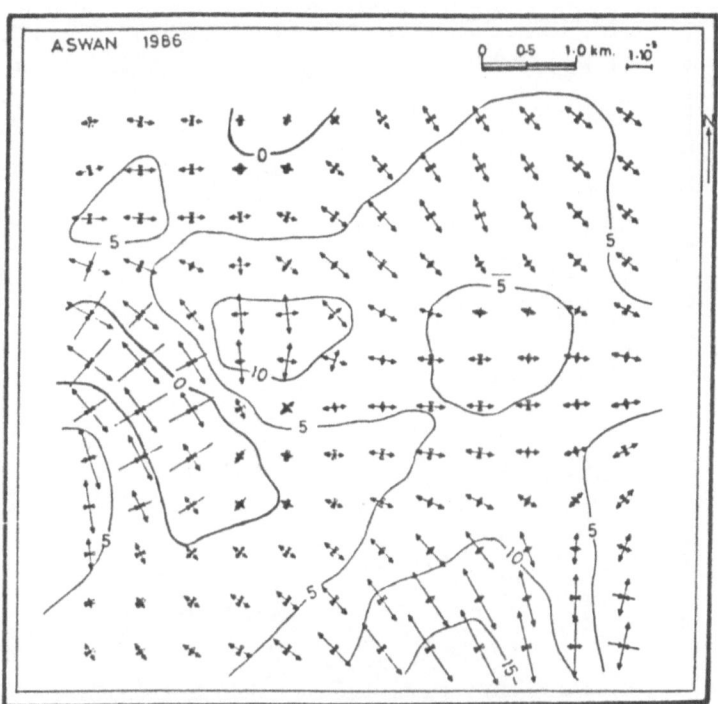

FIG.11.Zones of horizontal deformation
fields computed for the epoch from
December 1984 to February 1986.

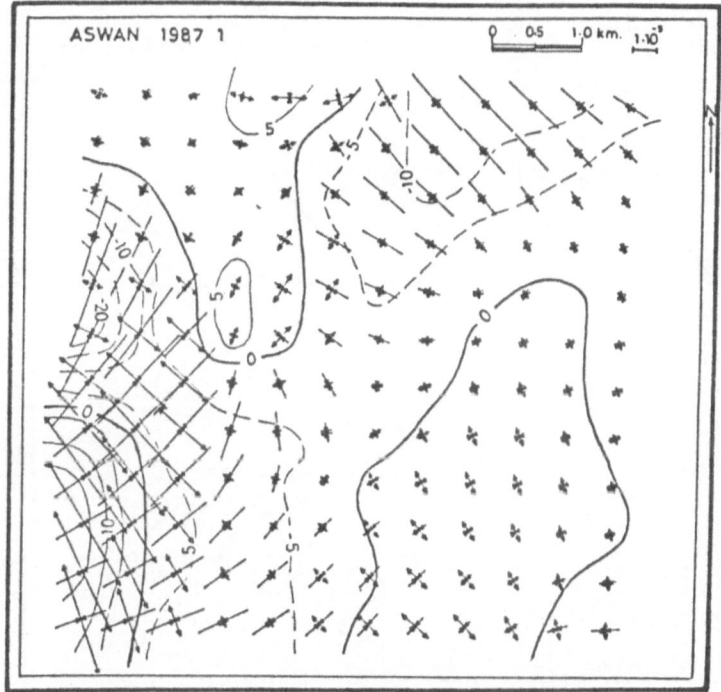

FIG.12.Zones of horizontal deformation
fields computed for the epoch from
December 1984 to January 1987.

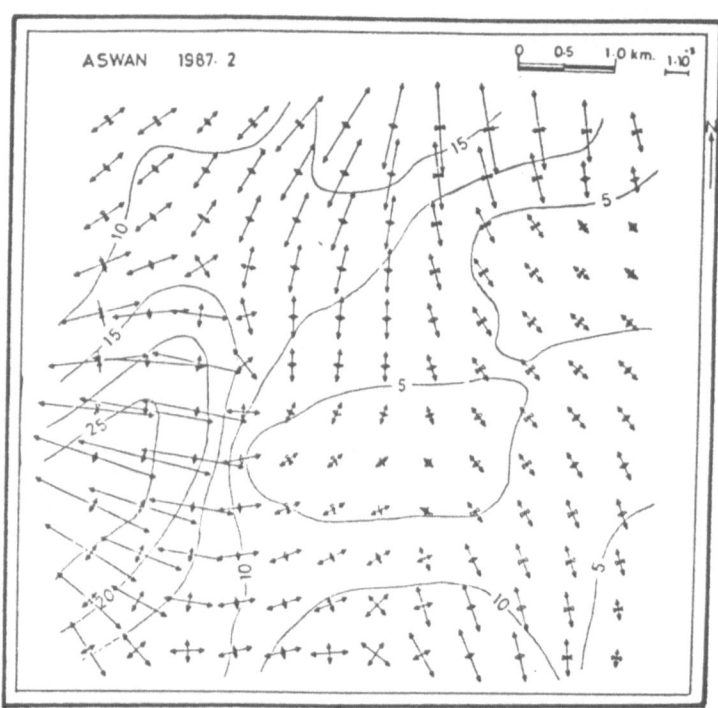

FIG.13.Zones of horizontal deformation
fields computed for the epoch from
December 1984 to September 1987.

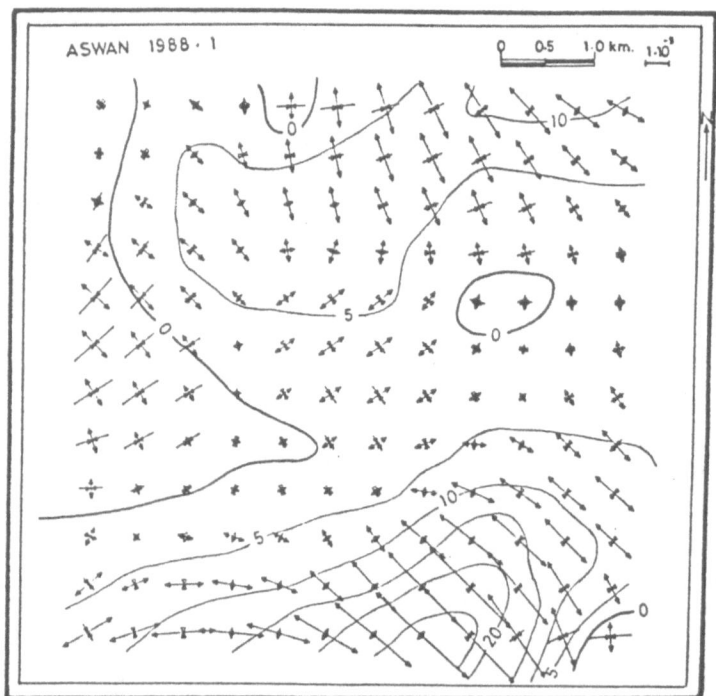

FIG.14.Zones of horizontal deformation
fields computed for the epoch from
December 1984 to January 1988.

REFERENCES

Issawi,B. (1978) . Geology of Nubian west area, Western Desesrt. Annals of the Geological Survey of Egypt.

Kebeasy,R.M., Maamoun,M. and Ibrahim,E.M. (1982) . Aswan lake induced earthquakes. Bull. Inter. Inst. of Seismology and Earthquake Engineering 19, Tokyo.

Kebeasy,R.M., Maamoun,M., Ibrahim,E.M., Megahed,A., Simpson, D. and Leith,W. (1987) . Earthquake studies at Aswan reservoir. Journal of Geodynamics 7, 173-193.

Mahmoud,S.M. (1988) . Crustal deformation measurements in the southwestern part of Egypt based on observations of local seismic and geodetic networks. M.Sc. Thesis, Faculty of Science, Ain-Shams University, Cairo.

Vyskočil,P. and Tealeb,A. (1985) . Report on activity by monitoring recent crustal movements at Aswan in period 1983-1985. Bull. Helwan Inst. of Astronomy and Geophysics V, Ser. B, 33-41.

Vyskočil,P., Tealeb,A., Kebeasy,R.M. and Mahmoud,S.M.(1987). Recent crustal movement studies along the western bank of Lake Nasser, Egypt. Proceedings of the 5th Annual Meeting of the Egyptian Geophysical Society, Cairo, Egypt.

Vyskočil,P., Kebeasy,R.M., Tealeb,A. and Mahmoud,S.M.(1988). The present state of crustal movement studies at Kalabsha area, Aswan, Egypt. Proceedings of the 6th International Symposium, Geodesy and Physics of the Earth, Potsdam, GDR.

COMPUTER AIDED DESIGN OF NETWORKS FOR MONITORING CRUSTAL TECTONIC ACTIVITY

M W Rayson and P A Cross
Department of Surveying
University of Newcastle upon Tyne

Abstract

Recent developments in terrestrial and satellite surveying techniques have led to the establishment of a number of new campaigns to determine recent crustal motions. In the light of this there is an increasing need for an interactive computer aided design program to design suitable measurement schemes.

The problem can be summarised thus: Given the approximate coordinates of a series of possible survey control points and a geological/geophysical model for the region, find the optimal set of (terrestrial and/or space) measurements that need to be made, and the optimal manner in which to make them, in order to recover the precision and reliabilty of the geophysical parameters (rates of motion, direction of motion etc.) to a satisfactory level.

A program has been written that can cope with a number of geological models including transform faults, regional rotation, divergent/convergent motion and multi-block rotation. Various statistical measures, computed from the functional and stochastic models allow the user to check on both the precision of the recovered parameters and the internal and external reliability of the observations. If the network quality is inappropriate, then experiments can be carried out with the addition or removal of both stations and observations until a satisfactory recoverable precision and reliability are achieved.

Introduction

Over the last thirty five years the framework of geophysical research has increased dramatically as a result of the plate tectonic theories and the launch of numerous artificial satellites. The furtherance of our knowledge and understanding of the nature of plate tectonic motions, especially continents and their margins (Brown and Reilinger 1983), involves the fundamental requirement of accurate measurements to provide relative motions between selected points on various tectonic plates. Near-earth satellites have greatly influenced and increased this ability by providing accurate positioning. However for satellites to fulfil their potential requires increased knowledge of

262

global gravity modelling, which in turn requires a better understanding of earth rotation, polar motion, earth tides and ocean loading. The solution to this jigsaw encompasses a broad spectrum of studies involving a variety of dynamical processes of the earth and is achieved by recent developments in both terrestrial and space based surveying techniques.

The Global Positioning System (GPS), Satellite Laser Ranging (SLR), Very Long Baseline Interferometry (VLBI) and Electonic Distance Measurements (EDM) have made, and are making, significant contributions to advancements in all the fields of study mentioned above, the space techniques having provided accurate spatial positioning of terrestrial points in both regional and global contexts.

Classical geodetic terrestrial measurements have for many years been used to monitor strain rates (Walcott 1984) in geologically active regions over a period of a decade of so with limited success. However the ability to determine spatial positioning to high levels of accuracy using space based techniques has made it a real possibility to study crustal motion and deformation within relatively short time scales. The use of GPS has meant campaigns oan be established to measure total strain of a region and obtain results in as little as five years.

Many such campaigns have been and are being performed and WEGENER-MEDLAS is one such example. WEGENER is an acronym for Working group of European Geoscientists for the Establishment of Networks for Earthquake Research, and one of its major projects has been MEDLAS (MEDiterranean LASer ranging) which has led to fourteen SLR stations being established in the eastern Mediterranean to measure rates of plate motion. In conjunction with this project several GPS campaigns have also been observed.

Other crustal motion studies carried out include Iceland (Foulger et. al. 1986), California (Sauber 1989), Central Asia (Chen 1988) and Alaska (Beutler et.al. 1985a). In all cases monitoring is performed using SLR, GPS, VLBI and EDM.

In the light of these projects it was considered that there was an increasing need for an interactive computer aided design program which would assist in designing suitable terrestrial and/or space networks for future measurement schemes. A program of this nature can answer many questions. For example, how many stations and observations are needed?, what are the best types of observations to make?, what type of instruments should be used to satisfy precision and reliability requirements ?, what time interval should be left between epochs?, etc. It may well occur that several solutions oan be found for the same problem, in which case, what network can be implemented at the cheapest cost? For example, if one wished to monitor the motion along a transform fault, what would be the most suitable configuration of stations and observations to use.

Though such a program can assist a network design and answer questions from a mathematical point of view there are other questions and factors that cannot be modelled. The sorts of problems addressed here are weather conditions, travel-times to and from the sites, machine failure, operator illness, etc. However, with good reconnaissance and common sense these kinds of problems can be allowed for in a design scheme. The design must also take into account instrument limitations. For example the requirement of intervisibility between stations for terrestrial slope distance and angle observations. Though satellite data does not require intervisibility and may be recorded in just about any weather conditions, and at any time of the day, there must be good all-round visibility in the immediate vicinity of the station and the site be clear from obstacles capable of multi-pathing effects. If this is not the case such obstacles as trees will mask satellite signals reaching the antenna. It is also extremely useful to try and locate stations as close to roads and paths as possible, as stations with easy access are a tremendous advantage.

The general design problem can be divided into four categories suggested by Grafarend (1974) :

   i) Zero order design : Where the design of the network reference system is considered.
  ii) First order design : Design of the network configuration, i.e.what geometric combination of observables (design matrix A) will yield the desired solution.
 iii) Second order design : Design of the observation weights.
  iv) Third order design : Design of network alterations to strengthen an existing network.

A further category, the so-called fourth order design, has also been considered, (Schmitt 1982) and considers the time that should elapse between epochs of measurements being made. The identification of these orders is useful for most practical problems, and a combination of several of the above will usually be solved for during any particular design procedure.

The first stage of network design is to consider the type of geological motion and parameters that are to be determined and to specify the precision and reliability with which these parameters are required. Once specified, it is necessary to determine the optimal observation scheme which satisfies these requirements. Various approaches to this problem are discussed in Grafarend and Sanso (1985). In general the approaches can be divided into two categories namely computer simulation (the method used in this program) and analytical methods. Computer simulation has been around for approximately twenty years and is well documented in the literature (Cross 1981, Mepham 1893).

Program implementation

A 3-D interactive graphics computer program called Netwrk3d has been specially developed to investigate the problem of network design for crustal motion monitoring. The program is a menu driven software written in Pro-pascal and implemented on a microcomputer. The main aims of the program were to make it user friendly (easy to use and learn) and interactive (with a series of graphics routines). The graphics allow a mouse to be used for the input of stations and observations and for their addition and deletion, and gives the user the advantage of being able to see on the screen, at any time, the network so far designed.

The program currently recognises all the coordinate systems, observation types and geological models discussed in the paper. The various statistical measures, illustrated later, allow the user to determine the precision of the geophysical parameters and coordinate differences, and the reliability of the observations.

The program is also interfaced with a printer to allow hard copies of the network details (i.e. observations, station coordinates and quality analysis), and screen dumps of the network, to be made.

Coordinate systems

For modern 3-D geodesy three coordinate systems are required, namely the Earth Centred Cartesian (ECC), Ellipsoidal and a Local Topocentric system (LT) the uses of which will become obvious throughout the paper. Although several systems may be used simultaneously, when computations and/or adjustments are performed, all data, regardless of its origin, must be in just one system.

Observation types

In crustal deformation monitoring, the raw positional data are observations of slope distance, horizontal and vertical angles, horizontal azimuths, height differences and coordinate differences derived from satellite observations. All observations are modelled in the ECC system and both terrestrial and space observations can be used simultaneously in the same network.

It is well known that the least squares estimate of the normal equations is described by

$$\hat{x} = (A^t WA)^{-1} A^t Wb \qquad (5)$$

and it is shown, by Cross (1983) for example, that the covariance of $\hat{x}$, $C\hat{x}$ is given by:

$$C\hat{x} = (A^t WA)^{-1} \qquad (6)$$

Notice that (6) does not depend on the vector b in (5). The covariance matrix $C\hat{x}$ is used to assess the quality of the parameters and if the A and W matrices are known a network design can take place without any knowledge of the actual values of the observations themselves.

Geological modelling

Up to date a number of geological motions, including transform faults (sinistral and dextral) divergent and convergent motion, faulting (normal and reverse), folding (synclinal and anticlinal), regional rotations and multiple block rotations have been mathematically modelled. All have characteristic parameters that describe the models i.e. rate of motion, azimuth of motion, angle of uplift, rate of rotation etc.. These parameters are termed geophysical parameters.

To model individual geological motions requires relationships to be developed between the geophysical parameters and the coordinate differences of the stations. These relationships are non-linear and because we wish to estimate the geophysical parameters and their quality using least squares it is necessary to linearise them. This is carried out in the standard way (by partial differentiation) leading to Jacobian matrices which are generally known as the "functional models". The models are initially developed in the LT system and then transformed to the ECC system. For example what follows is the development of the model for a simple transform sinistral motion.

266

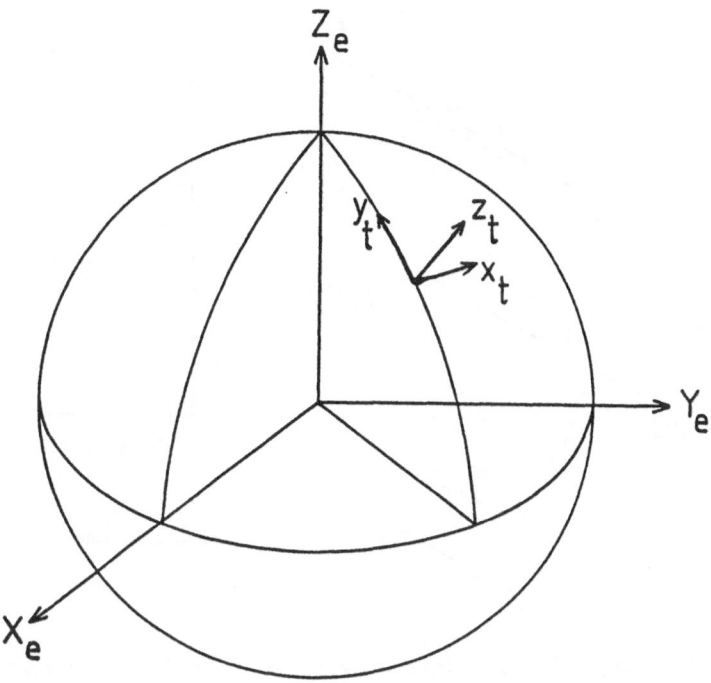

Fig. 1   Relationship between the EEC system and the LT system.

The right handed ECC system is extensively used for satellite positioning systems and its relationship to the LT system is shown by fig. 1. The transformation between the two is easily achieved using two rotations, one about the x-axis and one about the z-axis. The rotations are described by the following transformation matrix:

$$\begin{bmatrix} X_t \\ Y_t \\ Z_t \end{bmatrix} = \begin{bmatrix} -\sin\lambda & \cos\lambda & 0 \\ -\cos\lambda\,\sin\phi & -\sin\lambda\,\cos\phi & \cos\phi \\ \cos\lambda\,\sin\phi & \sin\lambda\,\cos\phi & \sin\phi \end{bmatrix} \begin{bmatrix} X_e - X_e{}^s \\ Y_e - Y_e{}^s \\ Z_e - Z_e{}^s \end{bmatrix} \quad (1)$$

The transformation of LT coordinates to the ECC is simply achieved by multiplying the LT coordinates by the transpose of the transformation matrix.

For an ellipsoidal coordinate (ϕ,λ,h) there exists a corresponding set of ECC coordinates (X_e,Y_e,Z_e). The relationship is given by the standard formulae below.

$$X_e = (V + H)\ \cos\phi\cos\lambda \qquad (2)$$
$$Y_e = (V + H)\ \cos\phi\sin\lambda \qquad (3)$$
$$Z_e = (V\ (1 - e^2) + H)\ \sin\phi \qquad (4)$$

The reverse process is also given by standard formula which are given in, for instance, (Bomford, 1980).

267

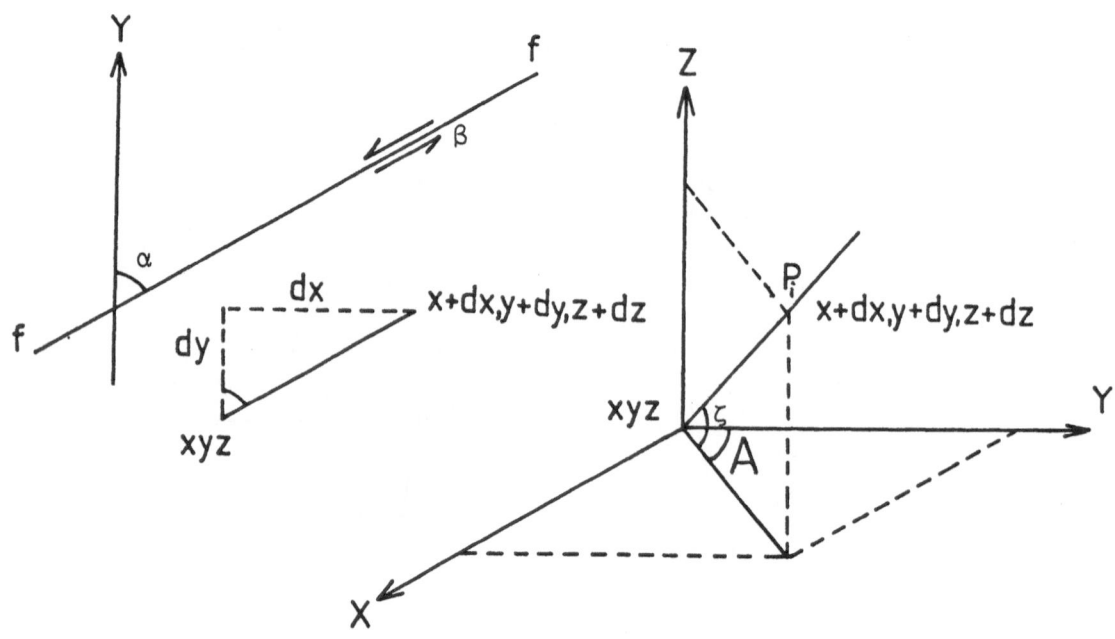

*Fig 2. a) Plan view of transform sinistral motion
b) transform motion described in 3-D using
vector Pi.*

Transform sinistral motion model

The transform motion model, represented in plan view by
fig 2a and in 3-D by fig 2b, can be totally described by
three parameters. These are the rate of motion (β), the
azimuth of motion (A) and the angle of vertical motion (ζ),
and they can be related to the coordinate differences by the
functions:

$$f_1 \ (A,\beta,\zeta,\Delta t,dX) = 0 \qquad (7)$$
$$f_2 \ (A,\beta,\zeta,\Delta t,dY) = 0 \qquad (8)$$
$$f_3 \ (A,\beta,\zeta,\Delta t,dZ) = 0 \qquad (9)$$

where
Δt = time difference between observation epochs $(t_2 - t_1)$.

In fig. 2a the coordinates (X,Y,Z) and (X+dX,Y+dY,Z+dZ)
represent the position of a station at times t_1 and t_2
respectively where the change in coordinates is given by

$$dX = \beta'\Delta t \ \sin A \ \cos \zeta$$
$$dY = \beta'\Delta t \ \cos A \ \cos \zeta$$
$$dZ = \beta'\Delta t \ \sin \zeta$$

To express the coordinate changes in the ECC system the
tranformation matrix, given in equ. (1), is applied. The
resulting coordinate differences are expressed by

268

$$dX_e = -\sin\lambda \ \beta'\Delta t \ \cos\zeta \ \sin A \ - \sin\phi\cos\lambda \ \beta' \ \Delta t \ \cos\zeta \ \cos A \qquad (10)$$
$$+ \cos\phi \ \cos\lambda \ \beta' \ \Delta t \ \sin\zeta$$

$$dY_e = \cos\lambda \ \beta'\Delta t \ \cos\zeta \ \sin A \ - \sin\phi \ \sin\lambda \ \beta' \ \Delta t \ \cos\zeta \ \cos A \qquad (11)$$
$$+ \cos\phi\sin\lambda \ \beta'\Delta t \ \sin\zeta$$

$$dZ_e = \cos \ \phi\beta' \ \Delta t \ \cos\zeta \ \cos A \ + \sin\phi \ \beta' \ \Delta t \ \sin\zeta \qquad (12)$$

which are general expressions of the form

$$\Delta \hat{x} \ = \ \Phi(\phi)$$

where ϕ is a vector of geophysical parameters.

Equations (10) to (12) are then partially differentiated with respect to the three geophysical parameters to form the elements of the Jacobian functional model. The matrix elements are then numerically represented by assigning provisional values for each of the geological parameters in the model. Similar linearised models are formed for the other geological motions mentioned.

Measures of Quality

It is essential for geophysical network design that the quality of both the coordinate differences and the geophysical parameters can be assessed. Unless accompanied by an assessment of their quality they can be considered to be of very little value. Two measures of quality are computed: precision and reliability. Precision is used to describe quality with respect to the random errors whilst reliability is with respect to gross errors.

Precision of the geophysical parameters

An extremely useful precision criteria exists where the precision of relevant (geophysical parameter) quantities can be directly derived from the previously estimated parameters ($C_{\hat{x}}$ matrix).

In general if a vector of geophysical parameters ϕ is to be estimated by least squares from

$$\Phi(\hat{\phi}) \ = \ \Delta\hat{x}$$

then their precisions can be calculated from

$$C_{\hat{\phi}} \ = \ (f^t \ Q f)^{-1} \qquad (13)$$

where

f = functional model of the selected geological motion and is given by,

$$f = \left| d\phi / d\Delta x \right|_{\Delta x = \Delta \hat{x}}$$

and Q = stochastic model (weight matrix) derived from the covariance matrix of the parameters i.e. $C_{\Delta \hat{x}}^{-1}$.

Deformation monitoring is usually achieved by making measurements at different epochs separated by a time interval Δt. Consider just two epochs of measurements. The coordinate differences between t_1 and t_2 can be given by:

$$x_{21} = \begin{bmatrix} I & -I \end{bmatrix} \begin{bmatrix} x_2 \\ x_1 \end{bmatrix}$$

by considering the covariance matrices from time t_1 and t_2 gives

$$C_{\Delta \hat{x}_{21}} = \begin{bmatrix} I & -I \end{bmatrix} \begin{bmatrix} C_{\hat{x}_2} & 0 \\ 0 & C_{\hat{x}_1} \end{bmatrix} \begin{bmatrix} I \\ -I \end{bmatrix}$$

which when multiplied out gives:

$$C_{\Delta \hat{x}_{21}} = C_{\hat{x}_1} + C_{\hat{x}_2} \qquad (14)$$

By assuming the observations at t_1 and t_2 will be the same equ. (14) reduces to

$$C_{\Delta \hat{x}_{21}} = 2C_{\hat{x}_1} \qquad (15)$$

By definition a weight matrix is the inverse of the corresponding covariance matrix, so Q in (13), the weight matrix of the coordinate differences, is given by

$$Q = C_{\Delta \hat{x}_{21}}^{-1}$$

$$= (2C_{\hat{x}})^{-1} \qquad \text{(from 15)}$$

$$= 1/2 \ A^t W A = Q \qquad \text{(from 6)} \qquad (16)$$

270

By selecting the appropriate geological model, f, and using the a priori matrix Q, the covariance matrix of the geophysical parameters can be computed by substituting (16) into (13). From the resulting matrix the precision of the individual geophysical parameters of the model can be derived.

Measures of Reliability

Reliability is defined as the ability or ease with which outliers or gross observational errors may be detected. Two measures of reliability can be identified, namely internal and external reliability. The former is a measure of the size of gross errors that could remain undetected in each observation and latter gives a measure of the effect the undetected gross errors will have on the estimated parameters. Pelzer (1977), amongst others, has shown that high internal reliability automatically leads to high external reliability and vice versa.

Internal Reliability

In order to assign numerical values to the internal reliability the symbols τ and Δ_i^u are introduced. Δ_i^u is known as the marginally detectable error and represents the upper bound on the gross error that can be detected with a probability α when the level of significance used in the testing is β. τ is called the tau factor and it is shown by Cross (1983) for example that it is given by:

$$\tau = 1.0 \ / \ \{ e_i{}^t \, C_l \, e_i \, e_i{}^t \, W C_{\hat{v}} W e_i \} \tag{17}$$

Where

$$C_{\hat{v}} = W^{-1} - A C_{\hat{x}} A^t \tag{18}$$

For the more simple case where C_l (and hence W) is a diagonal matrix the tau factor can be expressed by

$$\tau = \sigma_i \ / \ \sigma_{\hat{v}i} \tag{19}$$

where

$\sigma_{\hat{v}i}$ = the standard error of the ith least squares residual.

Note, it can be easily shown that (for a diagonal W)

$$1 < \tau < \infty \tag{20}$$

Once calculated the tau factor can be used to compute the marginal detectable error for the corresponding observation using (Cross 1987):

$$\Delta_i{}^u = \delta_i{}^u / \{ e_i{}^t W C_{\hat{v}} W e_i \}^{1/2}$$

which in the case of uncorrelated observations reduces to

$$\Delta_i{}^u = \delta_i{}^u e_i{}^2 / (\text{diag}\{ C_{\hat{v}} \})^{1/2}$$

which is often written as

$$\Delta_i{}^u = \delta_i{}^u \tau_i \sigma_i \tag{21}$$

where

$\sigma_i =$ standard error of the ith observation
$\delta_i^u =$ a function of the particular pdf where its value is given by the values selected for α and β, and is a unitless number.

When $\alpha = 0.05$ and $\beta = 0.10$ then $\delta_i^u = 3.85$. In the case of a distance measurement with a standard error of 0.01m, $\tau = 1.245$ and $\Delta_i^u = 3.85$ the marginally detectable error would be 0.0479m. This means that with the α and β used, there would be a 10% chance of a gross error of 0.0479m remaining undetected. Note observations are not required to compute either the tau factor or the marginally detectable error.

External Relibility

In general the internal reliability may not be the most useful picture of network reliability. We may wish to know, not just the size of the marginally detectable errors, but also their effects on the geophysical parameters and quantities computed from them. To do this it is necessary to define $\Delta \hat{x}_i^u$ as a vector which describes the effects on $\hat{x}$ of the marginally ith detectable error in the ith observation, this being given by

$$\Delta \hat{x}^u{}_i = \{ C_{\hat{x}} A^t W e_i \} \, \Delta^u{}_i \tag{22}$$

where

$$e_i = [\ 0,0,0\ \dots\ 1\ \dots\ 0,0,0]$$

That is to say, ei is a null column vector but for a value of unity in the ith element. It then follows that

$$\Delta \hat{\phi}^u{}_i = \{ (f^t Q f)^{-1}\ f^t Q \} \ \Delta \hat{x}^u{}_i \tag{23}$$

where f is the functional model describing the geological model and $\Delta\hat{x}_i^u$ is the effect of the marginally detectable error in the ith observation on the geophysical quantity.

Testing

Testing the program can be divided into two groups, namely correctness and usefulness. It is imperative before using the program for any investigative work that all algorithms are thoroughly tested to ensure the computations are correct. This has been achieved using hand calculation and comparison against existing software.

Limited testing the program for usefulness has also been carried out. The geometry of the observations, shape of the network and time difference between epochs all influence the values of precision obtainable for the geophysical parameters. Naturally observations, of all types, are of significantly more value when the change in observed values between t_1 and t_2 is greatest. Achieving the largest possible difference relies heavily upon the orientation of the observation with respect to the direction of motion. The case of a distance observation is illustrated with respect to the transform sinistral and divergent motions in fig 3. In both cases observations made in the direction of the dashed lines are more beneficial than their solid counterparts as a result of the greater component of motion being observed in each case.

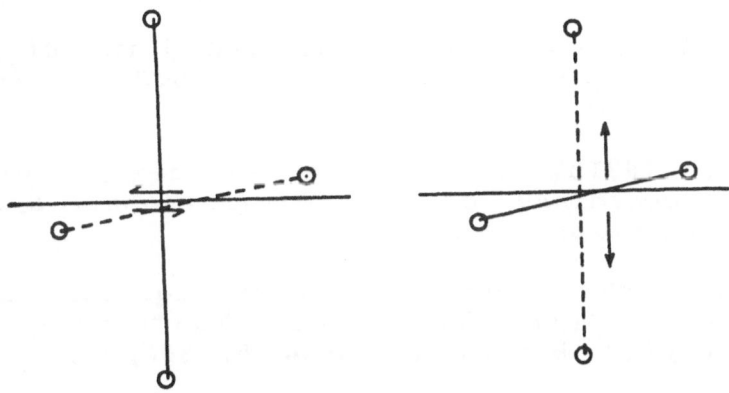

Fig 3. a) Transform sinistral motion b) Divergent motion.

The shape of a network also has a marked influence. Consider a simple four stations network. With all slope distance observations the more elongated a network is the more efficient it is. Hence, when measuring a transform sinistral motion or divergent motion elongated networks orientated to the directions of motion prove best. However for horizontal angle observations a square configuration is better.

The time difference between t_1 and t_2 as an individual factor has perhaps the most significant effect on the precision recoverable and plays an important part in the design.

Acknowledgements

The authors would like to express their thanks the Natural Environmental Research Council (NERC) and the Science and Engineering Research Council (SERC) for jointly funding the work under the Geodesy Initiative.

References

Beutler, G., W. Gunter, M. Rothacher, T. Schildknecht, and I. Bauersima. Evaluation of the 1984 Alaska GPS campaign with the Bernese Second Generation Software (Abstract). *Eos Trans. AGU, 66, 858.*

Bomford, G. 1980. Geodesy 4th Edition. *Clarendon Press, Oxford, 855pp.*

Brown, L, and R. Reilinger, 1983. Crustal Movement. *Rev. Geophys., 21(3), 553-559.*

Chen, R, 1988. Plate tectonics at the joint of Pamirs and Tien-Shan in Central Asia. *Report 88:3 of the Finnish Geodetic Institute.*

Cross, P.A, 1981b. Computer aided design of geodetic networks. *Proceedings of VI International Symposium on Geodetic Computations, Munich 12pp.*

Cross, P.A, 1983. Advanced least squares applied to position-fixing. *North East London Polytechnic, Department of Land Surveying. Working paper no.6: 205pp.*

Cross, P.A, 1987. CAD, geodetic networks and the surveyor. Land and Mineral Surveyor.

Foulger, G, Bilham, R, Morgan, W.J, and Einarsson, P, 1987. The Iceland GPS Geodetic Field Campaign 1986. *Eos, Vol 68, No. 52, 1817-1818.*

Grafarend, E.W, 1974. Optimisation of Geodetic Networks. *Bolletino di Geodesia a Science Affini. Vol. 33, No. 4 pp351-406.*

Grafarend, E.W, and Sanso, F, 1985. Optimisation and design of geodetic networks. *Springer-Verlag, 606pp.*

Mepham, M.P.A, 1983. Computer aided survey network design. *M.Sc. thesis, University of Calgary, 98pp.*

Pelzer, P, 1977. Criteria for the reliability of Geodetic Networks. *International Symposium on Optimization of design and computation of control networks, Sopron, Hungary. 10pp.*

Sauber, J, 1989. Geodetic Measurement of Deformation in California. *NASA Technical Memorandum 100732.*

Schmitt, G, 1982. Optimization of geodetic networks. *Review of Geophysics and Space Physics, vol. 4.*

NEW MODEL OF THE ENVIRONMENTAL ERROR SOURCES IN EDM DEVICES

Claudio Marchesini

Istituto di Scienze della Terra , Università di Udine , Italia

ABSTRACT

From an analysis of the environmental and instrumental error sources, the contribution of each source to the total dispersion of the distance measurements is determined. In particular, to evaluate the variance of the mean value of the temperatures along the base, use is made of a stochastic process, autoregressive of second order.

The calculated variances give the weights for the adjustment of a geodetic network and for the realization of the solution of the SOD problem.

1. OBJECT OF THE RESEARCH

We want to investigate the errors both instrumental and environmental concerning the distance measurements with EDM devices.

In modern high-precision distancemeters the global uncertainty of measurement depends on the instrument error to a small extent.

The major part of the uncertainty in the measurement is due to imperfect knowledge of the refractive index of the air along the path.

A reliable valutation of the mean square error σ_d of the distance measurement is essential to determine the weights in the adjustment of a geodetic network. In fact the number of repetitions of single measurements is normally too small to be considered significant from the statistical point of view and correct for the estimation of the measurement variance σ_d^2.

2. FUNCTIONAL DEPENDENCE OF A DISTANCE MEASUREMENT

The distance d between two points, measured with an electro-optic device is given by

$$d = \frac{1}{2}\frac{c}{f\,n} + k \tag{2.1}$$

with the following meaning of the symbols:

f : modulation frequency

l : number of half-wave-lengths contained between the end points, it is normally a not integer number and includes the phase measurement. Only in the Mekometer ME 5000 the modulation frequency is adjusted in order to obtain an integer value of l in any case (Meier-Loser, 1986)

c : light speed in vacuum

k : additive term, which includes the discrepances in the centering of the distancemeter and the prisms and in the position of the their optical centers

n : refractive index of atmosphere in the conditions of temperature and pressure at time of measurement

$$n = 1 + (n_s - 1) \frac{p}{q} \frac{1 + p(h - gt)10^{-6}}{1 + \alpha t} - le \qquad (2.2)$$

where

t : temperature in °C

p : pressure in hP

e : partial pressure of water vapor in hP

$q = 960.954$

$g = 0.0133$

$h = 0.817$

$l = 0.0404 \cdot 10^{-6}$

$\alpha = 1/273.15$

n_s: refractive index of standard atmosphere (temperature = 15°C, pressure = 1013 hP), in this work we use the formulae proposed by Edlen (1966)

$$n_s = 1 + \left(v + \frac{b}{m - 1/\lambda} + \frac{w}{y - 1/\lambda^2} \right) 10^{-8} \qquad (2.3)$$

where

b = 2406030

w = 15997

λ : carring wave-length of the distancemeter

$v = 8342.13$

$m = 130$

$y = 38.9$

3. VARIANCE COMPONENTS OF A DISTANCE MEASUREMENT

According to the law of error propagation the variance of a distance measurement is given by

$$\sigma_d^2 = \Sigma_i \left(\frac{\partial d}{\partial x_i} \right)^2 \sigma_{x_i}^2 \qquad (3.1)$$

where x_i are independent variables and $\sigma_{x_i}^2$ the corrisponding variances
From the previous formulae it result that we have seven independent variables l, f , k , λ , t , p , e ; the first four ones are connected with the distancemeter. In table 1 are listed the partial derivatives and the m.s.e. values if we use a Mekometer.

TABLE 1
INSTRUMENTAL COMPONENTS

x_i	$\dfrac{\partial d}{\partial x_i}$	σ_{x_i} (for ME 5000)	$\dfrac{\partial d}{\partial x_i} \cdot \sigma_{x_i}$ (in m)
l	$\dfrac{c}{2fn}$	$1.5 \cdot 10^{-11}$	$4.5 \cdot 10^{-11}$ (negligible)
f	$-\dfrac{d}{f}$	$1 \cdot 10^{-7} \cdot f$ (f = 500 MHz)	$-1 \cdot 10^{-7} \cdot d$
k	1	$0.2 \cdot 10^{-3}$	$0.2 \cdot 10^{-3}$
λ	$d\left[\dfrac{b}{(m\lambda-1)^2} + \dfrac{2w\lambda}{(\gamma\lambda^2-1)^2}\right]10^{-8}$	$1 \cdot 10^{-4} \cdot \lambda$ (λ = 0.6328 μm)	$2.9 \cdot 10^{-13}$ (negligible)

The last three variables are environmental parameters and precisely the average values of temperature, pressure and humidity along the path of measurement.

In practice it is not possible to measure the environmental values at all points along the path. Their values are measured only at the two end points of the line and in exceptional cases at some points in the middle (for instance with sounding balloons). It is possible to evaluate the variance of the mean of environmental parameters if their behaviour along the path is know.

Because of our poor knowledge of the laws of the microclimate, the best approach is the search for a stochastic process.

4. STOCHASTIC PROCESSES

In order to describe the behaviour of the temperature (or pressure, or humidity) along the path, the line of measurement d is divided into N intervals of size Δ, so that it is possible to obtain a set of N temperature data equally spaced.

The suitable stochastic process and parameters are searched for by means of various sets of data recorded alongside cableways. The temperature measurements are obtained using electronic sensors installed underneath the cablecars, which transmit the data by radio to a receiving unit.

We have taken into consideration the stochastic processes of Box and Jenkins (1970) and from the recorded data it comes out

TABLE 2

STOCHASTIC PROCESSES OF BOX-JENKINS

type	parameters	law
AR(1)	$\Phi = 0$	$t_i = a_i$
AR(1)	$\Phi = 1$	$t_i = t_{i-1} + a_i$
ARI(1,1)	$\Phi < 1$	$w_i = \Phi w_{i-1} + a_i$
ARI(2,1)	$\Phi_1 + \Phi_2 < 1$	$w_i = \Phi_1 w_{i-1} + \Phi_2 w_{i-2} + a_i$

t_i = term of the set of temperatures

$w_i = t_i - t_{i-1}$ first difference

a_i = random number

AR(1) : autoregressive of first order, the general law is $t_i = \Phi t_{i-1} + a_i$

ARI(1,1) : autoregressive integrated of order (1,1)

that the processes listed in table 2 are suitable to describe the behaviour of the environmental data.

5. VARIANCE OF THE MEAN VALUE IN A STOCHASTIC PROCESS

The mean value of a set of temperatures is given by

$$t = \frac{1}{N} \sum_1^N {}_i t_i \qquad (5.1)$$

According to the law of error propagation, the variance of t becomes
$$\sigma_t^2 = \underline{N} Q \underline{N}^T \qquad (5.2)$$
with $\underline{N}^T = | 1/N \quad 1/N \ldots 1/N |$
and Q = variance-covariance matrix of t

If the t_i terms are not correlated to each other Q is a diagonal matrix, furthermore we can assume that all t_i have the same variance σ_a^2, consequentely we can put
$$Q = \sigma_a^2 U \qquad (5.3)$$
where U is the unit matrix.

Putting the (5.3) into (5.2), we obtain

$$\sigma_t^2 = \frac{\sigma_a^2}{N} \qquad (5.4)$$

To describe better the real situation, we put $N \to \infty$, so we have
$$\lim_{N \to \infty} \sigma_t^2 = 0 \qquad (5.5)$$
If instead the terms are correlated, we can put $\underline{t} = F \underline{a}$ (5.6)
where F is the functional link, $\underline{t}$ the vector of t_i and $\underline{a}$ the resulting uncorrelated terms, that are random numbers. The variance is now
$$\sigma_t^2 = \underline{N} F Q F^T \underline{N}^T \qquad (5.7)$$
and Q is again a diagonal matrix, using the (5.3) definition we have
$$\sigma_t^2 = \sigma_a^2 (N F)(N F)^T \qquad (5.8)$$
The variances of the a_i terms are supposed to be all equal, they

depend also on the length of the sample interval $\Delta = d/N$.

To find an analytical description of this dependence let us consider a unit interval u contained an integer number r of times in Δ. The random increment a_i is decomposed in r increments b_k, that.is

$$a_i = \sum_{1}^{r} k \, b_k \qquad\qquad (5.9)$$

In the hypothesis that the variances of the b_k terms are all equal, according to the law of error propagation it follows that

$$\sigma_a^2 = r \, \sigma_b^2 \qquad\qquad (5.10)$$

where $r = \Delta/u$ and $\Delta = d/N$.

Replacing these terms in (5.10) it turns out that

$$\sigma_a^2 = \frac{d}{N} Z^2 \qquad\qquad \text{where } Z^2 = \sigma_b^2 / u \qquad (5.11)$$

The value of Z^2 is obtained from experimental data, as explained in the next paragraph. The F matrices, the $\lim_{N \to \infty} \sigma_t^2$ functions calculated on the grounds of the (5.8) and (5.11) with various stochastic processes are listed in table 3.

TABLE 3

type	F matrix	$\lim_{N \to \infty} \sigma_t^2$
AR(1) $\Phi = 1$	$\begin{vmatrix} 1 & 0 & 0 & 0 & \dots \\ 1 & 1 & 0 & 0 & \dots \\ 1 & 1 & 1 & 0 & \dots \\ 1 & 1 & 1 & 1 & \dots \\ \dots \end{vmatrix}$	$\dfrac{dZ^2}{3}$
ARI(1,1)	$\begin{vmatrix} 1 & 0 & 0 & 0 & \dots \\ 1 & 1 & 0 & 0 & \dots \\ 1 & 1 & 1 & 0 & \dots \\ 1 & 1 & 1 & 1 & \dots \\ \dots \end{vmatrix} \cdot \begin{vmatrix} 1 & 0 & 0 & 0 & \dots \\ \Phi & 1 & 0 & 0 & \dots \\ \Phi^2 & \Phi & 1 & 0 & \dots \\ \Phi^3 & \Phi^2 & \Phi & 1 & \dots \\ \dots \end{vmatrix}$	$\dfrac{dZ^2}{3(1-\Phi)^2}$
ARI(2,1)	$\begin{vmatrix} 1 & 0 & 0 & 0 & \dots \\ 1 & 1 & 0 & 0 & \dots \\ 1 & 1 & 1 & 0 & \dots \\ 1 & 1 & 1 & 1 & \dots \\ \dots \end{vmatrix} \cdot \begin{vmatrix} 1 & 0 & 0 & 0 & \dots \\ \Phi_1 & 1 & 0 & 0 & \dots \\ \Phi_1^2 + \Phi_2 & \Phi_1 & 1 & 0 & \dots \\ \Phi_1 + 2\Phi_1\Phi_2 & \Phi_1^2 + \Phi_2 & \Phi_1 & 1 & \dots \\ \dots & \dots & \dots \end{vmatrix}$	$\dfrac{dZ^2}{3(1-\Phi_1-\Phi_2)^2}$

6. TEMPERATURE

In order to find the most achievable stochastic model we have collected 8790 temperature data in total from 127 sets on 19 different locations. Some sets are measured with different carriers: aircraft, funicolar, etc. The above mentioned stochastic modells were applied to every set of temperatures and, if the autocorrelation of the residuals is statistically zero (test of Box-Pierce, 1970) the modell is accepted. The results are summarized as follows in table 4. The modell AR(1), $\Phi = 1$, the so-called "random walk" proposed in Crosilla and Marchesini (1988) is acceptable only as first step in the solution of the problem.

TABLE 4

type	parameters	accepted sets
AR(1)	$\Phi = 0$	0
AR(1)	$\Phi = 1$	50%
ARI(1,1)	$\Phi = 0.27$	72%
ARI(2,1)	$\Phi_1 = 0.36$	91%
	$\Phi_2 = 0.15$	

FIGURE 1

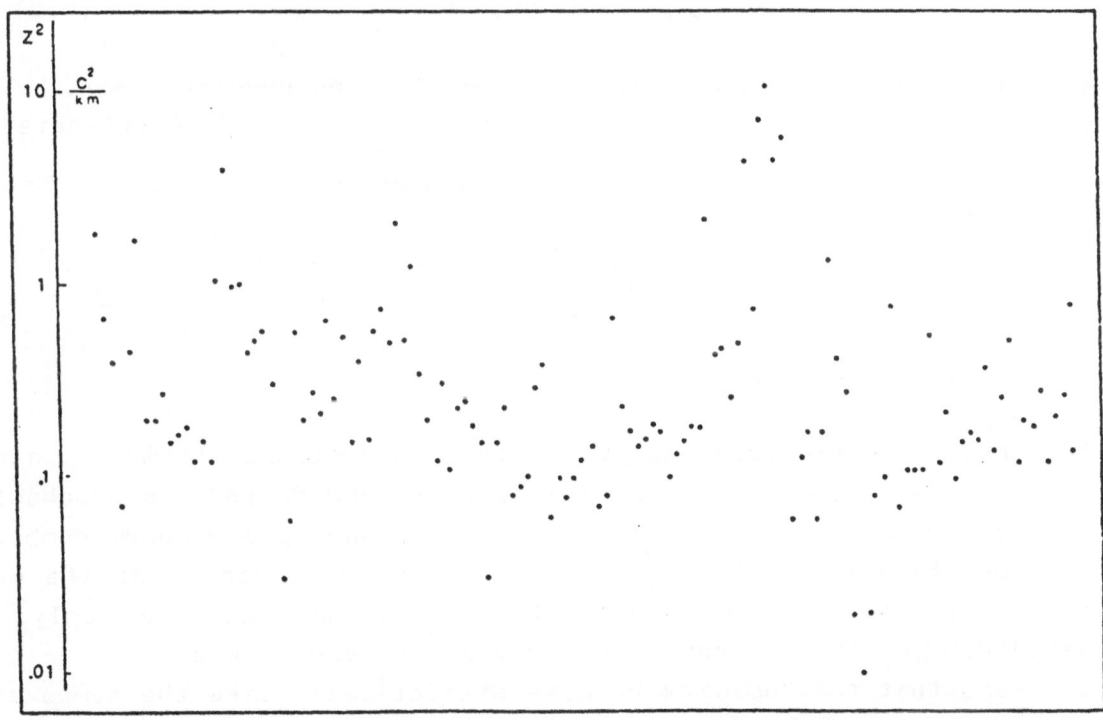

Z^2 values obtained from 127 sets of experimental data

The distribution of the Z^2 values is very scattered (see fig. 1), that is regular because it is a statistical variable and the process ARI(2,1) is quasi-stationary. The chosen value $Z^2 = 0.4$ C^2/km corresponds to the 68th percentile of all the available data which means a probability of 68% that the expected value of Z^2 will be less than the chosen value, i.e. 1 σ.

With those parameters we can calculate the component of the variance of the not measured temperatures along the path. We must now consider the variance component of the measured temperatures; we have three cases according to the number of points at which the temperature is measured

1. at one end point T_A (variance σ_T^2)
2. at both the end points of the measured line T_A, T_B (variance σ_T^2)
3. at the end points and at M intermediate equispaced points T_A, T_B (variance σ_T^2), $T_1,\ldots,T_M$ (variance σ_M^2)

In the first case the variance of T_A is added to the previous term

$$\sigma_t^2 = \sigma_T^2 + \frac{d\ Z^2}{3\left(1 - \Phi_1 - \Phi_2\right)^2} \qquad (6.1)$$

In the second case the set of temperatures is divided into two sub-sets of each $N/2$ data; after calculation we obtain

$$\sigma_t^2 = \frac{1}{4}\left[2\,\sigma_T^2 + \frac{d\ Z^2}{3\left(1 - \Phi_1 - \Phi_2\right)^2}\right] \qquad (6.2)$$

with a remarkable improvement compared with the previous case.

In the third case the set is divided into $2 + 2M$ sub-sets of each $\dfrac{N + M}{2\left(q + 1\right)}$ data, we obtain in the same way

$$\sigma_t^2 = \frac{1}{4\left(M+1\right)^2}\left[2\,\sigma_T^2 + M\,\sigma_M^2 + \frac{d\ Z^2}{3\left(1 - \Phi_1 - \Phi_2\right)^2}\right] \qquad 6.3)$$

7. PRESSURE AND HUMIDITY

The pressure fluctuations from the theoretical formulae along the path are very small (normally 0.2 hP) and follow the stochastic process AR(1) with $\Phi = 0$, that means they are random numbers. Consequently we have $\lim\limits_{N\to\infty} \sigma_p^2 = 0$, σ_p^2 is the variance of the mean of all pressures along the path, and we must consider only the variance σ_P^2 of the pressures measured at the end points.

It seems that the humidity behaves statistically like the temperature. Because of its lower effect on the distance measurements we have not yet investigated on the humidity behaviour.

8. CONCLUSIONS

Tables 1 and 5 report the complete formulae to compute σ_d with the list of the constants used when the distance is measured with a ME 5000. Figure 2 reports the m.s.e. σ_d of a measured distance with M points in which the temperatures are measured and for comparison also the classical linear trend : $\sigma_d = \pm(0.2 \text{ mm} + 1 \text{ mm/km})$. It is interesting to note that the behaviour of σ_d against the distance is not linear and decreases strongly when the number M of temperature points increases

ACKNOWLEDGEMENT

The author wisches to express his gratitude to Prof. Dr. Fabio Crosilla for helpful discussions on this paper.

TABLE 5

x_i	$\dfrac{\partial d}{\partial x_i}$	$\sigma^2_{x_i}$
		$\sigma^2_T + \dfrac{d\,Z^2}{3\left(1-\varPhi_1-\varPhi_2\right)^2}$ (one end)
t	$\dfrac{l\,c}{2f}(n_s-1)\dfrac{p}{q}\cdot\dfrac{p(g+\alpha h)}{(1+\alpha t)^2}10^{-6}+\alpha$	$\dfrac{1}{4(M+1)}\left[2\sigma^2_T+M\sigma^2_M+\dfrac{d\,Z^2}{3\left(1-\varPhi_1-\varPhi_2\right)^2}\right]$
p	$-\dfrac{l\,c}{2f}(n_s-1)\dfrac{1}{q}\cdot\dfrac{1+2\,p(h-g\,t)\,10^{-6}}{1+\alpha t}$	$\dfrac{\sigma^2_P}{2}$ $\qquad \sigma^2_P$ (one end)

$$\sigma_T = 0.3 \text{ C} \qquad \varPhi_1 = 0.36$$
$$\sigma_M = 0.3 \text{ C} \qquad \varPhi_2 = 0.15$$
$$\sigma_P = 0.2 \text{ hP} \qquad Z^2 = 0.4 \text{ C}^2/\text{km}$$
$$(n_s - 1) = 2.7652 \ 10^{-4}$$

FIGURE 2

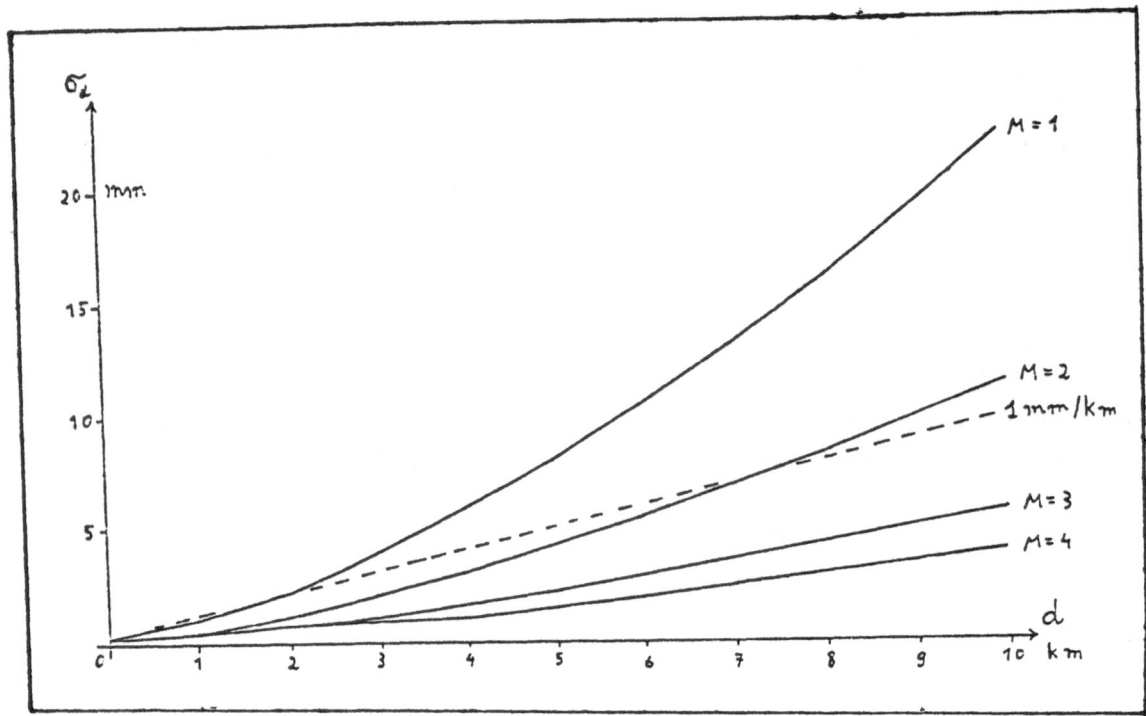

Mean square error σ of a measured distance with M points in which the temperatures are measured

REFERENCES

Box and Jenkins (1970). Time Series Analysis, Forecasting and Control. Holden-Day, San Francisco.

Box and Pierce (1970). Distribution of residuals autocorrelations in autoregressive integrated moving average time series models. Jour. Amer. Stat. Assoc. 65, 1509 - 1526.

Crosilla, F. and Marchesini, C. (1988). Realization of the second order design by modelling the environmental and instrument error sources in distance measurements. Manuscripta Geodaetica, 13, 210 - 217.

Edlen, B. (1966). The refractive Index of Air. Metrologia 2,71 - 80.

Meier, D. and Loser, R. (1986). Das Mekometer ME 5000 - Ein neuer Präzisionsdistanzmesser. All. Verm. Nachr., 5, 182 - 190.

COMPARATIVE RESULTS OF REPEATED DISTANCE MEASUREMENTS

H. Mizuno

Department of Earth Sciences, Kagawa University

l-l, Saiwai-cho, Takamatsu, Kagawa, 760 Japan

ABSTRACT

A simple statistical approach is applied to the repeated EDM of 1,383 lines, the average time interval of which is about 8 years, to obtain knowledge about the accuracy of EDM and possible accumulation of strain. The ratio d of a measured distance change to the distance is computed for each of these lines. This relative change in distance, d, shows a trend of decrease in magnitude with increasing line length. Then, the data of d are classified into three distance ranges, $D \leq 10km$, $10km < D \leq 20km$ and $D > 20km$. The histograms of d illustrated for each of the distance ranges suggest that d has a normal distribution with non-zero mean. The χ^2 test applied to the data of d provides support to this view. The standard deviation of d for a shorter distance range is larger than that for a longer distance range. The expression of random error in EDM divided by distance D; $(a + bD)/D$, is applied to describe the dependence of the standard deviation of d on D. The constant a is found to be about 10mm or more. This gives error in the phase difference determination of EDM. The quantity b is obtained as 1mm/km or less. This should provide information about distant-dependent random error and possible accumulation of strain. But it can be fully understood only in terms of error in atmospheric correction. The rate of horizontal crustal deformation must be much smaller in general than that so far estimated, even though the average interval between repeated measurements is short.

1. INTRODUCTION

Crustal strain has been obtained from repeated and adjusted triangulation surveys as well as trilateration surveys. However, estimation of the accuracy of strain thus obtained is rather complicated. The reliability of strain thus obtained is not intuitively clear. The PAG-U program has been used by the Geographical Survey Institute(GSI) to compute strains. The PAG-U Program was supplemented by Harada(1983) with the function for computing standard deviations of strain components. Though Harada's treatment was mathematically strict in expressing the complicated propagation of errors, yet some results which were obtained by applying it to repeated distance measurements seemed to be qualitatively open to question. It is still an important problem to evaluate the reliability of strain computed from geodetic measurements.

In the present paper, a simple statistical approach has been applied to electro-optical distance measurements to obtain knowledge about errors in them as well as the possible strain accumulation.

2. DATA

Suppose that distance measurements between two points have been made twice at dates T_1 and T_2. These results are denoted by D_1 and D_2. The difference between D_1 and D_2 consists of errors and deformation. The deformation cannot be discriminated from errors only on the basis of a single pair of distance measurements. But, it will be possible to obtain knowledge about them from the statistical viewpoint, if we have pairs of such measurements made over a lot of lines. The Geographical Survey Institute of Japan has conducted a systematic distance survey over the first order precise geodetic net since 1974. The survey had covered the whole of Japan for the first time by 1984, and the second survey is now under way. We can make use of distance measurements repeated at an interval of several years over a lot of lines.

In this analysis, measurements of the second survey made in 1984 and 1985 combined with the preceding measurements over the corresponding lines are used. Crustal strains computed for each of triangles of the net based on these repeated distance measurements have been reported by the Geographical Survey Institute in Report of the Coordinating Committee for Earthquake Prediction, whenever a survey over a certain regional extent was completed. The regions covered under the respective survey plans in 1984 and 1985 are given in Table 1. Letters A through H indicate these respective regions. Fig.1 roughly shows locations of these regions, which were defined in view of convenience for planning a survey. The preceding survey was not necessarily made in a year. In the column headed by T_1 of Table 1, the years over which the respective first surveys were carried out are given. Some of the surveys ranged over two years or more. The shortest time interval between the first and the second measurements is six years while the longest interval is as long as eleven years. For each of these regions, a strain distribution was reported by the Geographical Survey Institute (1985a,1985b,1986). In this analysis, distance measurements in each of these survey regions form a sample. The statistical examination is made for every sample.

Table 1. Survey regions and years of the first and second surveys, T_1 and T_2, respectively.

	Region	T_2	T_1
A	Sagami	84	75
B	Chubu, Kinki	84	74, 75, 76, 77
C	Fuji, Kofu	85	74, 75
D	Kinki	85	76, 77, 78
E	Chugoku	85	78, 79
F	Ibaraki	85	79
G	Kazusa	84	75, 76
H	Izu	84	78

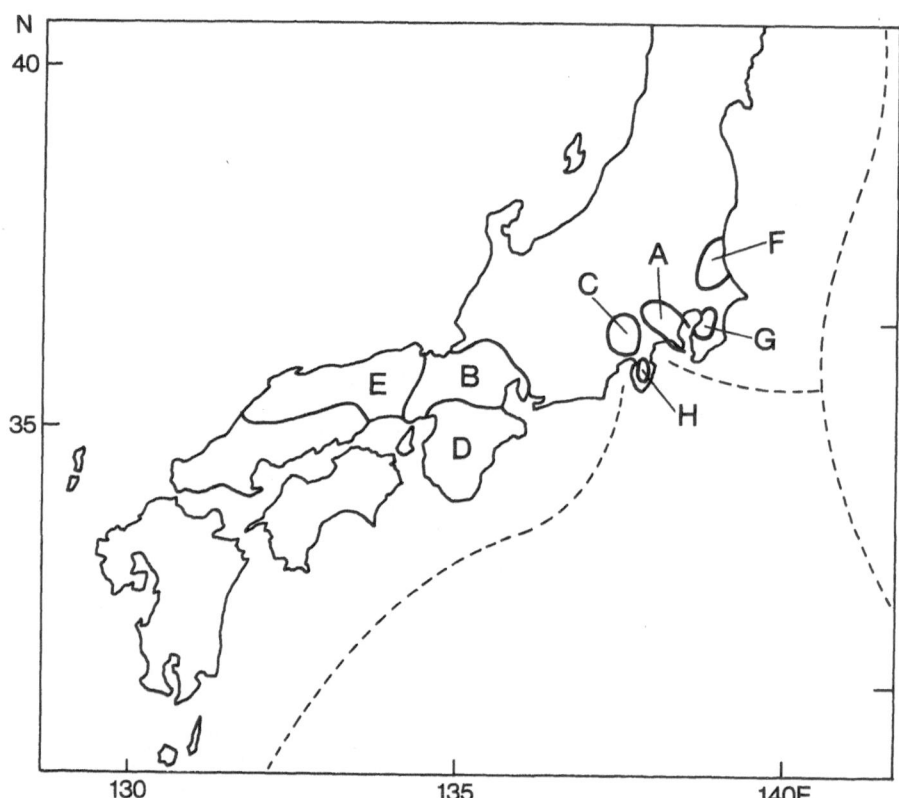

Fig.1. Location of the survey regions.

Table 2. Frequency distribution of the relative change in distance, d, in units of 10^{-6}. D_m is the average distance in km. Letters A through H are as given in Table 1.

	D_m	d	7.5	6.5	5.5	4.5	3.5	2.5	1.5	0.5	-0.5
A	7.8		0	0	0	1	1	0	6	3	6
A	13.7		0	0	0	0	1	2	2	2	6
B	7.8		4	0	3	8	10	17	24	54	67
B	11.9		0	0	0	1	0	4	18	31	43
C	7.9		3	3	3	6	7	5	5	4	9
C	12.3		0	1	3	1	3	3	6	8	7
D	7.9		1	0	0	1	1	3	8	14	28
D	12.2		0	0	0	0	0	0	4	18	12
E	7.9		0	0	4	3	4	8	14	19	27
E	12.3		0	1	0	2	4	1	6	21	20
F	16.1		0	0	0	0	1	2	4	6	4
F	24.9		0	0	0	0	0	2	1	1	5
G	7.2		3	1	2	2	4	5	6	8	6
H	7.8		0	0	0	0	1	1	3	0	2

	D_m	d	-0.5	-1.5	-2.5	-3.5	-4.5	-5.5	-6.5	-7.5	
A	7.8		6	4	8	6	1	1	0	0	0
A	13.7		6	5	7	2	1	0	0	0	0
B	7.8		67	66	54	31	23	7	6	5	1
B	11.9		43	43	48	28	7	2	2	0	1
C	7.9		9	6	1	4	2	4	2	1	3
C	12.3		7	8	3	3	0	0	1	0	1
D	7.9		28	34	29	29	14	8	4	2	0
D	12.2		12	31	19	14	10	4	0	0	0
E	7.9		27	13	4	8	0	0	0	0	0
E	12.3		20	13	3	0	0	0	0	0	0
F	16.1		4	7	3	1	0	0	1	0	0
F	24.9		5	3	0	0	0	0	0	0	0
G	7.2		6	10	3	9	3	0	4	0	3
H	7.8		2	3	4	2	2	0	1	0	0

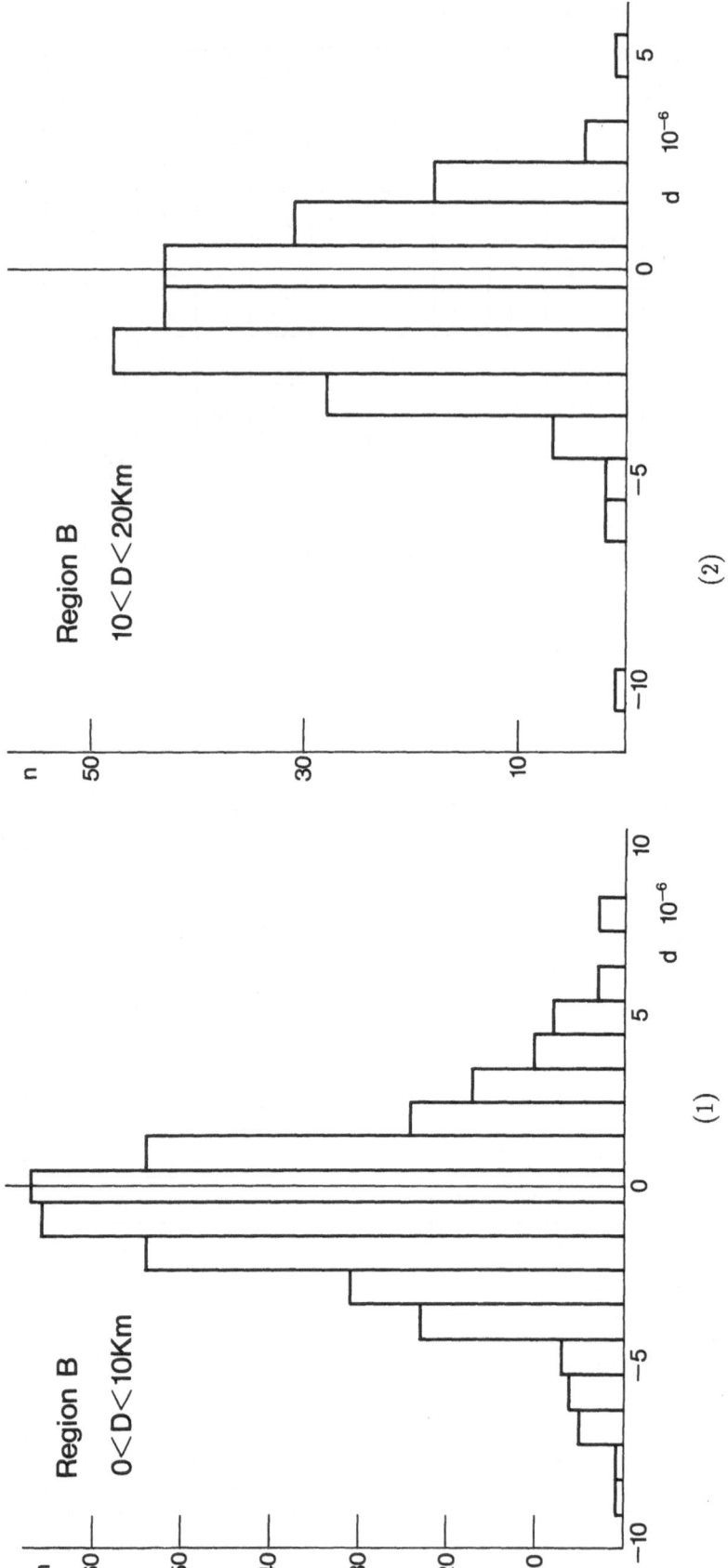

Fig.2 Histogram of the relative change in distance, d, for Region B.

3. FREQUENCY DISTRIBUTION OF THE RELATIVE CHANGE IN DISTANCE

Let the ratio of the difference $D_1 - D_2$ to D_1 be denoted by d.

$$d = \frac{D_1 - D_2}{D_1} \qquad (01)$$

This ratio d gives strain if the measurements are free of errors, but this is not the case in practice. The data of this relative change in distance has been statistically analyzed.

The data of the relative change in distance, d, shows that it has the general trend of becoming larger in magnitude as the distance reduces. The data of d in each of the survey regions have been divided into three groups according to distance ranges. The first group contains those from lines of which length is shorter than 10km, the shorter distance group, the second from lines with a length between 10km and 20km, intermediate distance group, and the third from lines with a length over 20km, longer distance group.

The frequency distribution of d for each of these distance groups in each of survey regions is given in Table 2. As a typical example, the histograms of d for region B is shown in Fig.2. The relative change in distance, d, is given in units of one part in million. The histogram for the shorter distance group is given in Fig.2(1). The total number of determinations of d is 382. This is suggestive of a normal distribution. It is characterized by a frequent occurrence of d with

Table 3. Standard deviation σ and mean value d_m of the relative change in distance, d, in units of 10^{-6}. D_m is the average distance in km. Letters A through H are as given in Table 1. N gives the number of measured lines available.

	N	d_m	σ	D_m	Remark
A	37	-0.67	2.25	7.8	
A	28	-0.50	1.88	13.7	
B	382	-0.52	2.72	7.8	*
B	228	-0.88	1.84	11.9	
C	68	0.89	4.27	7.9	
C	48	0.68	2.94	12.3	
D	176	-1.47	2.22	7.9	
D	112	-1.28	1.68	12.2	*
E	104	0.82	2.13	7.9	
E	71	0.69	1.73	12.3	*
F	29	0.18	2.01	16.1	
F	12	0.50	1.57	24.9	
G	69	0.01	4.24	7.2	
H	19	-0.75	2.59	7.8	

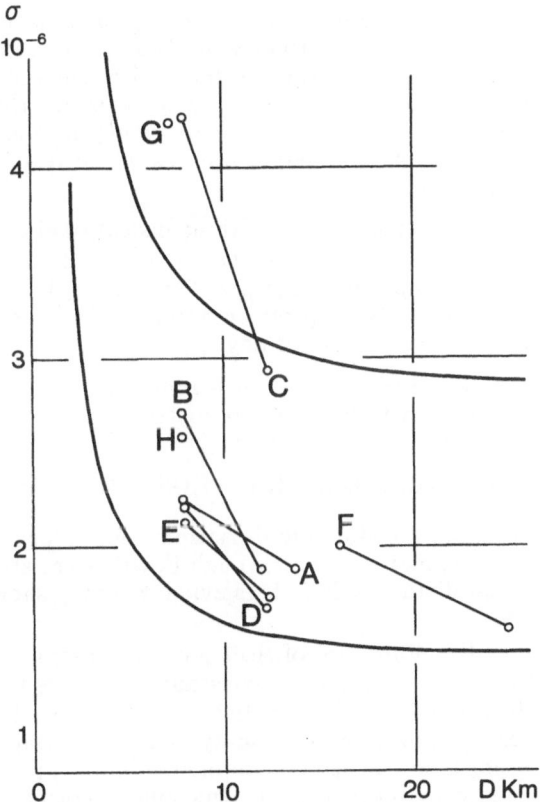

Fig.3. Standard deviation of the relative change in distance, d, plotted against the mean distance D_m.

small magnitude whereas the frequency rapidly decreases with the increase in the magnitude of d. The mean of d may be a small negative value. Fig. 2(2) is the histogram of d for the intermediate distance group of the same survey region. The total number of the determinations of d is 228. This has the same characteristics as those of the preceding one. These figures suggest that the relative change in distance, d, has a normal distribution with non-zero mean.

4. NORMAL DISTRIBUTION OF THE RELATIVE CHANGE IN DISTANCE

The standard deviation σ of d and its mean value, d_m, are obtained. Table 3 summarizes the results. The number of repeated measurements is given by N and the mean value of N measured distances is given by D_m. For regions A through E, the average d and the standard deviation σ are given for the shorter and intermediate distance groups. With regard to region F, they are given for the intermediate and longer distance groups, however, the number of data available from the longer distance group is only 12. For regions G and H, the results only from the shorter distance group are available.

By applying χ^2 test, we test the hypothesis that these samples of the relative change in distance, d, have a normal distribution with the standard deviation σ and the mean d_m. The longer distance group of region F is not tested because of its small number of measurements. The results show that we cannot reject the hypothesis of a normal distribution choosing a risk level of 5%, with the exception of these three cases, which are indicated by asterisks in the column of remark in Table 3.

The mean of the relative change in distance, d, expressed by d_m can be interpreted either as dilatation or as a systematic error. As can be seen in Table 3, it can be either positive or negative. In the cases, however, where the mean d_m of relative change in distance, d, is obtained for different distance groups in a given survey region, these d_m have the same sign and similar magnitude. This indicates that d_m has some definite meaning. But it is inconceivable that either uniform extension or contraction of the ground is produced over a region which has been arbitrarily bounded. Its magnitude is fairly large for real dilatation. This must be a systematic error which is proportional to the measured distance and is commonly introduced in measurements obtained under a given survey plan. The cause for this error can not be readily specified. Inaccuracy of the calibration of standard frequency is considered to be a possible factor. But the magnitude of d_m seems to be larger than expected from this factor. Further investigation is required in this respect.

With regard to the three exceptional cases indicated by asterisks in Table 3, the standard deviation σ is not necessarily large. Though Fig.2(1) is suggestive of a normal distribution, the assumption of normal distribution is rejected, as shown in Table 3. But this is due to comparatively frequent occurrence of smaller d. This implies that the real ground deformation must be very small, if any.

Taken all together, it is suggested that the quantity, $d - d_m$, results mostly from error, which is expected to be normally distributed.

5. RANDOM ERROR OF EDM AND CRUSTAL DEFORMATION

In Fig.3, the standard deviation σ for respective areas is plotted against D_m. We have six survey regions, marked by A through F, where the standard deviations have been determined for different mean distances D_m. It is evident that σ shows a decrease with increasing distance D_m in every case.

Random error of electro-optical distance measurements is conventionally represented by $a + bD$, where a and b are constants and D is the distance. The constant a is a part of error which does not depend on a distance. It is attributable to uncertainty in readings of a resolver. It results also from uncertainty in the centering of the instrument over a survey monument. On the other hand, the second term stands for a part of error linearly dependent on a distance and it is caused by inaccuracy in atmospheric corrections.

The atmospheric correction to distance measurements carried out by the GSI is made by using end-point temperature and pressure measurements. Random errors in distance measurements corrected by using the end-point measurements are described by $a = 10 - 5mm$ and

290

$b = 2 - 1mm/km$ with the distance D in km. The upper curve in Fig.3 shows the standard deviation of the relative change in distance, computed by using the pair of larger constants, $a = 10mm$ and $b = 2mm/km$. The lower curve is illustrated by taking another pair of the constants, $a = 5mm$ and $b = 1mm/km$, which gives the highest accuracy nominally expected.

These curves show a sharp rise as the distance D decreases to nearly zero. This is due to resolver error. The standard deviation σ of d for a shorter distance range is larger than that for a longer distance range without exception, as is seen in Table 3. This reveals resolver error in the distance measurements.

Let the standard deviation of d, denoted by σ, be written as

$$\sigma^2 = \frac{2(\alpha^2 + \beta^2 D_m^2)}{D_m^2} \tag{02}$$

where the constants α and β correspond to a and b, respectively. The constants α and β can be obtained from the determinations of σ for two different mean distances. The results are given in Table 4.

With regard to α, the values obtained are generally large. Four estimates out of six are over 10mm, which is the maximum of nominally given range for the constant a. The value of α for region C is especially great, reflecting the remarkable dependence of σ on a distance. There must have been something wrong with this survey. The standard deviation σ from region G for the shorter distance range is as great as that from region C. In region G, measurements only for shorter distance group are available, therefore, the dependence of σ on a distance cannot be obtained. But the circumstances must have been the same as those for region C. The value of α for region F is also large but we should also remark the small number of measurements available in this region.

The constant β contains information both about distance-proportional random error and ground deformation. Of course, the deformation cannot be exactly discriminated from error only on the basis of the results given here. The analysis of data which contains longer and intermediate distance measurements will give us more information.

Table 4. The constants α and β in equation (2) obtained for the regions A through F. α in mm and β in mm/km.

	A	B	C	D	E	F
α	8.4	14.6	22.6	10.6	9.1	18.7
β	1.18	0.43	0.98	0.81	0.98	0.82

In any case, the constant β so far obtained from the measurements is very small. Five determinations out of six are below one part in million. In general, we find good agreement between these values of β, considering that they are rather roughly estimated.

6. CONCLUSION

In a case where β is large enough, ground deformations of random nature can be expected. But the results of β obtained are very small. They are below the smallest nominal value of the constant b. Therefore, it is quite possible to understand that β is entirely due to errors proportional to a distance. The normal rate of horizontal deformation must be very low, though the time intervals between repeated measurements are as short as several years or more. The normal rate of vertical displacement is large enough to be revealed by repeated geodetic leveling (Mizuno, 1985a, 1985b, 1988). Vertical displacements should be dominant in comparison with horizontal deformations, so far as geodetic measurements of Japan are concerned. This is , of course, a view on the gradual deformation which is usually taking place. Sudden and remarkable displacements associated with an earthquake or a volcanic eruption should be considered separately from the problem under investigation here.

This approach does not provide the crustal strain for individual triangles. But it should be an effective way to get an intuitive view of the problem in hand. It is necessary to locate lines of various lengths in a survey region in such a way to fit this purpose and to repeat distance measurements over the lines systematically for many years, if we want to estimate errors in practice and to have information about the horizontal crustal deformation. And this will enable us to re-examine the availability of the triangulation surveys for the study of crustal deformation.

ACKNOWLEDGMENT

The author is deeply grateful to Mr. K. Kitada, a member of the staff of the Geographical Survey Institute, ret., for his kind help in processing and arranging the data.

REFERENCES

Geographical Survey Institute (1985a). Crustal movement in the Izu peninsula, (in Japanese), Rep. Coord. Comm. Earthq. Predict. 33, 236-257.

Geographical Survey Institute (1985b). Crustal movement in the southern Kanto, Chubu and Kinki districts, (in Japanese), Rep. Coord. Comm. Earthq. Predict. 34, 138-156, 170-178 and 346-357.

Geographical Survey Institute (1986). Crustal movement in the Kanto, Kinki and Chugoku-Shikoku districts, (in Japanese), Rep. Coord. Comm. Earthq. Predict. 36, 102-127, 333-354 and 355-364.

Harada,T. (1983). Crustal strains expressed with their standard deviations in the improved universal program (PAG-U), (in Japanese), J. Geod. Soc. Japan 29, 124-129.

Mizuno,H. (1985a). Leveling results in the Tokai region and the estimation of the accuracy of leveling based on them, Earthq. Predict. Res. 3, 425-440.

Mizuno,H. (1985b). Stochastic characteristics of vertical movements of the ground as revealed by repeated geodetic leveling, J. Geod. Soc. Japan 31, 46-58.

Mizuno,H. (1988). A study of random vertical displacements and estimates of the accuracy of leveling based on weekly measurements, Manuscripta Geodaetica 13, 275-289.

LOCAL CRUSTAL STRAINS OBSERVED WITH HOLOGRAPHIC INTERFEROMETRY

Shuzo Takemoto
Department of Geophysics, Faculty of Science
Kyoto University, Sakyo-ku, Kyoto 606, Japan

INTRODUCTION

In observational studies of crustal deformations, various types of strainmeters such as rod, wire and laser strainmeters have been developed over the last half of this century (Agnew, 1986). The rod strainmeters, the early development of which was due to Benioff (1935, 1959), are still employed in many countries because of the simplicity of construction and easiness of maintenance. The rod strainmeter uses, in its common form, a length of solid material consisting of the fused-quartz tube or super-invar bar to measure a relative displacement between two piers fixed into the bedrock. Disadvantages considered are the effects of deformations in the solid material used and frictional effect between the material and its supports (Takemoto, 1979, 1989a). On the other hand, the laser strainmeter is free from these problems because it can detect the small displacement quantitatively in terms of the wavelength of laser light without using a length of the solid material. In any case, however, the conventional strainmeter can detect only one component of the Earth's motion, i.e. a linear strain between two points.

We thus developed a new measurement system based on holographic interferometry (Takemoto, 1986, 1989b). This system enables two- or three-dimensional small strains to be measured quantitatively in terms of the wavelength of laser light, without any contacting the object. We report observational results of holographic strain measurements carried out in two observation tunnels at Amagase and Iwakura Observatories, Kyoto, Japan (Fig. 1).

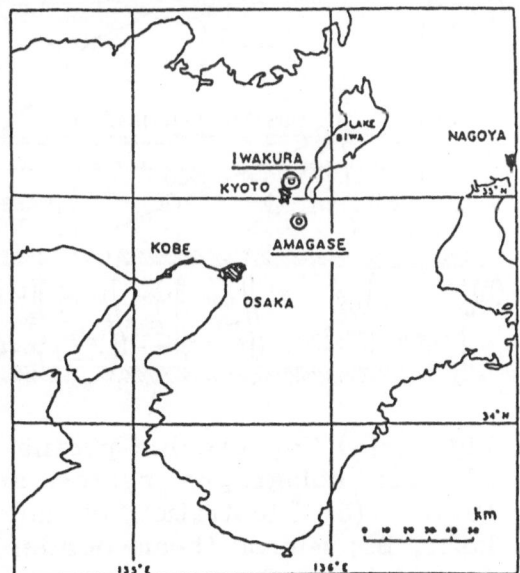

Fig. 1. Location map around Amagase and Iwakura Observatories in the southern part of Kyoto Prefecture, Japan.

RECORDING SYSTEM AND OBSERVATION SITES

A laser holographic recording system comprises a high power He-Ne laser source (30 ~ 50 mW) and associated optical elements. The system is installed in a tunnel where temperature change is negligible and artificial noise does not exist. A white-painted area of the tunnel wall within a section of 1 ~ 2 m in diameter is used as an object for holographic interferometry. A coherent light beam emitted from the laser source is divided into two beams, i.e. the object and reference beams. The two waves, reflected from the object wall and the reference mirror, respectively, are superimposed on a photographic plate set on the plate-holder. After an exposure of 1 ~ 3 min, the plate is developed, fixed and washed. The developed plate, on which the 'hologram' is recorded, is then exactly put back in its position where the hologram was taken. By again illuminating the plate with the same reference wave, we can see through the plate the real scene of the tunnel wall and its holographic virtual image overlapping each other. In this procedure, if the plate is slightly inclined from the original position, a number of dark and bright lines ('fringes') superimposed on the holographic image can be seen. The changes in the interference fringe pattern are continuously monitored with a video camera and a time-lapse video recorder. Video signals stored on a video-cassette tape are reconstructed with a re-recording video recorder and then analyzed with an image processing system.

In 1984, the first holographic recording system comprising a 50 mW He-Ne laser was installed in the tunnel at the Amagase Observatory where various types of strainmeters and tiltmeters were installed and continuous observations of crustal deformations have been carried on since 1967. As shown in Fig. 2, the tunnel is 1830 m long and has a gradient of 1/1300. The section of the tunnel has a horseshoe shape, with a diameter of about 6 m. The holographic recording system was

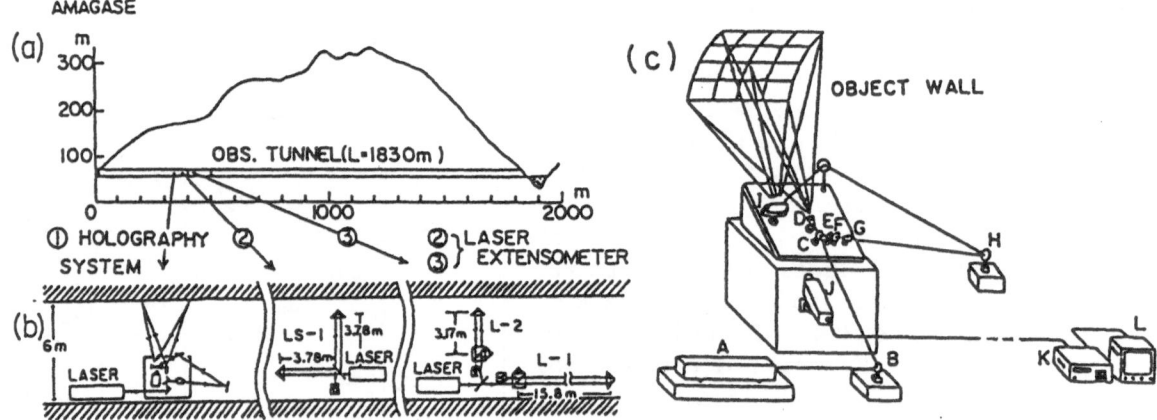

Fig. 2. (a) Topographic profile of the Amagase tunnel. (b) Arrangement of laser holography system and laser extensometers installed in the tunnel. (c) Illustration of holographic recording system. A ; He-Ne laser, B ; mirror (beam-bender), C ; beam-splitter, D ; beam-expander, E; pin-hole, F; neutral-density filter, G ; collimator, H ; reference mirror, I ; photographic plate and plate-holder, J ; video camera, K ; time-lapse video recorder, L ; monitor TV.

installed 320 m from the entrance of the tunnel and 130 m below the surface. An annual variation of temperature at the observation site is about 0.2 °C, and the daily variation is less than 0.05 °C. The amplitude of microtremors measured in the tunnel does not exceed 20 nm/s at 0.1 ~ 10 Hz. Simultaneous observations with the holographic recording system and laser strainmeters in the Amagase tunnel enable the tunnel deformation to be measured precisely. Detailed discussion will be made in the following section.

We have then developed a portable holographic recording system comprising a 30 mW He-Ne laser and associated optical elements. Using the system, holographic measurements of tunnel deformation were executed at the Iwakura Observatory during the period from July 22 to November 21, 1988. As shown in Fig. 3, the Iwakura tunnel is 33 m long and has a cross-section of 1.8 x 1.2 m. In addition to the entrance door, there are partition doors 17.1 m and 27.4 m from the entrance of the tunnel. In the innermost vault, three components of super-invar bar strainmeters and two components of horizontal pendulum tiltmeters were installed, and continuous observations using these instruments have been carried on since 1964. Annual variations of air temperature at positions of 15 m, 20 m and 30 m from the entrance of the tunnel are measured to be 1.5 °C, 0.5 °C and 0.3 °C, respectively, and that of ground-water temperature is smaller than 0.2 °C. The holographic recording system was installed at 20 ~ 25 m inside the second and third doors. The tunnel wall within a section about 1 x 1 m was used as the object of holographic interferometry. Because the maximum path difference from any point on the object to the recording position of holographic interference patterns should be within the coherence length (about 30 cm) of the laser light, the object wall was made hollow to be a concave shape by cutting the crude rock surface of the tunnel wall.

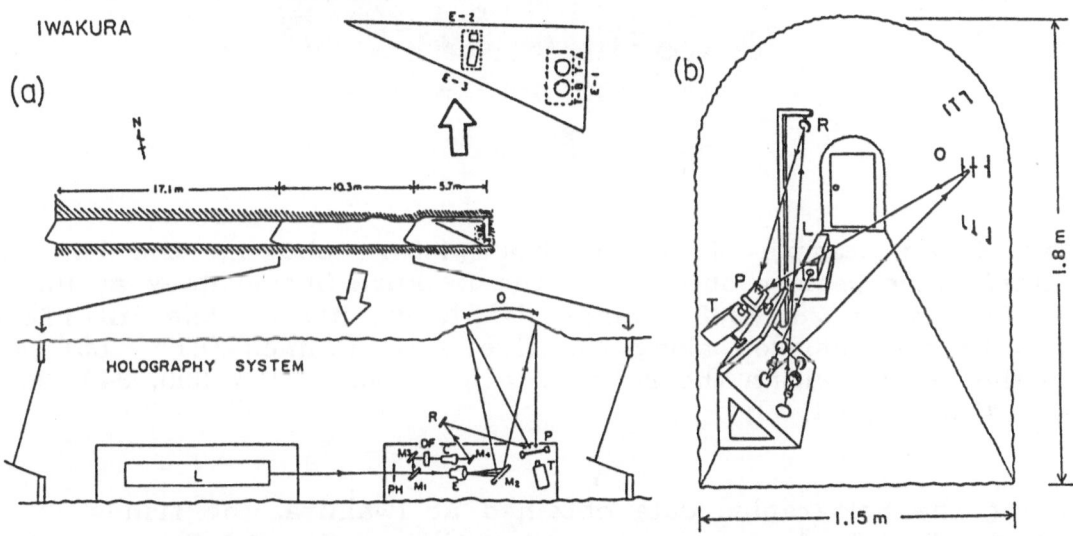

Fig. 3. (a) Arrangement of laser holographic recording system together with super-invar bar strainmeters (E - 1 ~ 3) and tiltmeters (T - A, B) in the Iwakura tunnel. L ; 30 mW He-Ne laser, PH ; pin-hole, M_1 ~ M_4 ; mirrors, E ; beam-expander, DF ; neutral-density filter, C ; collimator, R ; reference mirror, P ; photographic plate and plate-holder, T ; Video camera, O ; object wall. (b) Schematic view of holographic recording system.

EXAMPLES OF DATA ANALYSIS

Analysis of the holographic interference pattern is carried out using the image-processing system. At first, video signals stored on a video tape are reconstructed with a re-recording video recorder which is the same type as that used for recording the fringe patterns at the observation site. The video output is then fed to an image processor and converted to the 8-bit digital data of 768 x 256 pixels per field. Access to the image processor is controlled by a personal computer.

Figure 4 shows an example of the re-recording monitor picture printed out on thermal printing paper using a TV-printer. The original record of the picture was obtained at the Iwakura Observatory on September 28, 1988. In this figure, more than 10 dark and bright fringes can be seen, together with the standard mark '+' drawn with black paint on the white-painted, highly reflective area of the object wall.

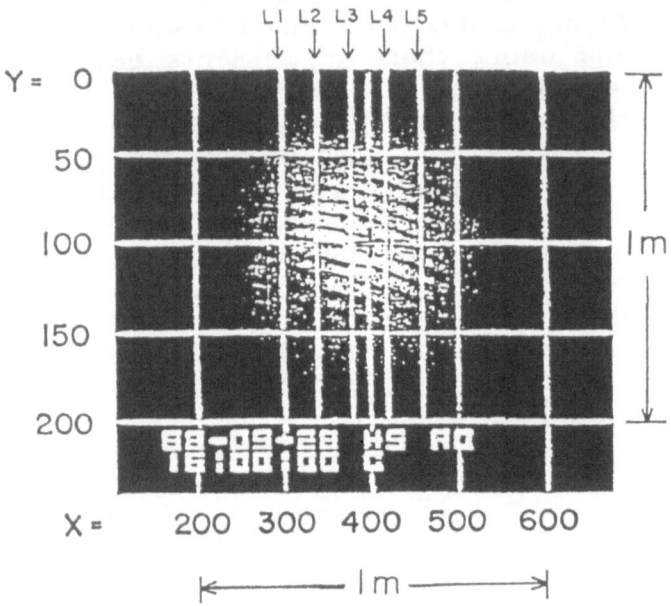

Fig. 4. An example of the monitor picture showing the holographic interference pattern obtained at the Iwakura Observatory at 16: 00: 00 on September 28, 1988. Fringe displacements in the interference pattern are analyzed along the five vertical lines (L1 ~ L5) parallel to the Y-axis, within the range Y = 0 ~ 200, at X = 300, 340, 380, 420 and 460, respectively.

Using the holographic data obtained at Iwakura, the fringe displacements in the interference pattern were then analyzed. Figures 5 (a) and 5(b) show distributions of the brightness values (B) and its derivative (dB/dY) along L1 ~ L5 lines (see Fig. 4). The brightness value is the 8-bit grey shade level of each pixel and shows that the left side is black (B = 0) and the right side is white (B = 255). Figure 5 (c) shows the changes of dB/dY during the period from 13 h on September 28 to 12 h on October 05, 1988. These curves are obtained by printing out the every 5-min data which are stored on a floppy disk as a form of binary image ('0' or '1') : if dB/dY has a negative value or zero, it should be '0', while if it has a positive value, it should be '1'. In these curves of fringe displacements, we can clearly recognize the sinusoidal changes which indicate the tidal deformation of the tunnel. Similar changes can be seen in the strain records obtained from super-invar bar strainmeters installed in the same tunnel.

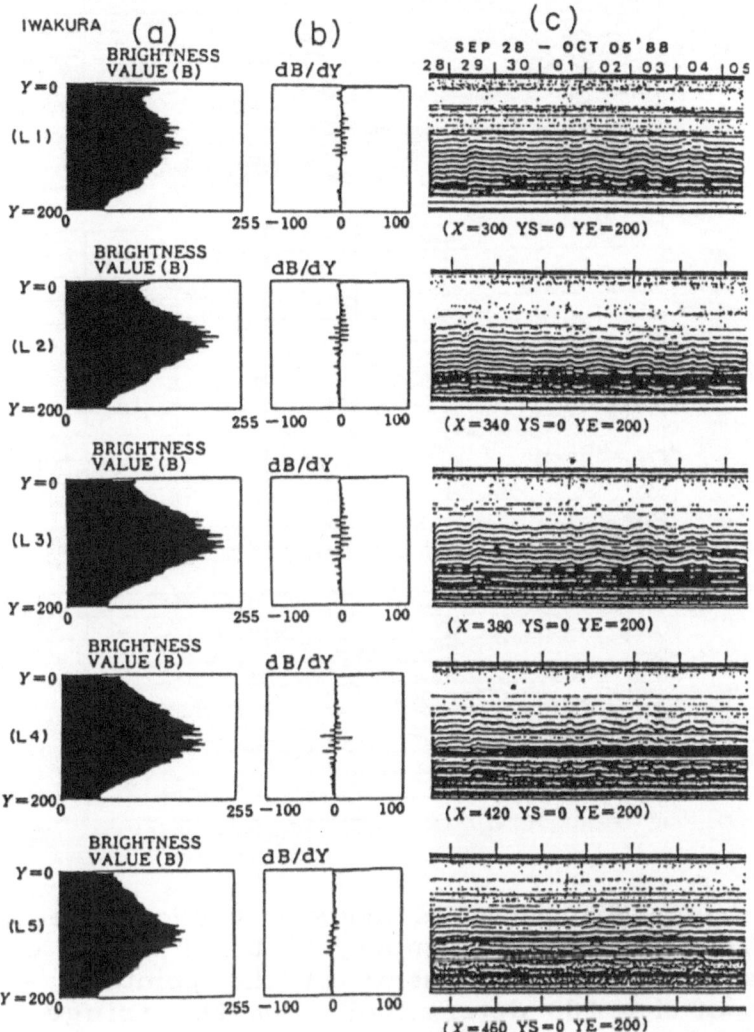

Fig. 5. Examples of image processing of interference patterns obtained at Iwakura. (a) Brightness value (grey shade level of pixels) of the interference fringe pattern along L1 ~ L5 lines (see Fig. 4). (b) Derivatives of B (dB/dY). (c) Changes of dB/dY during the period of one week from September 28 to October 05, 1988.

More detailed analysis of holographic fringe displacements was carried out using data observed at Amagase. As an example, the fringe displacements along a vertical line ($X = 300$, $Y = 0 \sim 200$) during the period of one week are shown in Fig. 6 (a), which was obtained by the same procedure as that used for Fig. 5 (c). The displacement of an arbitrary

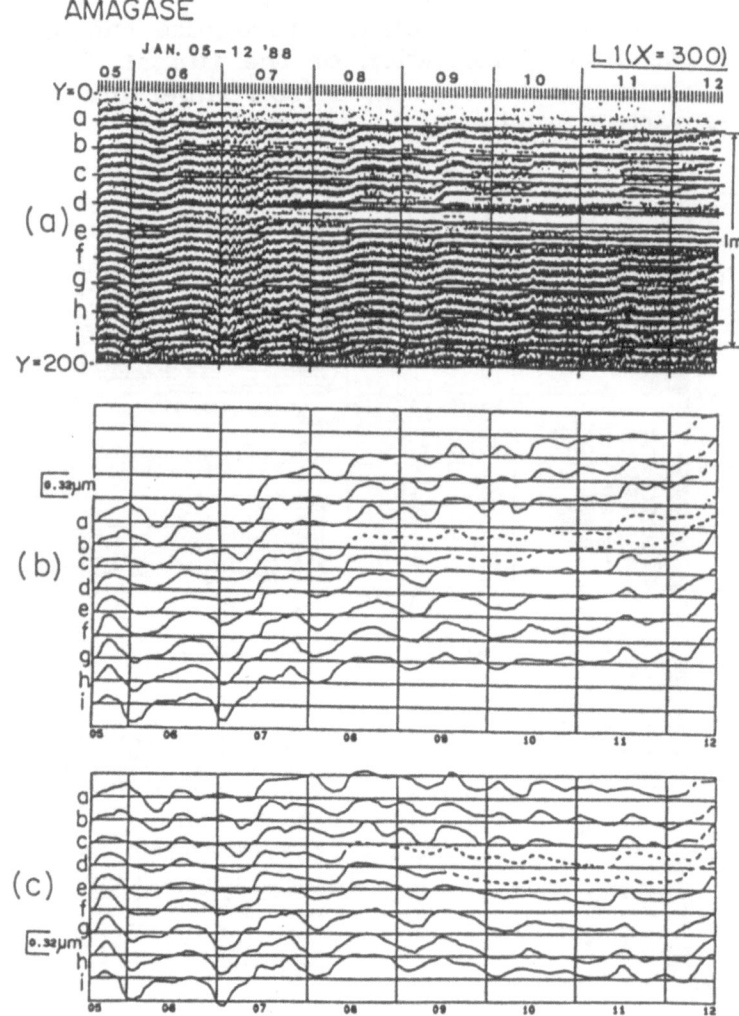

Fig. 6. (a) Fringe displacements along a vertical line (X = 300) for the period from 15: 00: 00 on January 05 to 13: 00: 00 on January 12, 1988 at Amagase. (b) Displacements of the points a ~ i along the vertical line of X = 300 determined from the fringe displacement curves (a). (c) Drift-eliminated curves of (b).

point P on the tunnel wall can then be determined by examining the changes of the fringe pattern at the corresponding point P' (X, Y) on the monitor picture : one cycle of dark-bright-dark change at the point P' is equivalent to the out-of-plane displacement of 0.32 μm at P, if the reference light path is considered to be invariant during the time of observation. In the Amagase tunnel, the reference mirror was installed on the floor in the direction along the the tunnel axis at the distance of nearly equal to the light path of object wave (see Fig. 2). Because the displacement along the tunnel is far smaller than that across the tunnel, the change in the length of the reference light path is negligible compared with that of object wave which travels across the tunnel. Thus, the change in the reference light path is negligible and the fringe displacements indicate the back and forth movements of the tunnel wall in its cross-section. In Fig. 6, (b) shows the tunnel wall displacements determined from the changes of fringe patterns at the 9 points ('a' ~ 'i') along the vertical line, and (c) is the drift-eliminated curves, in which linear trends obtained from the least squares fitting were subtracted from the original curves of (b). In these figures, sinusoidal changes of the order of one fringe (= 0.32 μm) can be seen, but peak and trough positions are significantly different along the vertical line. For example, the first trough in the curve at the point 'a' (X = 300, Y = 20) appears at 08 h on January 06, while that of the point 'i' (X = 300, Y = 180) appears about 01 h on the same day, nevertheless the distance between the points 'a' and 'i' corresponds to the length of 1 m on the tunnel wall. The later peaks and troughs show a similar pattern. We then compared the tunnel wall displacements determined from the holographic method with the strain records obtained from laser strainmeters in the same tunnel.

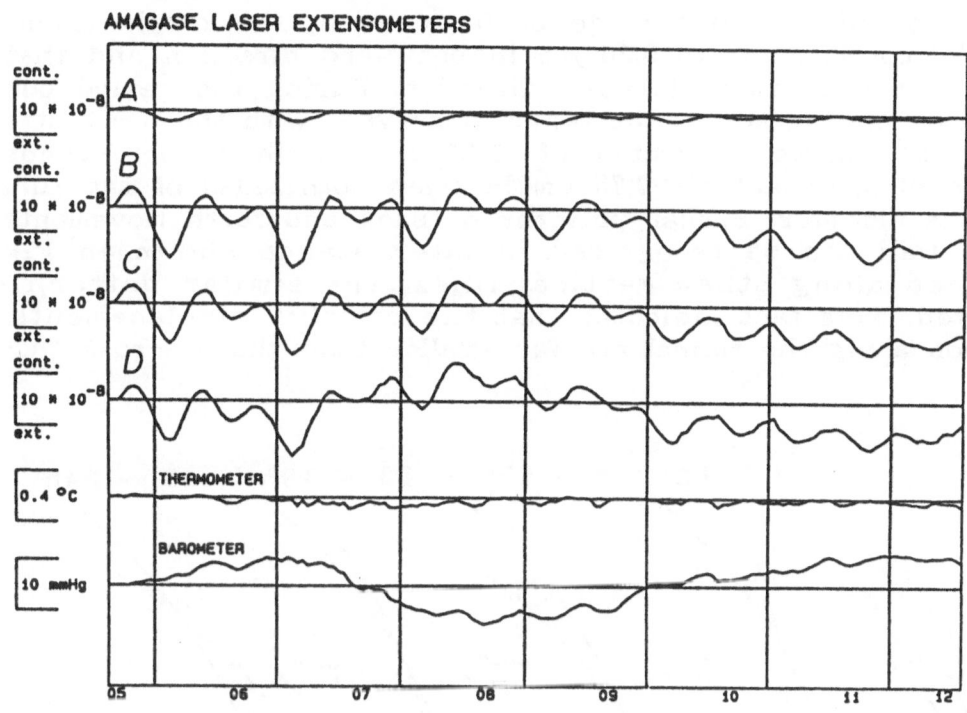

Fig. 7. Strain changes observed with laser extensometers at Amagase (A ~ D : see text), together with thermometric and barometric changes for the same period of Fig. 6.

Figure 7 shows strain changes observed with laser strainmeters together with thermometric and barometric changes in the Amagase tunnel for the same period as that of Fig. 6. In this figure, A and B are the linear strain components measured by L - 1 and L - 2 laser strainmeters, respectively, the former is oriented along the tunnel and the latter is across the tunnel. The curve C shows the difference between these two components $((L - 2) - (L - 1))$. The curve D shows the strain change measured by the LS - 1 strainmeter, which detects the difference between two-axial linear strains, i.e. along and across the tunnel. Thus, the curve C should be equal to D. In any case of A ~ D, meteorological disturbances are very small and tidal strains are clearly recorded. It can be seen, however, that the tidal strain along the tunnel is far smaller than that across the tunnel. The difference was predicted from the stress-strain estimation around the tunnel considering the azimuthal dependence of tidal strains at Amagase and including the 'cavity effect' (Takemoto, 1981). The result of calculation shows that the amplitude of B should be about seven times larger than that of A. Thus, it is reasonable that the tidal strains in curves C and D are nearly equal to that of B.

Comparing Figs. 6 and 7, the displacements of the tunnel wall at its lower portion (e.g. 'h' and 'i' in Fig 6. (c)) are fairly well consistent with the strain changes across the tunnel (Fig. 7; B). However, the displacements of the upper portion of the tunnel wall (e.g. 'a' and 'b') are out of phase. In Fig. 8, the tunnel wall displacements at the points a ~ i for the period of 24 hours on January 06, 1988 are illustrated. In the first stage of 01 ~ 08 h, the displacement of the point 'a' corresponds to 0.30 μm in backward direction and that of the point 'i' is 0.22 μm in forward direction. During the period concerned, the strain change across the tunnel observed with the L - 2 strainmeter (Fig. 7; B) shows contraction of 1.25×10^{-7}, which corresponds to the relative displacement of −0.75 μm between both sides of the tunnel wall 6 m in a diameter. Similar patterns of back and forth movements of the tunnel wall can be recognized in later stages shown in Fig. 8. If examined along other vertical lines, the similar pattern can be observed. This fact indicates that the relative displacements in the direction along the tunnel are far smaller than those across the tunnel.

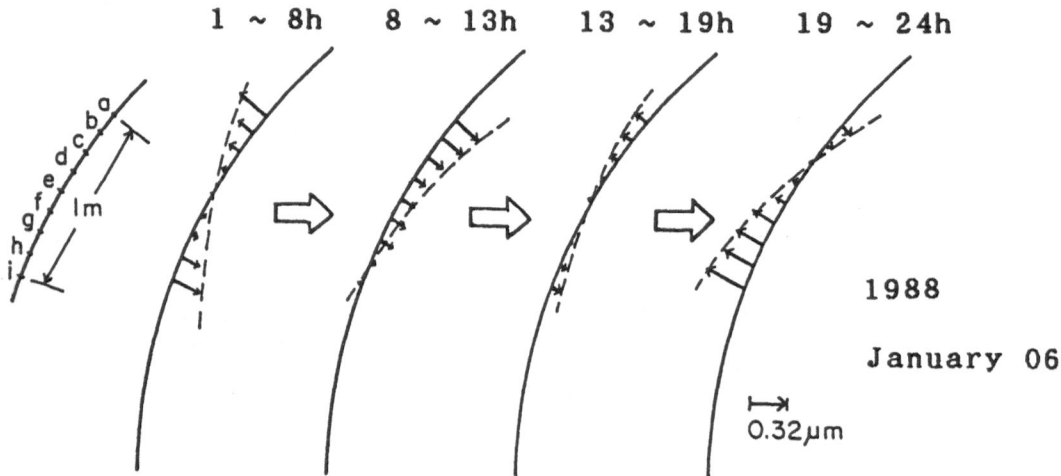

Fig. 8. Tunnel deformation determined from the holographic method.

CONCLUDING REMARKS

Using a new technique of holographic interferometry, local strain changes in the two observation tunnels at Amagase and Iwakura have been observed. The holographic method has an advantage of detecting two- or three-dimensional strains simultaneously in terms of the wavelength of laser light. In addition, this method allows small deformations to be measured directly without any physical contact. This obviates some of the problems inherent in conventional strainmeters where the effects of deformations in the solid materials used (e.g. the super-invar bar or fused-quartz tube) and the frictional effects between the lengths and their supports have to be considered.

In the Amagase tunnel, the laser holographic recording system comprising a 50 mW He-Ne laser source was installed in 1984. Using the system, detailed patterns of the tunnel deformations have been precisely investigated. The fringe displacements in holographic interference patterns have indicated the back and forth movements of the tunnel wall in its cross-section according to the tensile and compressive stresses associated with tidal forces. The results obtained are consistent with the strain changes observed with laser strainmeters which have been installed in the same tunnel. On the other hand, relative displacements or strains observed in the direction along the tunnel are very small. Consequently, local strain changes in tunnels have been successfully determined using the holographic method referring to laser strainmeters. Similar holographic measurements were executed at the Iwakura Observatory using a portable holographic recording system.

As further applications, we are now planning to use the holographic method to investigate the 'cavity effect' of more complicated tunnels. If a holographic recording system is installed at a corner or the dead-end of the tunnel, the resultant fringe patterns would clearly indicate the local inhomogeneity of the strain field caused by the 'cavity effect'.

The present system, however, has a margin for improvement to measure the long-term strain accumulation because the fringe pattern observed through the photographic plate gradually blurs over the course of time. Thus a continuous record cannot be obtained even over a week. Therefore, the photographic plate must be renewed every week for long-term observations. Two schemes are now considered for improving the system; one is the adoption of a thermoplastic recording system instead of the photographic recording system, and the other is the introduction of the technique of Electronic Speckle Pattern Interferometry (ESPI). We believe these improved systems will contribute to the continuous strain measurements for earthquake prediction and other geophysical studies.

Acknowledgments. The author wishes to thank Professor Ichiro Nakagawa of Kyoto University for helpful suggestions and encouragement. This work was partially supported by the Grant-in-Aid for Scientific Research from the Ministry of Education, Science and Culture of Japan (Nos. 62540292 and 63540304).

REFERENCES

Agnew, D. C. (1986) Strainmeters and tiltmeters, Rev. Geophys., 24, 576–624.

Benioff, H. (1935) A linear strain seismograph, Bull. Seism. Soc. Am., 25, 283–309.

Benioff, H. (1959) Fused-quartz extensometer for secular, tidal, and seismic strains, Bull. Geol. Soc. Am., 70, 1092–1132.

Takemoto, S. (1979) Laser interferometer systems for precise measurements of ground-strains, Bull. Disas. Prev. Res. Inst., Kyoto Univ., 29, 65–81.

Takemoto, S. (1981) Effects of local inhomogeneities on tidal strain measurements, Bull. Disas. Prev. Res. Inst., Kyoto Univ., 31, 211–237.

Takemoto, S. (1986) Application of laser holographic techniques to investigate crustal deformations, Nature, 322, 49–51.

Takemoto, S. (1989a) Continuous measurement of crustal deformation in a long tunnel for studying tectonics in the subduction zone, J. Geod. Soc. Japan, 35, 243 – 256.

Takemoto, S. (1989b) Real-time holographic measurement of crustal deformation, *Laser holography in geophysics*, S. Takemoto (ed.), John Wiley & Sons, Inc., New York.

EARTHQUAKE PREDICTION BY GEODETIC SURVEYS AND CONTINUOUS CRUSTAL MOVEMENT OBSERVATIONS IN JAPAN

Minoru Tanaka
Geographical Survey Institute
1 Kitazato, Tsukuba-shi,Ibaraki Pref. 305 Japan
kachishige Sato
National Astronomical Observatory
Hoshigaoka-cho, Mizusawa-shi, Iwate Pref. 023 Japan
Torao Tanaka
Disaster Prevention Research Institute, Kyoto Univ.
Gokasho, Uji, Kyoto-fu, 611 Japan

INTRODUCTION

Japanese Islands have been considered to be located at the eastern edge of Eurasian plate boundary to the Pacific and Phillipine Sea Plates. Recently, another plate boundary is proposed to divide northeastern from southwestern Japan(for example, Seno,1985), and by this the former is on the North American and the latter the Eurasian plate, respectively. Relative motions of these four plates control the tectonics of Japanese Islands, and consequently the occurrence of large earthquakes along these plate boundaries.

In order to obtain fundamental data for long-term earthquake prediction in Japan, the Geographical Survey Institute has repeated precise geodetic surveys over the main islands of Japan. Especially the south Kanto area has a high population and is considered to be attacked by a large earthquake in the near future. Accordingly the Institute has been conducting dense and frequent geodetic surveys in this area assigned as one of the "areas of intensified observation" in Japan.

In the south Kanto area, continuous subsidence at the southernmost tips of the three peninsulas, Boso, Miura and Izu have been detected from repetition of leveling surveys. The subsidence is caused by the drag of the subducting Phillipine Sea plate under the Kanto area and therefore is considered to relate with accumulation of earthquake generating energy within the plates.

A sinusoidal fluctuation with a period of about 20 years has been detected in the subsidence at Aburatsubo in the Miura peninsula by Tanaka and Gomi(1989). On the other hand Ishii et al.(1978,80) found a migrating change of maximum shear strains on a station array for observation of crustal movements in the northeastern part of the main island of Japan. With the purpose of explaining such a migrating crustal movement by a periodic shear force at the plate boundary generated by subduction of the Pacific plate, Sato(1989) carried out a viscoelastic finite element analysis by applying a periodic shear force at the plate boundary, but he could not obtain such a slow speed as 40km/year

303

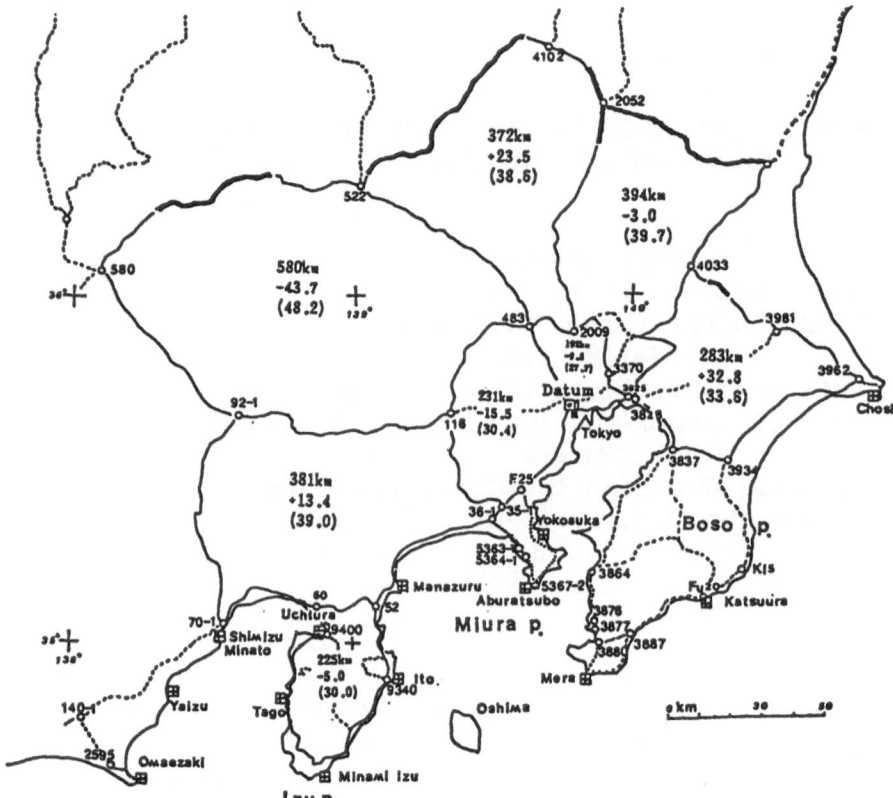

Fig.1. Leveling net in Kanto area for determining the height of Tokyo Datum. Solid line;leveling net adjusted for the period of about 60 years from 1923-26 to 1984. Thick solid line; leveling route fixed by stable leveling points (75 points selected). Numerals;closing circuit(km)(upper), closing error(mm) (middle), and limit of closing error(mm)(bottom). Mark ⊞ shows tidal stations.

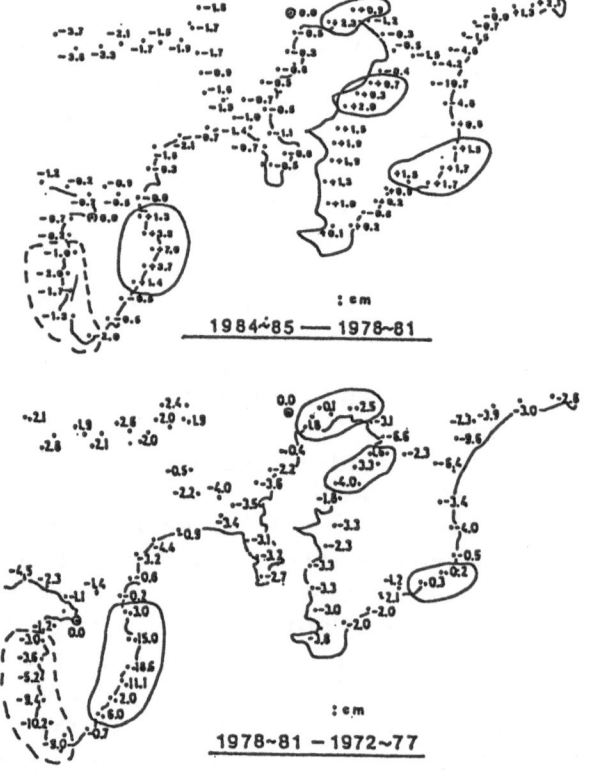

Fig.2. Vertical displacements in the south Kanto area. Tokyo Datum and Uchiura Tidal Station (mark ⊙) were fixed. Areas surrounded by solid and dashed curves indicate uplifting and subsiding areas to be noticed.

proposed by Ishii et al.(1978).

We present here a result from a similar finite element analysis by applying a shear force with a period of 20 years that gives a migrating velocity of about 80 km, which is of the same order as that observed by Ishii et al.

An eustatic change of the mean sea level is also estimated to be 1.9mm/year around Japanese Islands by the results from leveling surveys repeated in Kanto area in the last 80 years.

LEVELING SURVEYS IN KANTO AREA

In Kanto area leveling surveys have been frequently repeated along the route shown in Fig.1. In Fig.2 vertical displacements from three surveys in 1984/85 - 1978/81 and 1978/81 - 1972/77 are given. Persistent pattern of uplift and subsidence is remarkable. Especially relative subsidence of the tips to the central parts of Boso and Izu peninsulas, and subsidence of Miura peninsula are to be noted. The subsidence of Boso peninsula is shown more clearly in Fig.3. The speed of subsidence seems to be accelerated since 1983 toward the tip of the peninsula. A similar change in the trend of linear strain also has been found on the record from a strainmeter of NW-SE component at Tateyama close to the tip of Boso peninsula(Tanaka and Gomi,1989). We consider that this subsidence

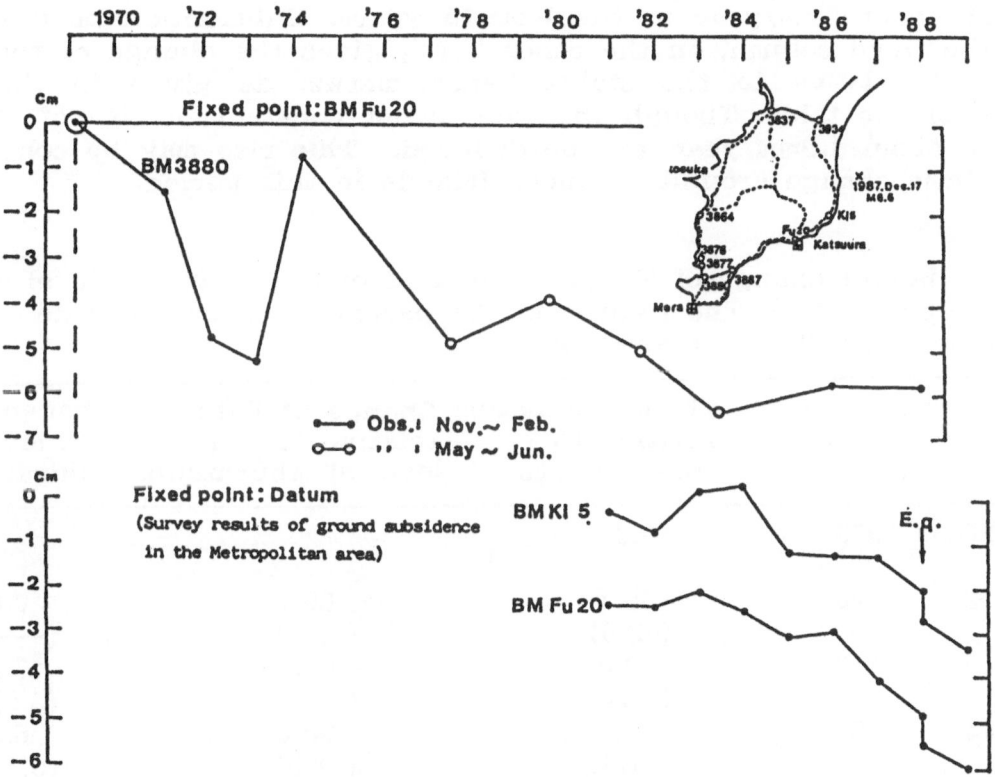

Fig.3. Vertical displacements of the Boso peninsula.

should be carefully monitored from the viewpoint of future occurrence of large earthquakes caused by subduction of the Philippine Sea Plate under the Kanto area.

The route from Tokyo Datum to Aburatsubo Tidal Station at the southern tip of the Miura peninsula has frequently been surveyed, especially every year in this 20 years, and the height change of Tokyo Datum relative to the mean sea level at Aburatsubo has obtained as shown in Fig.4. In this figure a fluctuation from linear subsidence is clearly seen and the period is estimated to be about 20 years in average.

SEA LEVEL CHANGE IN KANTO AREA

From the results in the previous section it is possible to estimate the sea level change itself at Aburatsubo, if the height change of Tokyo Datum could be estimated referring to some stable bench marks. We have selected a number of bench marks along the route shown by thick lines in Fig.1 in the northern Kanto area by taking their small vertical crustal movements, tectonic circumstances and local geology into consideration. Adopting the mean height of these bench marks as the fixed height reference, the height change of Tokyo Datum was calculated by adjusting the local leveling networks. The third column in Table 1 gives the change of the Datum thus obtained and the forth column the change of the Datum relative to the mean sea level at Aburatsubo. Taking the mean of these two columns, it is apparent that Tokyo Datum is uplifting at a rate of about 2mm/year in the last 15 years. Subtraction of the forth from the third column, on the other hand, gives the change of the mean sea level relative to the stable bench marks, as given in the fifth column of the table. Though the rate is not constant, a rise of the sea level of about 1.9mm/year may be deduced. This rise may be considered as eustatic change around Japanese Islands in this period.

Table 1. Height change of Tokyo Datum and of the mean sea level at Aburatsubo (in mm). The figures in the parentheses are the rate (mm/year) from the first leveling.

Survey year	Stable point number	Change of Datum relative to stable points	Change of Datum relative to sea level at Aburatsubo	Change of sea level at Aburatsubo
1923–26	203	0.0		
1949–52	120	−65.7 (−2.5)	−65.7 (−2.5)	0.0 (−−−)
1965–68	113	−59.0 (−1.4)	−106.1 (−2.5)	47.1 (2.9)
1972–75	94	−84.0 (−1.8)	−90.2 (−1.9)	6.2 (0.3)
1980–81	77	−58.6 (−1.0)	−105.3 (−1.9)	46.7 (1.5)
1984	75	−34.2 (−0.6)	−101.7 (−1.7)	67.5 (1.9)

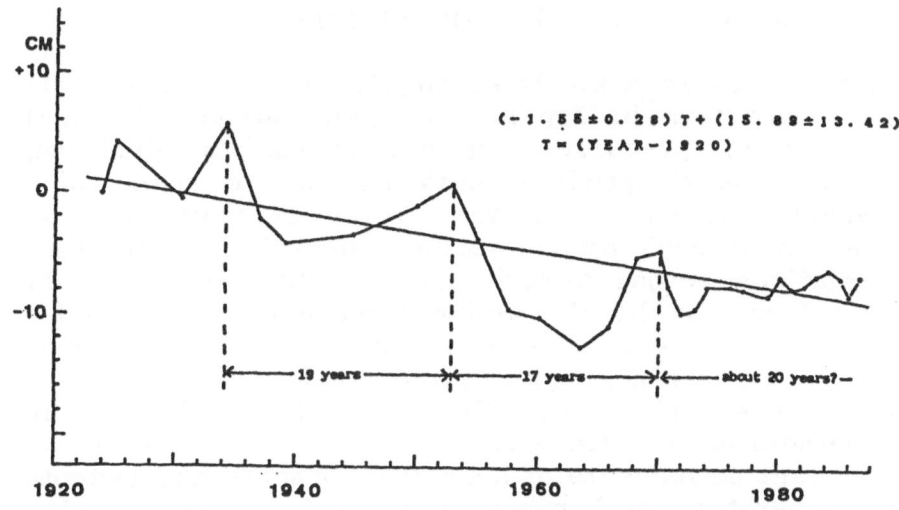

(−1.55±0.28)T+(15.89±13.42)
T=(YEAR−1920)

Fig.4. Height change of the Tokyo Datum relative to the mean sea level at Aburatubo.

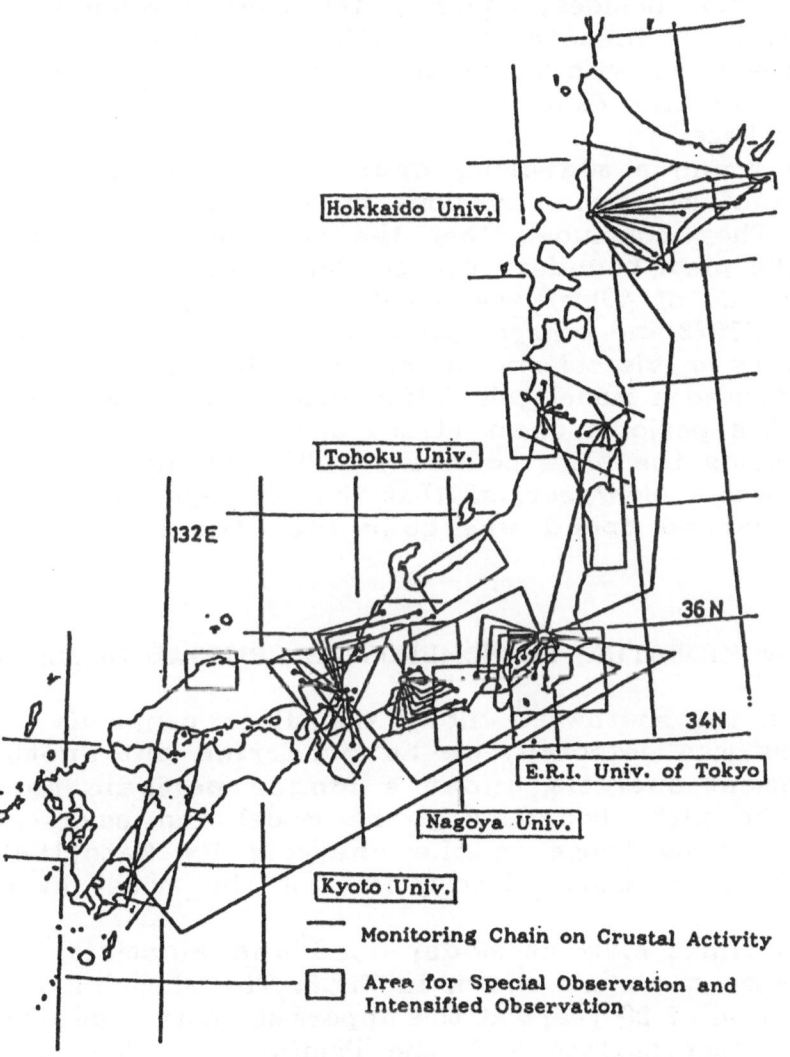

Fig.5. Stations of "Monitoring Chain on Crustal Activity " and the network for data exchange by Universities. (Ishii,1988)

CONTINUOUS MONITORING OF CRUSTAL MOVEMENTS

Crustal movements have also been monitored continuously with strainmeters and tiltmeters in Japan. Its main purpose is earthquake prediction by detecting precursory crustal movements which might be overlooked in the case of geodetic surveys due to their fairly long repetition interval. Seven "Monitoring lines of crustal activities" were constructed in 1979-85 by connecting several stations for crustal movement observations across tectonically important areas in Japanese Islands and have been run by five universities ever since. Each station is equipped with strainmeters, tiltmeters and meteorological sensors and some with long-period seismographs, proton precession magnetometers, earth current meters and so on. The data from these lines are continuously telemetered to observation centers of each university and some of the important components of crustal movements and meteorological data such as temperature and precipitation being further telemetered routinely to Earthquake Prediction Data Center of the University of Tokyo. The total number of observation stations which consist of the seven monitoring lines reaches 78. The stations and telemetering network system are given in Fig.5. Besides, Japan Meteorological Agency and National Research Center for Disaster Prevention have networks of crustal movement observations with bore-hole type volume strainmeters and tiltmeters in and around Kanto area and are monitoring crustal movements on real-time base.

Ishii et al.(1978,80) found a migrating crustal movements on the records from an array of monitoring stations of Tohoku University in northeastern Honshu. They concluded that the maximum shear strain migrated inland from the plate boundary between the North American and Pacific plates with a speed of 40km/year. Similar migrating phenomenon was reported by Yamada(1973)and others. Sato(1989) tried to explain this migrating phenomenon as a viscoelastic deformation in the crust and upper mantle, and performed a model calculation using the finite element technique. In his model a periodic shear stress changing with a period of 5 years was given along the plate boundary. The obtained speed of migration in this case was 130km/year, so that the discrepancy between the calculated and observed speed was concluded too large to be overlooked as a trifle.

MODEL CALCULATION OF MIGRATING PHENOMENON ON CRUSTAL MOVEMENT

Since the fluctuation in the southward tilt of the Miura peninsula with the period of 20 years was detected, we have carried out another viscoelastic finite element analysis under a longer periodic shear stress change along the plate boundary. The model and calculation procedures are the same as those in the analysis by Sato(1989). Accordingly we present here simple descriptions of the model in the following.

The two-dimensional finite element model used and elasticity and viscosity assigned in each mesh are shown in Fig.6. The shearing force was applied with the period of 20 years at the uppermost part shown by a thick line along the contact surface with the Pacific plate. The model boundary at the left and bottom was fixed in the normal directions,

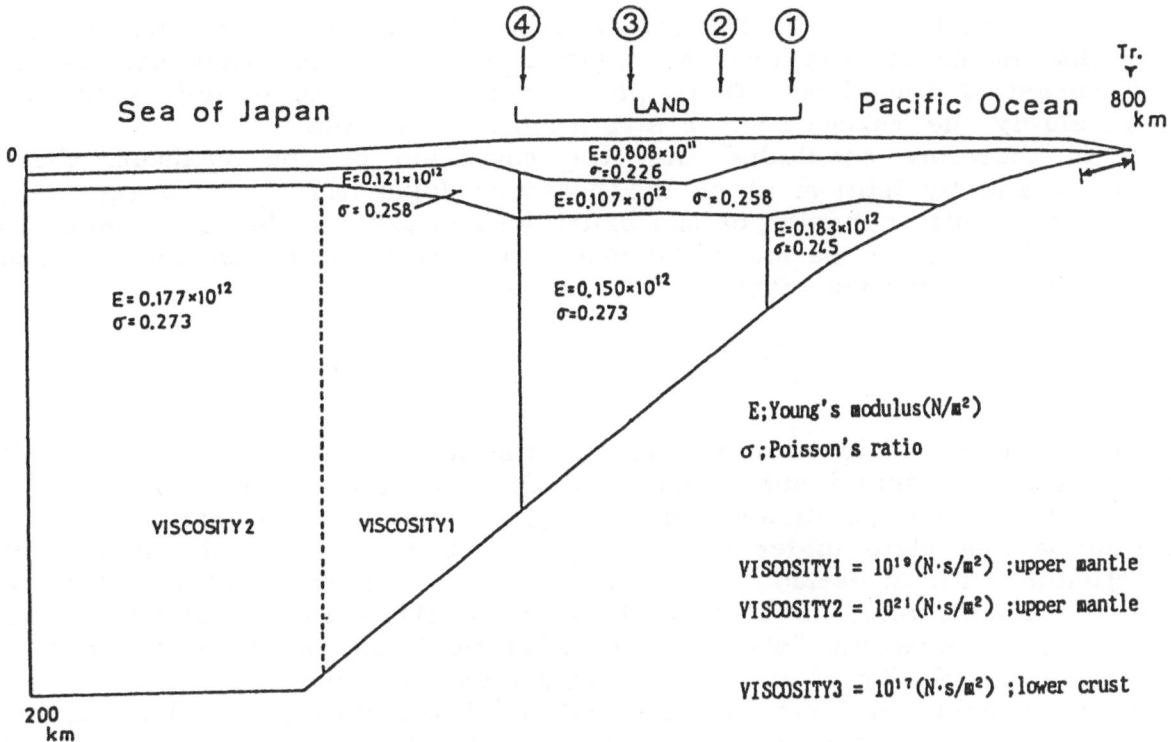

Fig.6. Elasticity and viscosity of the two-dimensional finite element model for the simulation. The numbers 1 to 4 indicate the points for which the change of the maximu shear strains is plotted in Fig.7.

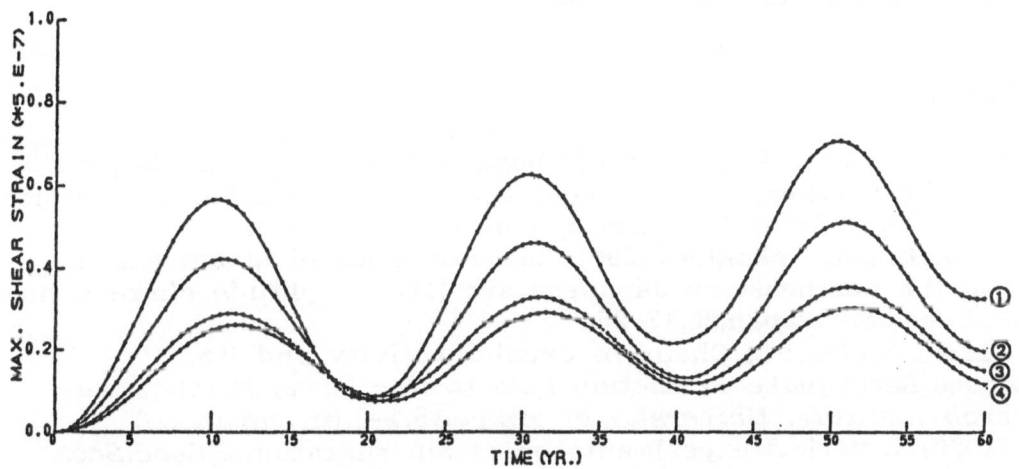

Fig.7. Maximum shears at the points 1 , 2 , 3 and 4 shown in Fig.6 obtained from the finite element analysis.

respectively. Deformations were solved assuming the Maxwell body. Obtained maximum shear strains at points 1 ,2 , 3 and 4 on the land are given in Fig.7. As seen in the figure the farther the point is located from the plate boundary, the later the maximum shear strain reaches due to the stress relaxation. It should be noted here that the apparent speed is obtained as 70–80km/year from the shift of the peaks and troughs. Moreover, the present analysis also indicates that the longer the period of the shear stress given along the plate boundary is, the more slowly the maximum shear strain migrates inland.

It is therefore concluded that the migration of the maximum shear strain found by Ishii et al. in northeastern Japan may be well explained as a viscoelastic behavior of the crust and upper mantle in response to a fluctuation with a period of 20 years or longer in the thrusting force due to the subduction of the Pacific plate.

CONCLUSION

From repetition of leveling surveys in the Kanto area, Japan a periodic fluctuation was found superposing on the continuous subsidence at the tip of the Miura peninsula. This suggests that the subduction of the Phillipine Sea plate under Kanto area is not constant in its speed but fluctuated with a period of about 20 years. It is concluded from a finite element analysis that such a fluctuation in subduction may generate an apparent "shear strain migration" on Japan Island with a propagation velocity of order of several tens km per year.

It is probable that periodic fluctuation of the speed of subduction is also closely related to the seismicity in Kanto and the neighboring region, and accelerated subsidence seen at the tip of Boso peninsula in the last 5 years should be carefully investigated from the view point of earthquake prediction in this area.

We have tried an estimate of eustatic sea level rise, and obtained the speed of 1.9mm/year by combining the data from the leveling and the tidal observation in the Kanto area.

REFERENCES

Ishii,H.,Sato,T. and Takagi,A.(1978).Characteristics of strain migration in the northeastern Japanese arc (I) –Propagation characteristics-,*Sci.Rep.Tohoku Univ.,Ser.5,Geophysics*,25,83–90.

Ishii,H.,Sato,T. and Takagi,A.(1980).Characteristics of strain migration in the northeastern Japanese arc (II) –Amplitude characteristics-,*J.Geod.Soc.Japan*,26,17–25.

Ishii,H.(1988).Monitoring Chain on Crustal Activity and its database,*Earthquake Prediction Data Center News, Earthquake Research Institute, University of Tokyo*,15,1–3.(in Japanese)

Sato,K.(1989).Numerical experiments on strain migration,*J.Geod.Soc. Japan*,35,27–36.(in Japanese)

Seno,T.(1985).Is northern Honshu a microplate?*Tectonophysics*,115,177–196.

Tanaka,M. and Gomi,T.(1989).Crustal movement observed from horizontal and vertical variations above the subduction zone,*J.Geod.Soc.Japan*, 35,187–206.

Yamada,J.(1973).A water-tube tiltmeter and its applications to crustal movement studies,*Spec. Bull. Earthq. Res. Inst. Univ. Tokyo*,10,1–147. (in Japanese)

INTEGRATED APPROACH TO KINEMATIC ADJUSTMENT AND SPATIAL ANALYSIS OF LEVELING CONTROL DATA OF THE ANCONA 82 LANDSLIDE

M. Crespi - B. Crippa - L. Mussio
Istituto di Topografia, Fotogrammetria e Geofisica
Politecnico di Milano - Italy

1. Introduction

In December 1982 a large scale landslide occured in the district of Ancona (Italy), including also some residential areas of the city.

A leveling network was established to control the settlement of the terrain (fig. 1) and eleven high precision spirit levelings were performed during the following two years (tab. 1).

Since the levelling network is wide, the execution of the measurements required generally a long time, so that the frequency of the repetition (about two months) was comparable with their lastings (15 - 20 days).

Thus an unique adjustment of all the repetitions has to be carried out by applying a suitable kinematic model (kinematic adjustments). The movements are very quiet so that they can be represented by a polynomial of the third order as shown by the behaviours of three representative benchmarks (fig. 2); at a fixed time it results:

$$z_k(T) = z_{0k} + z_k' \, T + z_k'' \, T^2 + z_k''' \, T^3 \tag{1}$$

The parameters z_{0k}, z_k', z_k'', z_k''' represents respectively the height at a reference time, the velocity, the acceleration and the impulse (the first derivative of the acceleration) of the benchmark k; they are estimated by the kinematic adjustment too.

2. Kinematic adjustment and spatial analysis

After the adjustment it is possible to perform a spatial analysis of each kinematic parameter. It can be analyzed as sample from a realization of a generally stationary, isotropic and ergodic stochastic process, a part of which can be previously removed e.g. by a bicubic

311

spline functions interpolation. The procedure consists of the three steps, which imply the solution of three systems of equations:

1) kinematic adjustment, which furnishes the kinematic parameters; the related system is formed by observation equations of differences of heights of two points (i,j) at a fixed time T:

$$\bar{z}_{0i} + \bar{z}'_i T + \bar{z}''_i T^2 + \bar{z}'''_i T^3 - \bar{z}_{0j} - \bar{z}'_j - \bar{z}''_j T^2 - \bar{z}'''_j T^3 - \Delta z^0_{ij}(T) =$$
$$= \bar{w}_{ij} \tag{2}$$

2) interpolation of the kinematic parameters by bicubic spline functions and computation of the residuals; the related system is formed by pseudo-observation equations:

$$\Sigma_{m,n} \, S_{m,n} \, \beta_{m,n}(k) - \bar{z}'_k('(')) = \bar{v}_k \tag{3}$$

where:

$\beta_{m,n}(k)$ is the value of the bicubic spline function with origin in (m,n) in the benchmark k;

$S_{m,n}$ is the coefficient of the $\beta_{m,n}$ spline function;

$z'_k('(')))$ is the velocity or the acceleration or the impulse of the benchmark k;

3) filtering of the residuals by collocation, after having estimated the covariance function of each set of residuals (fig. 3); the system related to the filtering is formed by the following equations:

$$\bar{s}_k + \bar{n}_k = \bar{v}_k \tag{4}$$

Therefore, note that in this approach the kinematic adjustment and the spatial analysis are completely separated. It can be substitute by the new integrated approach, which provides for the contemporary execution of the adjustment and the spatial analysis.

3. Integrated approach: kinematic adjustment together with spatial analysis

According to the computing point of view, the integrated approach requires all the systems of equations are solved simultaneously.

The equations (3) and (4) can be compacted in the following pseudo-observation equation:

$$\Sigma_{m,n} \, S_{m,n} \, \beta_{m,n}(k) - \bar{z}_k'('(')) - \bar{s}_k = \bar{n}_k$$

so that the unique system to solve takes the following form:

$$B\bar{s} - \bar{n} - \alpha_0 = 0 \tag{5}$$

where B is the design matrix, α_0 the observed quantities, $\bar{n}$ the vector of the residuals (noise) and $\bar{s}$ the vector of the parameters, which are both non-stochastic (incorrelated: $\bar{z}_{0k}$, $\bar{z}'_k$, $\bar{z}''_k$, $\bar{z}'''_k$, $\bar{S}_{m,n}$) and stochastic (correlated: $\bar{s}_k$).

Therefore the covariance matrix of the signal C_{ss} is assumed to be known a priori, consists of four blocks: two diagonal blocks containing the covariance matrices of the stochastic and non-stochastic parameters, and two off-diagonal blocks containing the crossvariance matrices between the two types of parameters, which are identically equal to zero.

The block of the stochastic parameters is determined by one or more auto and crosscovariance functions, which can be estimated empirically by the results of preceeding separate adjustments. The block of the non-stochastic parameters is a diagonal matrix, whose elements have to be chosen in balance with the covariances of the stochastic parameters, in such a way that the solution is not constrained too much to either type of parameters.

The general variance of the noise σ_n^2 also has to be known a priori; it is assumed equal to the estimate square sigma naught σ_0^2 of the separate kinematic adjustment.

The weights of the observationsare equal to the ones of the kinematic adjustment, while pseudo-observations have the same weights assumed equal to the variances of the residual noises derived from separated spatial analyses.

Therefore, before the integrated approach, it is necessary to perform separate adjustments for some reasons:

1) to compute the σ_n^2 (kinematic adjustment);

2) to build up the covariance functions (kinematic adjustment followed by bicubic spline functions interpolations);

3) to compute the variances of the residual noises (kinematic adjustment and separated spatial analyses).

The use of both observation and pseudo-observation equations causes the need to introduce a hybrid least squares criterium:

$$1/2 \begin{bmatrix} \bar{s}^t & \bar{n}^t \end{bmatrix} \begin{bmatrix} C_{ss}^{-1} & 0 \\ 0 & P/\sigma_n^2 \end{bmatrix} \begin{bmatrix} \hat{s} \\ \hat{n} \end{bmatrix} + \lambda^t(B\bar{s} - \bar{n} - \alpha_0) = min \qquad (6)$$

where P is the weight matrix and λ is a vector of Lagrange multiplicators.

4. Computational problems of the integrated approach

According to the expressions (5) and (6) the estimates for the signal and the noise become:

$$\bar{s} = C_{ss}B^t(BC_{ss}B^t + \sigma_n^2 P^{-1})^{-1} \alpha_0 \qquad (7)$$

$$\hat{n} = \alpha_0 - B\hat{s} \qquad (8)$$

The computation of these quantities requires the solution of a system with dimension m, equal to the number of observations. Since it is much more convenient to have analogous expressions which need the solution of a system with dimension n < m, equal to the number of parameters, the previous ones (7) and (8) can be trasformed, by using two theorems of linear algebra:

$$(Q \mp RST)^{-1} = Q^{-1} \pm Q^{-1}R(S^{-1} \mp TQ^{-1}R)^{-1} TQ^{-1}$$

$$Q^{-1}(Q^{-1} \mp S)^{-1} Q^{-1} = (Q \mp QSQ)^{-1}$$

into the following ones:

$$\bar{s} = (B^tPB)^{-1} B^tP\alpha_0 - \sigma_n^2[(B^tPB)C_{ss}(B^tPB) + \sigma_n^2(B^tPB)]^{-1} B^tP\alpha_0$$

$$\hat{n} = \alpha_0 - B\bar{s}$$

By starting from these new expressions, the law of variance propagation lets to compute the expression of the corresponding covariance matrices:

$$C_{\bar{e}\bar{e}} = \sigma_n^2(B^tPB)^{-1} - \sigma_n^4[(B^tPB)C_{ss}(B^tPB) + \sigma_n^2(B^tPB)]^{-1} B^t$$

$$C_{\hat{n}\hat{n}} = \sigma_n^2[P^{-1} - B(B^tPB)^{-1}B^t] + \sigma_n^4B[(B^tPB)C_{ss}(B^tPB) + \sigma_n^2(B^tPB)]^{-1} B^t$$

where $\bar{e} = s - \bar{s}$ represents the estimation error of the signal.

Note that the matrix (B^tPB) is sparse and the covariance matrix C_{ss} is sparse too when it is built up by finite covariance functions; therefore the product $(B^tPB)C_{ss}(B^tPB)$ is sparse and it has to be computed taking into account this characteristic.

A remarkable problem of the integrated approach is the fixing of the weights of the stochastic signal in comparison with the ones of the other observations. The question requires further investigations and, for the moment, it is possible to say only that by repeating the integrated approach, the incertainly about the weight ratios can be eliminated and suitable values for the weights can be established.

5. Results and conclusions

The sets of measurements coming from the eleven high precision spirit levelings were adjusted in three different ways:

1) separately, leveling by leveling (classical adjustment);

2) simultaneously, by the kinematic approach;

3) simultaneously, by the integrated approach.

The main results are shown in the tab. 2. Note that the integrated approach needs a very long computing time, even if an efficient algorithm and routines able to treat large sparse matrices are emploied; besides the obtained results are just the first one and require to be tested again after an adequate deepening of the discussed problem of weights.

6. References

Colombo L., Fangi G., Mussio L., Radicioni F. (1986): "Kinematic Processing and Spatial Analysis of Leveling Control Data of the Ancona 82 Landslide". In H. Pelzer and W. Niemeier (EDS.), Determination of Heights and Height Changes, pp. 703-719, Duemmler Verlag, Bonn, 1987.

Forlani G., Mussio L. (1986): "The Kinematic Adjustment of Leveling Networks". In M. Unguendoli (ED.), Proc. of the Meeting on the Modern Trends in Deformations Measurements, Bologna 26-27 June 1986, pp. 265-290, Bologna, 1986.

De Haan A., Forlani G. (1987): "Digital Height Variation Modelling by Least Squares Collocation". In O. Jacoby and P. Frederiksen (EDS.), Proc. of the Int. Colloquium Progress in Terrain Modelling, Technical University of Denmark 20-22 May 1987, pp. 113-126, Copenhagen, 1987.

Barzaghi R., Crippa B., Forlani G., Mussio L. (1988): "Digital Modelling by Using the Integrated Geodesy Approach". Int. Archives of Photogrammetry and Remote Sensing, 16th Congress of the ISPRS, Vol. 27, part B3, pp. 47-56, Kyoto, 1988.

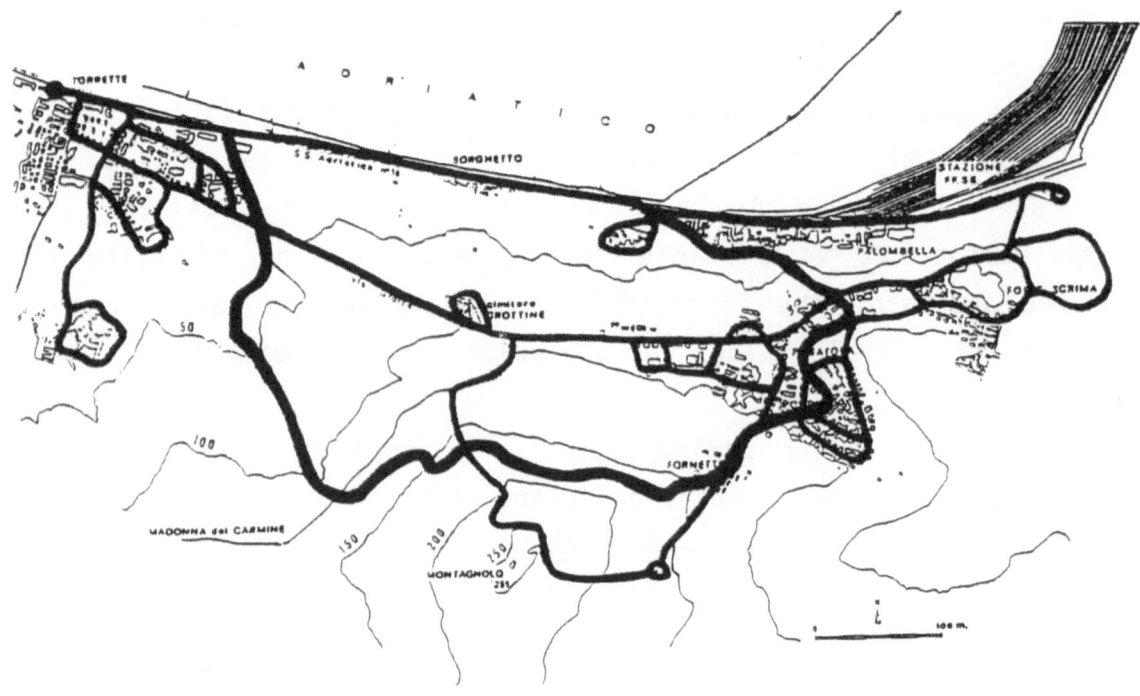

Fig. 1 - Configuration of the leveling network in the Ancona 82 landslide area.

Leveling number	Beginning date	Ending date	Lasting (days)	Net's length
1	22.02.83	21.03.83	28	13 Km
2	26.04.83	06.05.83	11	
3	30.06.83	04.07.83	5	
4	26.09.83	03.01.83	8	
5	02.12.83	19.12.83	18	23 Km
6	30.01.84	16.02.84	18	
7	05.04.84	18.04.84	14	
8	03.09.84	12.09.84	10	18 Km
9	17.12.84	22.01.85	37	
10	18.03.85	29.03.85	12	
11	01.06.85	24.06.85	24	

Tab. 1 - Measurements periods (T=913 days) of the eleven leveling campaigns.

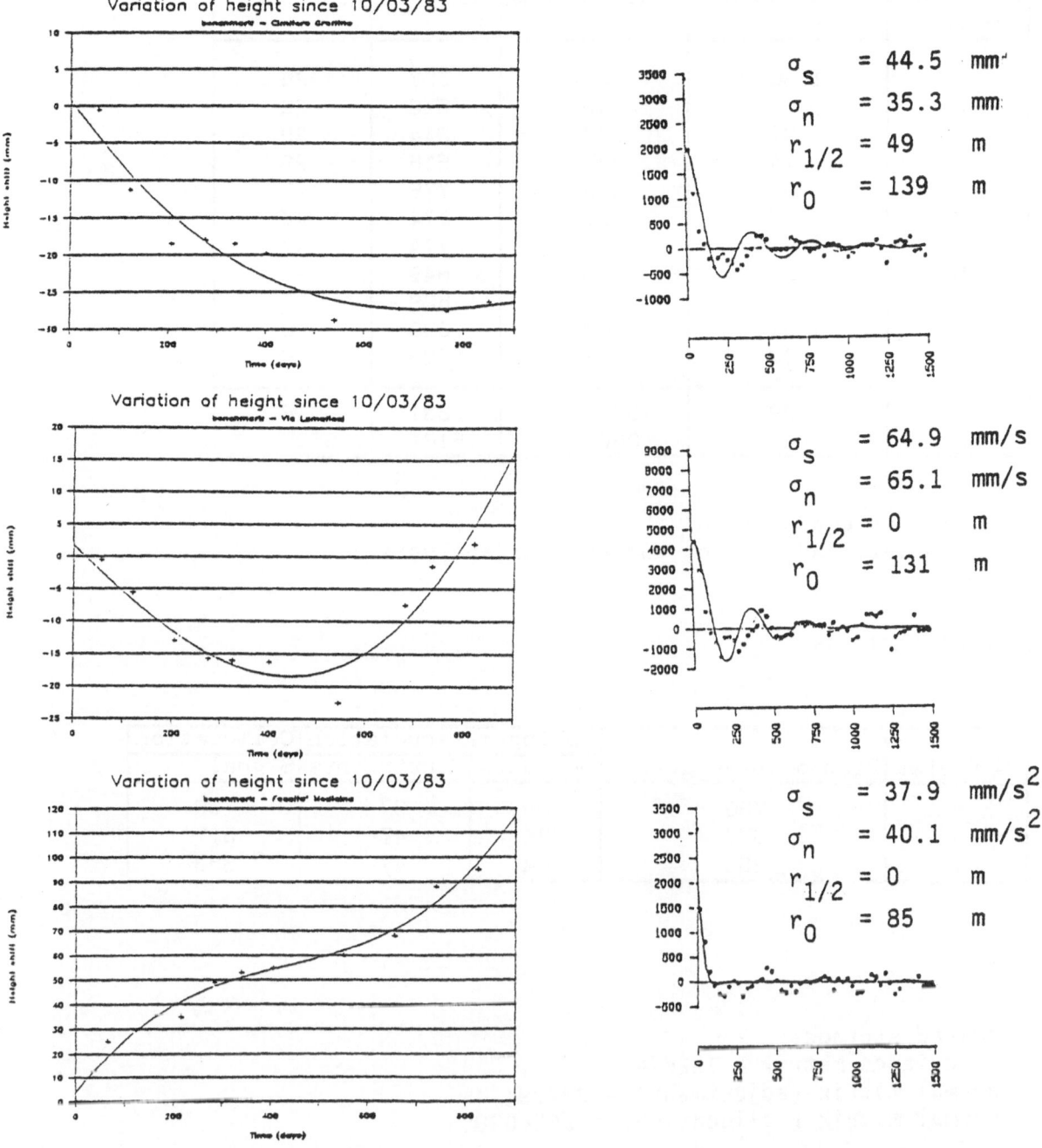

Fig. 2 - Representative movements
of the benchmarks in the
Ancona 82 landslide area
fitted with a polinomial
of the third order.

Fig. 3 - Autocovariance functions of
the kinematic parameters
(v, a, i) of normal-J_0 type
after the bicubic splines
functions interpolation.

Leveling number	Sigma-naught (mm)	Observations	Unknowns	Redundancy
1	1.36	533	507	26
2	.88	545	516	29
3	.97	543	514	29
4	.74	545	516	29
5	1.01	823	776	47
6	.72	821	774	47
7	.91	821	774	47
8	1.21	676	648	28
9	.81	656	629	27
10	1.01	660	630	30
11	1.00	661	632	29
K	2.99	6449	2831	3618
KS	5.43	8666	5161	3505

K : kinematic adjustment
KS : kinematic adjustment and spatial analysis

Spatial analysis:

Parameter	Pseudo-observations	Spline interpolation		Collocation
		Unknowns	Constraints	Signal
v	736	96	57	736
a	677	84	47	677
i	653	84	47	653

Required storage:
- covariance matrix = 152674;
- normal matrix (adjustment) = 93729;
- normal matrix (collocation) = 2426090.

Solution method: conjugate gradient with reordering and preconditioning.

Computing time to built up the normal system and to solve the least squares adjustment = $2^h 20^m$.

Computing time for solving the least squares collocation = $9^h 01^m$.

Tab. 2 - Main results.

KALMAN FILTERING IN LEVELLING : AN APPLICATION

Maurizio Barbarella   Fabio Radicioni
Dipartimento di Scienze dei Materiali e della Terra
Ancona University, Italy

INTRODUCTION

The present study deals with the analysis of repeated levelling networks set up to control geological deformation (such as subsidence or landslide phenomena). Vertical movements are determined by comparing repeated measurements in subsequent epochs on the area under study. For large networks, this comparison is not easy. The measurements can require a great deal of time, to the extent that this may be compared to the interval of time between two repetitions. This may lead to problems of loop closure in areas of fast movement and the obtained results would in this case be of very difficult interpretation. Moreover, even though the measurements involve the same area, they may concern a network that gradually modifies according the new data acquired. This means that in subsequent repetitions there are entire areas for which no data on movements are available. It is therefore essential to define a model which takes these additional characteristics of variability into consideration, so as to identify the evolution of the phenomenon correctly (VANICEK, 1979). The knowledge of the kinematic parameter i.e. velocity and acceleration is very important for the physical interpretation of the phenomena.

1. VERTICAL MOVEMENT ANALYSIS BY REPEATED SURVEYS

Several computational methods can be used for studying vertical movements by means of repeated levelling (HOLDAL, 1979). Each survey can be elaborated separately from the others if the observations are considered to be contemporary and uncorrelated with those of the other epochs (the so called "static adjustment"). The adjusted heights are then used to obtain the velocity of the benchmarks. Otherwise a particular model for the height variations can be hypothesized for each point with height, velocity and acceleration as the unknowns of an adjustment problem which takes into account the data of all the surveys: the kinematic parameters are directly estimated from the data (kinematic adjustment). Finally, it is possible to suppose that the kinematic parameters (height, velocity, acceleration) represent the "system status", and their values known at time t_i can be predicted at the subsequent epoch t_{i+1} and updated with the new measurements (Kalman filtering).

1.1 Static adjustment and height fitting

The height of all points for each epoch are estimated by least square adjustments (Gauss - Markov model). Since the point movements are not taken into account during the time survey, the adjustments should be considered as static. Subsequently the adjusted heights at the different epochs are compared and the vertical displacements are computed for each point. Dividing the movements by the time interval, the mean velocities of vertical displacement are obtained for each point (obviously a linear motion is assumed). To eliminate the rank deficiency of the adjustment and in order to compare the results, the height of the same reference point (at least one) is assumed as "fixed" for all the epochs. The estimated heights of each point are interpolated by polynomials and the kinematic parameters can be obtained with the polynomial coefficients. This simple method can be applied effectively when the time required for measuring is negligible with respect to the interval between the two repetitions.

1.2 Kinematic adjustment

In recent years, so called kinematic techniques have been developed which also take into account the temporal behavior of the point height during the survey. Several kinematic methods have been proposed, generally by associating an equation of motion for each point in the form of a simple polynomial. The height of a generic point A may be expressed as follows:

$$h_A^i = h_A^o + v_A^o \left[t^i - t^o\right] + \frac{1}{2\,!}\, a_A^o\left[t^i - t^o\right]^2 + \qquad (1)$$

where h_A^i is the height of point A at time t^i and h_A^o, v_A^o, a_A^o respectively are the height velocity acceleration of point A at time t^o, with t^o as the reference time (time of the first measuring campaign) and $t^i > t^o$ as time at the ith epoch. It is important to note that the rank deficiency, that is the non-invariance to temporal translations, in agreement with this observation equation, is three. The datum in the case of equation (1) consists in fixing the known height, the velocity and the acceleration constants, for a reference point. It is evident that there is no simple way to test this assumption, but it may be backed by geological investigation. Subsequent coefficients of equation (1) have no geometric interpretation, but they could be admitted if it were possible to estimate them and if they were statistically significant. The most important assumption that requires the

320

application of the motion equation (1) is that the movements of the single points should be continuous functions without irregular disturbances. In addition the subsequent adjustment provides the estimates of the kinematic parameters which refer to the initial time t^o but no estimate on the subsequent values. According to the motion equation (1), which is valid for the single points, each difference in height Δh^i_{AB} between the two points A and B should still be considered as a function of time t. The functional model of the height difference between the points A and B is given by:

$$\Delta h^i_{AB} + r^i_{AB} = h^i_B - h^i_A = \left[h^o_B + v^o_B (\Delta t^i_{AB}) + a^o_B \frac{1}{2!} (\Delta t^i_{AB})^2 \right]$$

$$+ \left[h^o_A + v^o_A (\Delta t^i_{AB}) + a^o_A \frac{1}{2!} (\Delta t^i_{AB})^2 \right] \quad (2)$$

where $\Delta h^i_{AB} = l^i_k$ is the kth height difference observed (between the points A and B) at the ith epoch, $k = 1, 2, , m^i$ (m^i : the number of observations at the i-th epoch) h^i_A, h^i_B height of point A, B at time i, $i = 0, ..p$; r^i_{AB} residual of observation ; $t^i_{AB} - t^o_{AB} = \Delta t^i_{AB}$: time interval between the ith epoch height and the initial measurements of the height difference Δh^i_{AB}. From equation (2), by ordering the unknowns, the following equation may be written (3):

$$\underline{l}^i_k + \underline{r}^i_K = [-1\ 1\ 0\ .0]\underline{h}^o + \Delta t^i [-1\ 1\ 0\ .0]\underline{v}^o + \frac{\Delta t^{(i)^2}}{2!}[-1\ 1\ 0\ .0]\underline{a}^o$$

For the m^i observations at the ith epoch:

$$\underline{l}^i + \underline{r}^i = A^i \underline{h}^o + T^i_1 A^i \underline{v}^o + T^i_2 A^i \underline{a}^o \qquad (4) \quad \text{with:}$$

$$T^i_1 = \text{diag}(\Delta t^i_k) ; \quad T^i_2 = \frac{1}{2!} \text{diag} \left[(\Delta t^i_k) \right], \quad k = 1, 2, ., m^i;$$

Equation (4) may be expressed in a more compact form:

$$\underline{l}^i + \underline{r}^i = A^i + T^i_1 A^i + T_2 A^i [\underline{h}^o, \underline{v}^o, \underline{a}^o]^T \qquad (5)$$

The coefficient matrix A is written on measurements m^i taken at the ith epoch, which may be different from the m^j measurements of a different j^{th} epoch, since the network can show a different graph. For all epochs e, the system becomes:

$$\begin{bmatrix} \underline{l}^1 \\ \underline{l}^2 \\ .. \\ .. \\ \underline{l}^e \end{bmatrix} + \begin{bmatrix} \underline{r}^1 \\ \underline{r}^2 \\ .. \\ .. \\ \underline{r}^e \end{bmatrix} = \begin{bmatrix} \underline{A}^1 T^1_1 & A^1 T^1_2 & A^1 \\ \underline{A}^2 T^2_1 & A^2 T^2_2 & A^1 \\ .. & .. & .. \\ .. & .. & .. \\ A^e T^e & A^e T^e & A^e \end{bmatrix} \begin{bmatrix} \underline{h}^o \\ \underline{v}^o \\ \underline{a}^o \end{bmatrix} \qquad (6)$$

which constitutes a GAUSS-MARKOFF mathematical model:

$$\underline{l} + \underline{r} = A \cdot \underline{x} \qquad (7)$$

(M.1) (M.1) (M.n) (n.1)

where: $M = \Sigma m$ total measurements on e repetitions; P = the

321

number of points; $n = (3 \cdot p) =$ unknowns in the model with second degree polynomial. Considering that the observations are the height differences measured, the following stocastic model is considered for each repetition:

$$\Sigma_{ll}^i = \sigma_o^{(i)^2} Q_{ll}^i \qquad \text{with:} \qquad Q_{ll}^i = \text{diag}\left[\frac{1}{d_k}\right],$$

$d_k = k^{th}$ distance, $k=1,2, ., m$. If there are no correlations for M observations at e epochs, then:

$$\Sigma_{ll} = \text{diag}(\Sigma_{ll}^i) = \sigma_o^2 \text{ diag}(Q_{ll}^i) \qquad i=1,2, .,e$$

It should be noted that the measurements do not require pretreatment since the height differences measured are adjusted directly. In addition it should be observed that the network must not be partitioned otherwise it would be necessary to define a point of reference for each disjointed section.

2. KALMAN FILTER

The disadvantages related to the very considerable dimension of the numerical problem, as well as the fact that the estimate of kinematic parameters refer to the initial time, can both be overcame by using recursive calculation techniques such as Kalman filtering (PELZER, 1987).
For each measuring campaign, this technique makes it possible to update the solution previously obtained, and provides the current estimate of kinematic parameters instead of the "historical " kinematic values for the first measuring time. An additional advantage of this technique, whose solution is calculated by successive phases of "prediction" and "up-dating", is that the future pattern of the phenomenon may be predicted. The unknown parameters are arranged in a vector $\underline{y}(t)$ which represents the "system state vector", and a "transiction matrix" T is introduced, which make it possible to transform the system state at a time t_i, to another system state at a subsequent time t_{i+1}. In the transformation, an additional term w_i which constitutes the "system noise" must be considered, so that the process is described by the following *transiction model*:

$$\underset{(n.1)}{\underline{Y}_{i+1}} = \underset{(n.n)}{T} \underset{(n.1)}{\underline{Y}_i} + \underset{(n.1)}{\underline{w}_i}$$

The system state is also observed at time t_i by suitable sensors, which provide the measurements l_i so that the system also follows the *measuring model* ,

$$\underset{(m.n)}{A_i} \underset{(n.1)}{\underline{Y}_i} = \underset{(m.1)}{\underline{l}_i} + \underset{(m.1)}{\underline{v}_i}$$

where A_i is the design matrix at time t_i and $\underline{v}_i$ the measurement error at time t_i. For the stochastic part of the model, the following is supposed:

$$E\left[\underline{w}_i \underline{w}_k^T\right] = \delta_{ik} Q_i \quad ; \quad E\left[\underline{v}_i \underline{v}_k^T\right] = \delta_{ik} R_i \quad ; \quad E\left[\underline{w}_i \underline{v}_k^T\right] = 0 \quad \forall i,k$$

322

At this stage it is possible to obtain the best signal value using a recursive linear estimator. The calculation procedure for obtaining the signal estimate is described in the following algorithms, known as "The Vectorial Kalman Filter" (GELB, 1974). The filtering process starts with a prior estimate of state vector $\underline{y}_i$ and its covariance matrix P_i^-. The best estimate is intended as the estimate which minimizes the estimate error of each signal component at the same time. The current time estimate is in this way given by a suitable linear combination of the previous time estimate and the new measurements at the current time. The covariance matrix P_i for updated estimate is then computed. The previous estimate is projected ahead and duly premultiplied by the transiction matrix to obtain the predicted estimate. The algorithm follows a succession of prediction and updating phases as shown in Fig.1

Fig. 1. - The vectorial Kalman filtering.

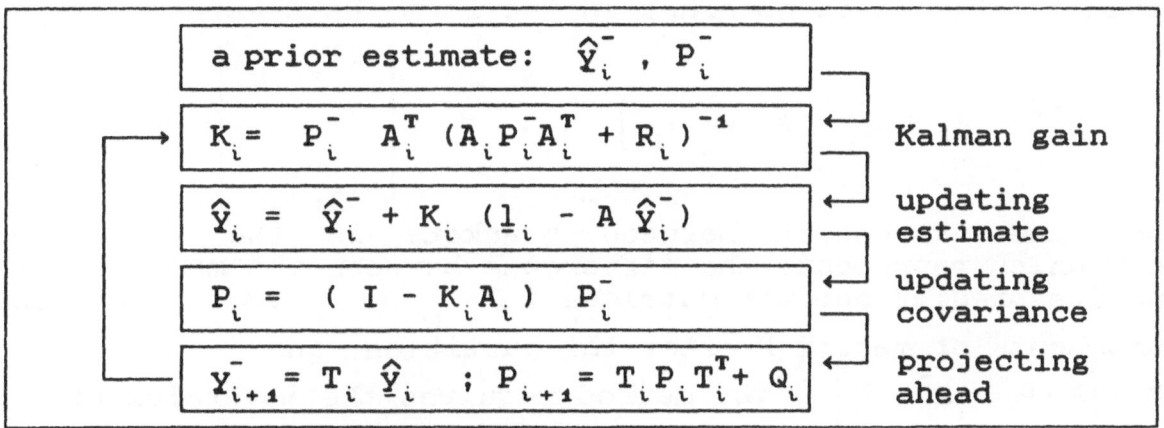

3. LEVELLING APPLICATION

The temporal behavior of the benchmarks can be described at each epoch by height, velocity and acceleration: the state vector at time t_i for p points is:

$$\underline{y}_i = [\ \underline{h}_i \ , \ \underline{v}_i \ , \ \underline{a}_i \ ]^T$$

where the (p.1) vectors $\underline{h}_i$, $\underline{v}_i$, $\underline{a}_i$ represent the heights, velocities, accelerations of the benchmarks respectively. Assuming that the accelerations have a constant value, apart from a white noise, the motion law for each point is : $\partial^2 h / \partial t^2 = w$. From this law it is possible to obtain the transiction matrix in the following form:

$$T = \begin{bmatrix} I & I\Delta t & \Delta t^2/2 \\ 0 & I & I\Delta t \\ 0 & 0 & 0 \end{bmatrix}$$

which depend only on the Δt difference between observation epochs. In the levelling networks only the height differences are observed, and neither the velocities nor the accelerations can be measured. It is therefore convenient to partition the design matrix into three parts:

$$\begin{bmatrix} A_l & | & 0 & | & 0 \end{bmatrix} \begin{bmatrix} (\underline{y}_i)_1 \\ (\underline{y}_i)_2 \\ (\underline{y}_i)_3 \end{bmatrix} = \underline{l}_i + \underline{v}_i \quad ; \qquad \Sigma = \begin{bmatrix} \Sigma_{ll} & 0 & 0 \\ 0 & 0 & 0 \\ 0 & 0 & 0 \end{bmatrix}$$

$(m. n) \qquad (n.1) \quad (m. 1) \; (m.1)$

with $\Sigma_{ll} = \text{diag}(1/d_k, \quad k=1,2 \quad ,m \quad) \sigma_o^2$.

This partition is very convenient since it reduces the dimension of the matrices involved in the Kalman algorithm. The two sub-matrices related to velocity and acceleration are zero, and each row of the first contains only two elements which are different from zero. It is therefore not necessary to memorize an A_l matrix in the computing program: two pointer vectors are sufficient. Using the partition , the updated estimate is obtained in the form (BARBARELLA, 1989):

$$\hat{\underline{y}}_i = \hat{\underline{y}}_i^- + K_i \left[\underline{l}_i - A_l (\hat{\underline{y}}_i^-)_1 \right] \qquad ; \qquad K_i = \begin{bmatrix} (P_i^-)_{11} \\ (P_i^-)_{21} \\ (P_i^-)_{31} \end{bmatrix} A_l^T \; M$$

where the expression between brackets is the *innovation* term which represents the difference between the measurements and its prediction; the matrices $(P_i^-)_{kl}$, $k,l = 1,2,3$ are the components of matrix P after the partition, and

$M = \left[A_l (P_i^-)_{11} A_l^T + R_i \right]^{-1}$. The method requires the inversion of a full triangular matrix of order m, equal to the number of the observations at time t_i. The updated value of the covariance matrix of the estimated state vector is:

$$P_{jk} = (K_i)_j A_l (P_i^-)_{1k} - (P_i^-)_{jk} \qquad ; \; j,k = 1,2,3$$

In the prediction phase it is not necessary to memorize the full matrix $T_i P_i T_i^T$ since its submatrices can be computed from those of P with suitable equations.

This form of the algorithm makes it possible computer memory; the major memory requirement is due to two triangular matrices, P ($3p.(3p+1)/2$ locations) and M ($m.(m+1)/2$ locations), and some auxiliary matrices of order m.p, p.p and some vectors. The inversion of the triangular full M matrix is required and may be performed using the Cholesky method.

The agreement of the data with the models can be verified by simple statistical test on the innovations ν (TEUNISSEN, 1988) for each epoch: i.e. the so-called "local overall model" for all the data, and the "local slippage" to test each observation :

$$\underline{\nu}_i \; \Sigma_\nu^{-1} \; \underline{\nu}_i \,/\, m_i \;\simeq\; \chi_m^2 \quad ; \quad w_j = \underline{e}_j \; \Sigma_\nu^{-1} \; \underline{\nu}_i \,/\, \left[(\Sigma_\nu^{-1})_{jj} \right]^{1/2} \simeq N$$

3.1. LEVELLING NETWORK

A program based on the described algorithm has been applied to the levelling network set up in 1983 to control the Ancona landslide (Italy). Of particular importance is the fact that the network design underwent strong changes. The number of points has changed from 530 to 820 according to the different epochs: new lines have been added, others have been eliminated. 15 repetitions were carried out in a period of 6 years (from 1983 to 1988);redundancy has varied from 26 to 47. The surveys lasted for 10 - 30 days while the time interval between two subsequent repetitions has often been only 2 - 3 months. This is a very difficult test - net for a first application of the Kalman filter !. To reduce the dimension of the problem, a program has been applied that automatically find the nodal points and the lines connecting them; the original data are substituted with the nodal points and height differences between these. The heights of intermediate benchmarks and their covariance matrix can be computed from the corresponding quantity for the nodal points in a static adjustment. The total number of unknown points is in this a way reduced to 90. The data were elaborated with the three forementioned methods.

3.1.1. Static adjustment. The first phase of the procedure consists in estimating the heights of the points by a traditional least squares adjustment for each epoch; the a-posteriori value of the sigma naught ranges from 0.8 to 1.4 . In a second phase a least squares polynomial fit of the estimated (in phase one) heights is performed for some points; for each epoch a weight inversely proportional to the estimated height variance is assumed. The orthogonal polynomial method is used to find the polynomial degree that performs the best fit. The effectiveness of an increase of the degree of the interpolating polynomial is evaluated by statistical tests on the sum of the squares of the residuals of the fit. The values of static adjustments and the best polynomial fit (computed on the first 11 epochs) are shown in Fig. 2. for point No. 364.

3.1.2. Kinematic adjustment. The kinematic adjustment is realized following the model shown in Par. 1.2 and involves 11 epochs, 6450 observations and about 2800 unknowns (the derivative of acceleration have also been taken into account); the results have been presented in a previous paper (COLOMBO et al 1987). The estimated initial values of the velocities $\underline{y}^{\circ}$ of all 800 points are shown in the surface plot (Fig. 3.), in which the lowered area is clearly visible at the top of the landslide while the bottom section is rising.

3.1.3. Kalman filtering. The present implementation of the computing program based on the discussed algorithm can not take into account strong changes in the network design. Since the first seven epochs present a fairly similar net, only 7 epochs were analyzed using the Kalman method.

Fig. 2. Static height and polynomial fit.

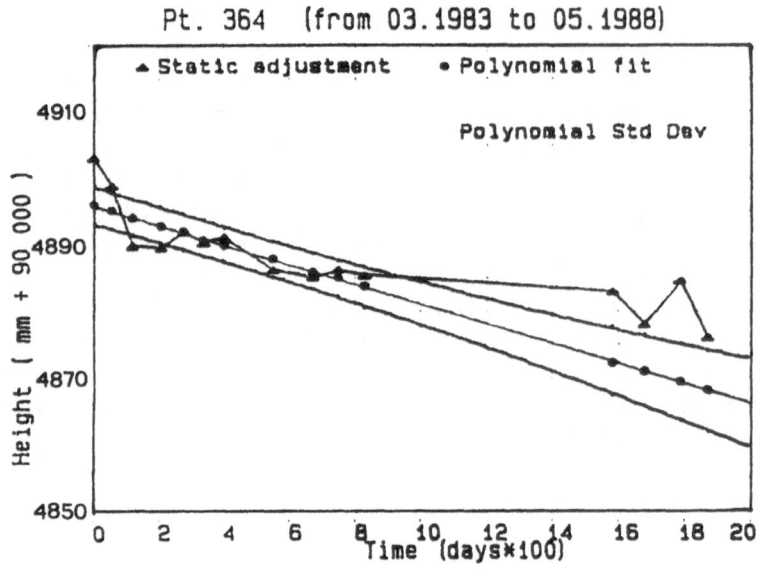

Fig 3. Initial velocity (kinematic adjustment).

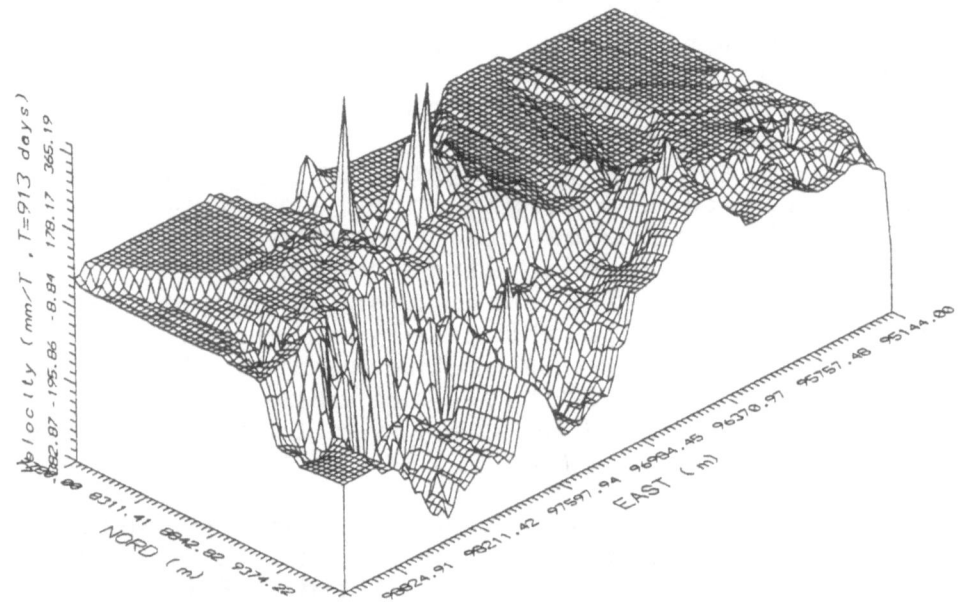

Obviously, the validity of the results could be affected by the very short time series analyzed. The *current* velocity of vertical displacement, updated to the 7th epoch, is shown in the surface plot of Fig.4 : the behavior of the velocity vector is quite similar to that shown on the previous plot, but with smaller values (positive for lowering).The Fig. 5 shows the Kalman prediction and updating for the motion of a typical point together with the height values obtained from the static adjustments. The assumed system noise variance matrix has the scalar form, $\sigma_v^2.I$, and the constant $\sigma_v^2 = 10^{-3}$:

the updating heights exactly reproduce the static values, while the predictions follow the trend of the previous updated values. The measurements assume a predominant role with respect to the transiction model. Fig. 6 is on the other hand obtained assuming $\sigma_v^2 = 10^{-6}$ for the same point : in this case, the system contribution is predominant with respect to the measurements (which define the static heights) The main problem in applying Kalman filtering to levelling involves the choice of the system noise variance matrix. Physically the system noise is due to various perturbations acting on the points, whose effects are very difficult to model; in order to solve this problem some adaptive filters may be applied.

Fig. 4. Current velocity (Kalman updating 7' epoch).

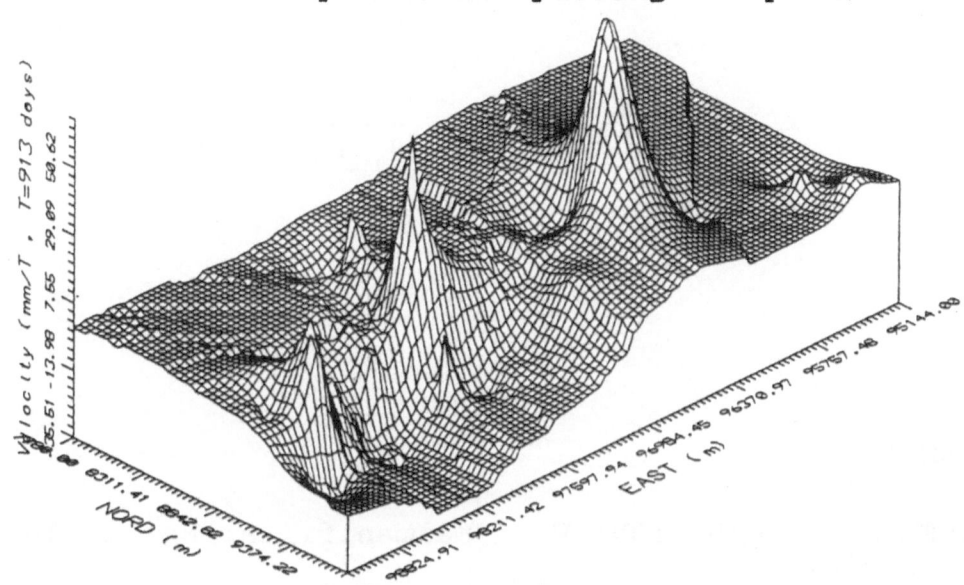

Fig. 5. Kalman prediction and updating $\sigma_v^2 = 10^{-3}$.

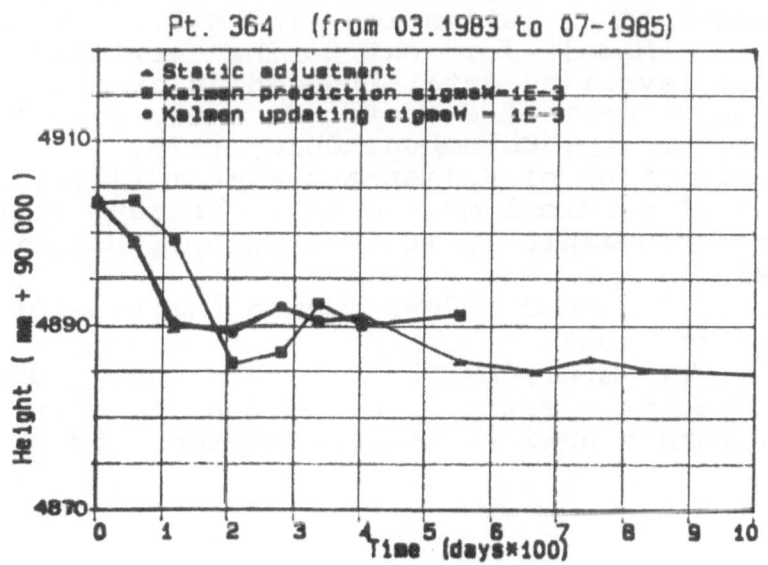

Fig. 6 Kalman prediction and updating $\sigma_{\tilde{v}}^2 = 10^{-6}$.

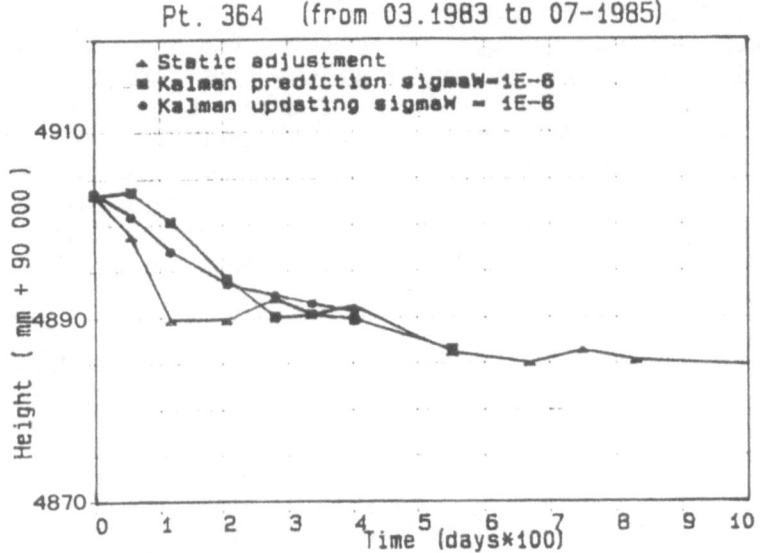

Pt. 364 (from 03.1983 to 07-1985)

REFERENCES

BARBARELLA M., RADICIONI F. : Kalman filtering in levelling. In printing.

COLOMBO L.,FANGI G. , MUSSIO L., RADICIONI F. :Kinematic processing and spatial analysis of levelling control data of Ancona '82 landslide. Presented paper on Recent Crustal Movements in Western Europe,Hannover,14-19 September 1986.

GELB A. :Applied optimal estimation. The M.I.T. Press, 1974.

HOLDAL S. R. : Model for extracting vertical crustal movements from levelling data. Proceeding of the 9th GEOP Conference, Department of Geodetic Science, Report n. 280, Ohio State University, Columbus, Ohio, 1979.

PELZER H. : Application of Kalman-and Wiener filtering on the determination of vertical movements . Presented paper on Recent Crustal Movements in Western Europe, Hannover, 14-19 September 1986.

TEUNISSEN P.J.G., SALZMANN : Performance analysis of Kalman filters. TUDelft, 1988.

VANICEK P.,ELLIOTT M.R.,CASTLE R.O. : Four dimensional modelling of recent vertical movements in the area of southern California uplift. Tectonophysics, 52,pp.287-300, 1979.

OBSERVATION OF CRUSTAL MOVEMENTS BY MEANS OF A LONG BASELINE WATER-TUBE TILTMETER AT SAGARA, SHIZUOKA, JAPAN

Mikio Satomura, Shigeki Kobayashi
Faculty of Liberal Arts, Shizuoka University, Shizuoka 422,JAPAN
Yasushi Hasegawa
Matsuyama Girls' High School,Saitama 355, JAPAN
Tadaaki Toyama
Japan Meteorological Agency, Tokyo 100, JAPAN
Takeshi Dambara
Japanese Association of Surveyors, Tokyo 112, JAPAN
Ryuichi Shichi
School of Science, Nagoya University, Nagoya 464-01, JAPAN
Morio Ino and Takayoshi Iwata
Shizuoka Prefectural Government, Shizuoka 420, JAPAN

ABSTRACT

It has been pointed out that a destructive earthquake will occur in the Tokai District, central Japan, by the interactive motions between the Philippine Sea Plate and the Eurasian Plate in the near future.

In order to detect precursory crustal movements of the earthquake as a earthquake prediction scheme, a long baseline (365m) water-tube tiltmeter was set in March, 1983, at Sagara-cho, which is the central part of the source region of the presumed earthquake. The direction of the water-tube is N60°W, which is the convergent direction of the Philippine Sea Plate to the Eurasian Plate. The effects of ground water and room temperature were investigated from the records obtained by the tiltmeter since June, 1983.

In order to check the recorded values, the leveling surveys have been repeated 11 times between the two observation sites at the both ends of the tiltmeter. The values obtained by the tiltmeter correspond well with those by the leveling surveys.

The ground water levels were measured near the northwest site, since December, 1985. The tiltmeter results corrected for the effects of ground water level and room temperature show that the ground is consistently tilting to northwest. This direction is reverse to that of the regional tilting change obtained from the leveling survey by the Geographical Survey Institute.

The local gravity high anomaly exists near the southeast site of the tiltmeter. The northwest tilting observed by the tiltmeter is possibly caused by a local folding, while the tilting change obtained by the leveling survey is possible the regional phenomenon by the plate motion.

1.INTRODUCTION

Destructive earthquakes have often occured in the Tokai Region, central Japan, by the subductive motion of the Philippine Sea Plate under the Eurasian Plate. Though Kanto Earthquake occurred in 1923 and Tonankai Earthquake in 1944 in the neighbor of the Tokai District, the destructive earthquake whose source region is in the Tokai District has not occurred since 1854. And moreover, it is found from the geodetic surveys that the contraction and ground subsidence have continued for more than hundred years by the motion of the Pilippine Sea Plate. From such circumstances, it is thought that a destructive earthquake shall occur in the Tokai District in the near future (Ishibashi,1981).

In order to detect the precursory crustal movements of the earthquake as a earthquake prediction scheme, a long baseline water-tube tiltmeter was set in March, 1983, at Sagara-cho, which is the central part of the source region of the presumed earthquake.

The data obtained with this tiltmeter were analyzed and we discussed the result in the present paper.

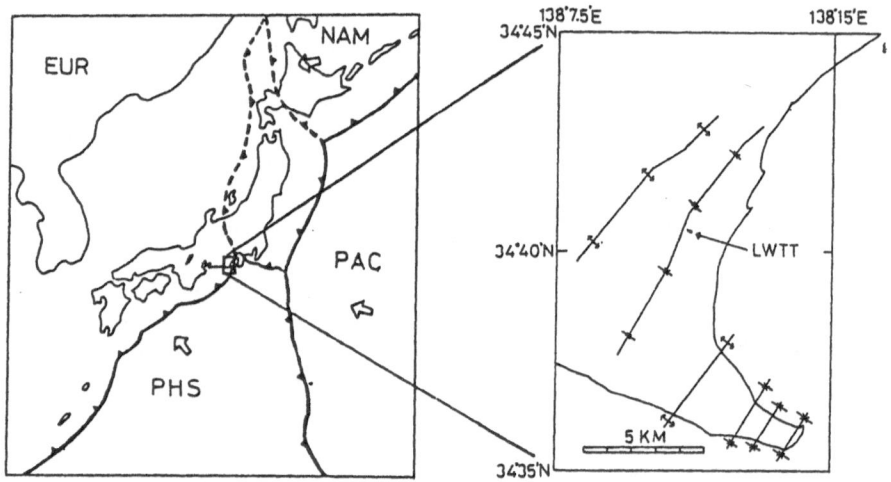

Fig.1 Location of the observation site. EUR,NAM,PAC,PHS and LWTT-
    Eurasian, North American, Pacific, Philippine, Sea Plate
    and long water-tube tiltmeter.

2.LOCATION AND INSTRUMENT

The tiltmeter was set in a tea farm. Its latitude is 34°40.5'N, its longitude is 138°11.0'E and its height is 105.0m.The location is shown in Fig.1. Only one component of the tilt is measured, and the direction is N60°W, which is almost the convergent direction of the Philippine Sea Plate to the Eurasian Plate. The baseline length is 365.0m, which is the longest water-tube tiltmeter in Japan (Satomura et al.,1987).

330

The water levels in two pots are automatically measured by using floats and magneticsensor (Shichi et al.,1980), the data obtained are sent to Shizuoka University through telephone cable, and they are stored in floppy disks there.

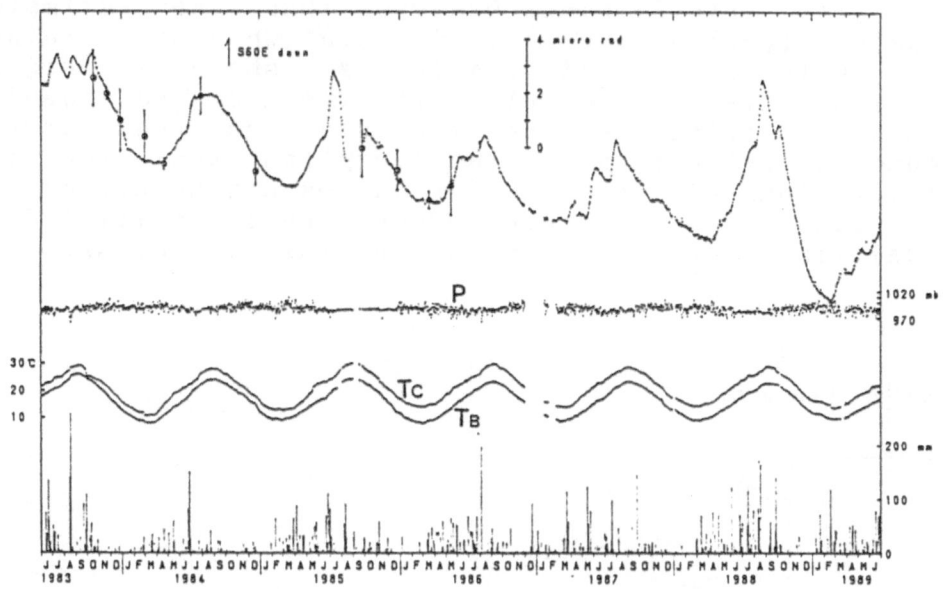

Fig.2 Daily mean values of the tiltmeter, air pressure(P) and room temperature (A,C). The circles with bars show the results of leveling survey between the both sites of the tiltmeter and the lowermost bars show the daily precipitation near the tiltmeter.

3.RESULTS OF THE TILTMETER

The daily mean values of the tiltmeter obtained from June 1983 to June 1989 are shown in Fig.2, as well as those of air pressure and room temperatures at northwest site(B) and central site(C) of the tiltmeter, and daily precipitation near the tiltmeter.

In order to check the obtained tiltmeter values, the leveling surveys have repeated 11 times between the two observation sites at the both end of the tiltmeter. Their results are also shown in Fig.2. The values obtained by the tiltmeter correspond well with those by the leveling surveys.

We can easily see that the tiltmeter's values have large seasonal variations which are related to room temperature and precipitation. The variations seem to be caused by the temperature change and ground water level change.

4.ANALYSIS AND DISCUSSION

We have tried to reduce the seasonal variations through the following two methods.
(a) We have been measured ground water level at a well near the northeast site (Fig.7) since Desember 1985. We estimated the ground water level of the whole duration when we obtained the tilt data with a simulation model as shown in Fig.3 from precipitation data, by comparing with the measured ground water level. Next, we calculated the effects of the room temperature and ground waterlevel under the asumption that the measured seasonal variatons of the tiltmeter values can be expressed by a linear combination of the room temperature and estimated ground water level. The reduced tilt values by this method are shown in Fig.4.

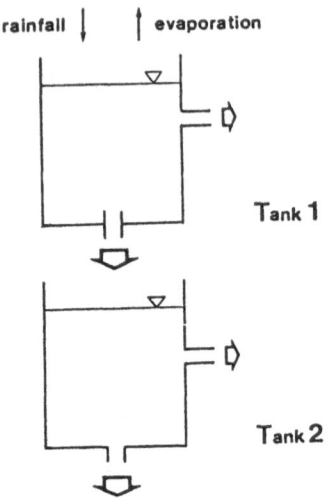

Fig.3 Simulation model for estimating

ground water level

(b) We estimated their effects directly from temperature and precipitation values with a linear transfer function as follow (Yanagisawa, 1980);

$$Y(t)=a_1+a_2 t+a_3 t^2+bT(t)+\sum_{i=0}^{50} c_i P(t-i) \qquad (1)$$

where $Y(t)$: daily mean values of the tiltmeter on t-th day,
$T(t)$: daily mean values of the room temperature on t-th day,
$P(t)$: daily mean values of the precipitation on t-th day.

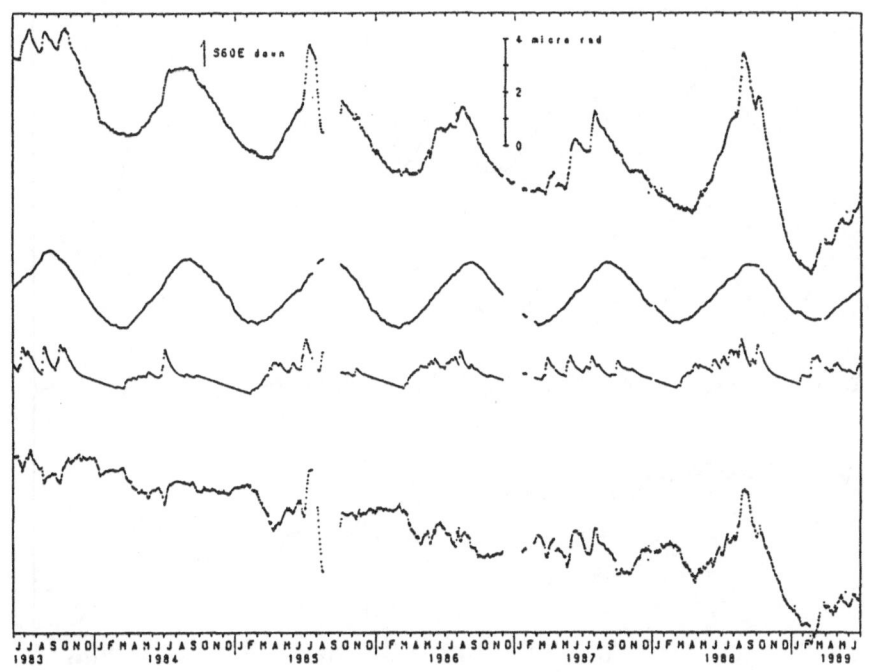

Fig.4 Daily mean values of the tiltmeter, effects of room temperature, effects of the ground water level estimated by using a simulation model, and the tilt values reduced their effects.

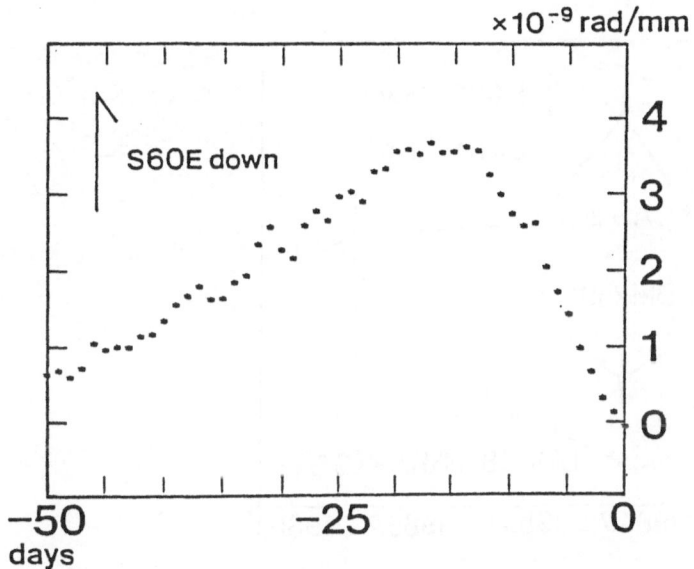

Fig.5 The coefficients Ci of the linear transfer function (1)

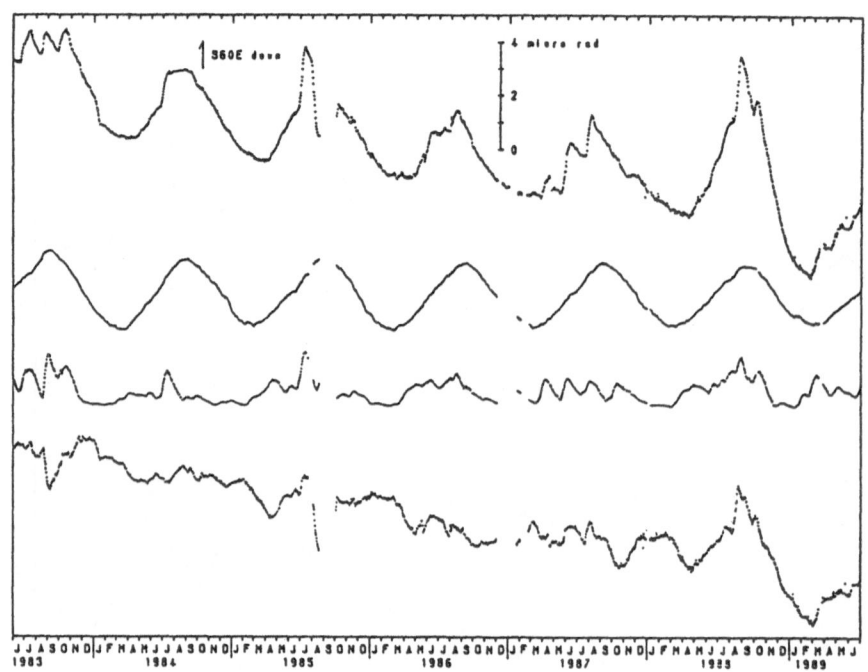

Fig.6 Daily mean values of the tiltmeter, effects of room
    temperatur, effects of the precipitation estimated by using
    a linear transfer function, and the tilt values reduced
    their effects.

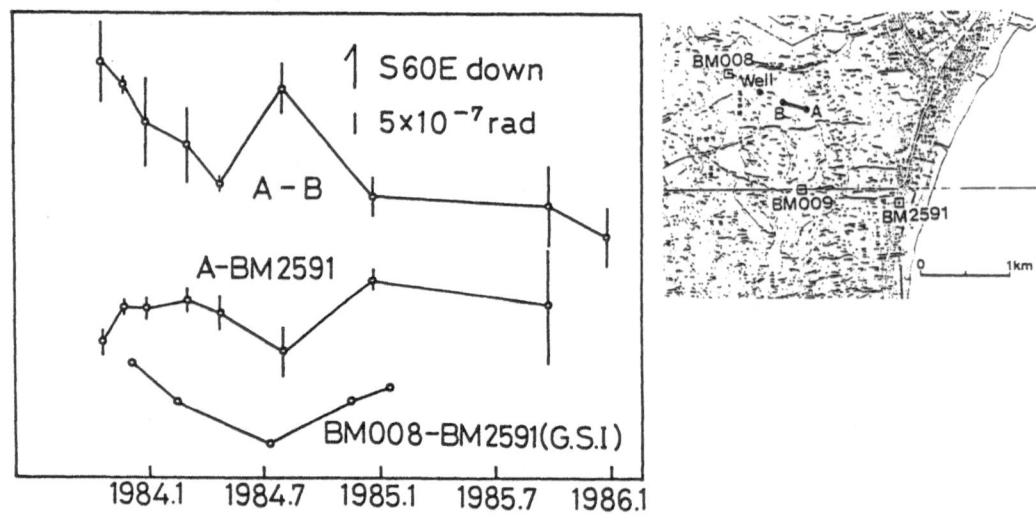

Fig.7 Results of the leveling surveys between both sites of the
    tiltmeter(A-B), that between the tiltmeter and the
    benchmark at the nearest coast and that between benchmark
    near the tiltmeter and the nearest coast.

The obtained coefficients c_i are shown in Fig.5. This result
shows that the effects of the precipitation continue for more
than 50 days. The tilt values reduced by using this function are
shown in Fig.6. Both results obtained through (a) and (b) show
that the ground is tilting continuously to northwestward. These
results are reverse to the regional tilt change including the
observation site observed through leveling surveys by
Geographical Survey Institute (G.S.I.,1989). We also carried out
another leveling survey between observation site (A) and the
bench mark near the coast (BM2591). The result shown in Fig.7
is also reverse to the change of tiltmeter's result.

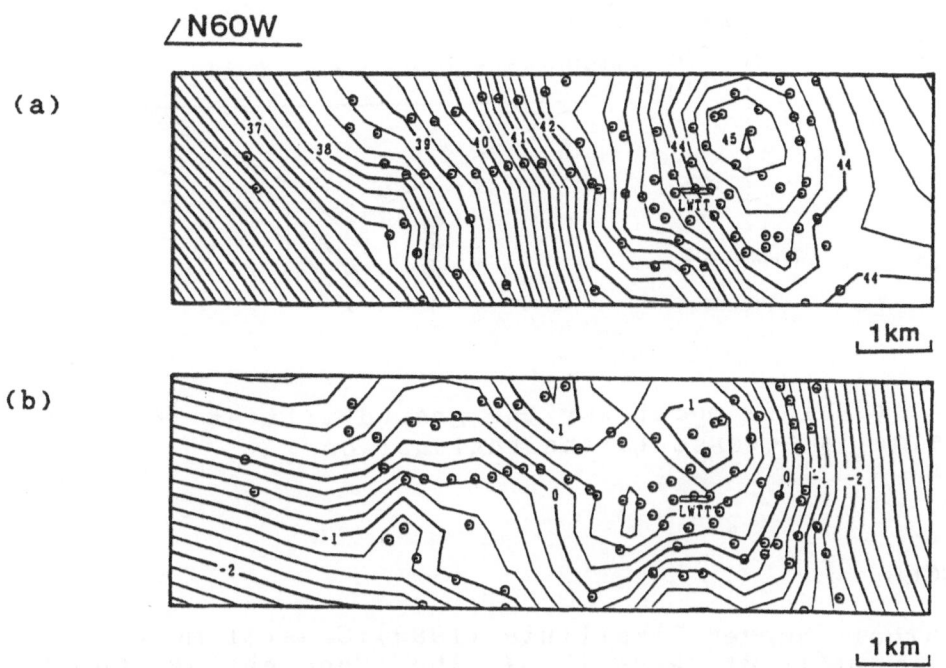

Fig.8 (a) Bouguer anomalies near the observation site(LWTT)
      (b) Residual gravity anomalies near the observation site.
      Assumed density is 2.35g/cm³. unit in mgal.

In order to study the reason why they show the reverse tilts, we
carried out a gravity survey near the observation site. Bouguer
anomaly map shown in Fig.8(a) with assumed density 2.35 g/cm³. As
the Bouguer anomaly values have a tendency that the northwest is
low and southeast high, we calculated the general tendency by
using a linear function and we obtained the residual anomaly
values by subtructing the general tendency. They are shown in
Fig.8(b). A local gravity high exists near the east site of the
tiltmeter. From the geological and geographical studies, it is
pointed out that there is a synclinal axis just near the
northeast site of the tiltmeter(Ujiie,1962), and also that there
are some small scale flexures.

From these investigations, we think that the crust of this region tilts continuously to southeastward by the subductive motion of the Pilippine Sea Plate, but that the tiltmeter shows local crustal movements by folding or flexuring, as shown in Fig.9. This local movements, however, must be related to the plate motion, and therefore the data obtained by the tiltmeter shall be a great use in detecting the precursory crustal movements of the assumed destructive earthquake.

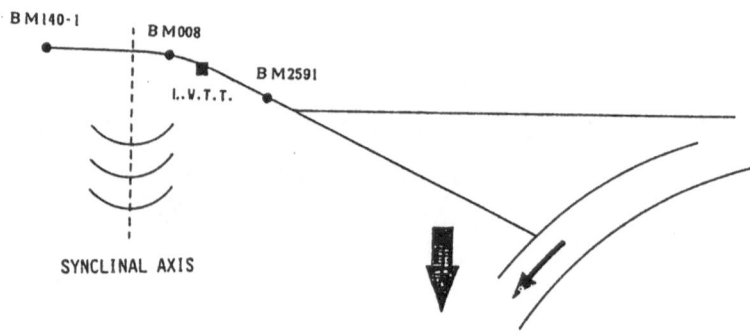

Fig.9 A possible model why the ground tilt at the observation
        site is reverse to the regional one.

REFERENCES

Geographical Survey Institute (1989).Crustal movements in the
        Tokai District, Report of the Coodinating Committee for
        Earthquake Prediction,vol.42,257-267. (in Japanese)
Ishibashi,K. (1981).Specification of a soon-to-occur seismic
        faulting in the Tokai District, central Japan, based upon
        seismotectonics, Earthquake Prediction, An International
        Review, D.W.Simpson and P.G.Richards(Eds.), Maurice Edwing
        Ser.,IV, Am. Geophys. Union, Washington,D.C.,297-332
Satomura,M., Dambara,T., Hasegawa,Y., Toyama,T., Shichi,Ry.,
        Samesima,T., Ino,M., Iwata,T. and Hagiwara,T.(1987).
        Descreption of Sagara and Kamisaka Observation station for
        Crustal Movements, Geosci.Repts. Shizuoka Univ.,157-164.
        (in Japanese)
Shichi,Ry., Okuda,T. and Yoshioka,S.(1980).A new design of moving
        float type water-tube tiltmeter, Jour.Goed.Soc.Japan, 26,1-16.
        (in Japanese)
Ujiie,H.(1962).Geology of the Sagara-Kakegawa Sedimentary Basin
        in central Japan, Sci.Rep.Tokyo Kyoiku Diagaku. ser.C., no.75,
        123-188.
Yanagisawa,M.(1980).Rainfall effect on the tilt observations at
        Usami, the Izu Peninsula, Jour.Geod.Soc.Japan,26,189-199.
        (in Japanese)

A REFINED ADJUSTMENT MODEL FOR THE SECULAR CHANGE OF GRAVITY ON THE FENNOSCANDIAN 63° LATITUDE GRAVITY LINE

Lars E. Sjöberg
The Royal Institute of Technology
Department of Geodesy
S–100 44 Stockholm, Sweden

Repeated precise gravity observations along the precise gravity lines in Fennoscandia covering the period 1966–1984 were published by Mäkinen et al. (1986). The observations of the most extensively measured gravity line at about latitud 63° were analysed by Ekman et al. (1987), Sjöberg et al. (1988) and Sjöberg (1989). These investigations show that the estimated secular change of gravity, of the order of –0.2 μGal/mm uplift, is very sensitive to the individual weighting of the data. In the present study the adjustment model is refined, e.g. by the inclusion of an iterative weight estimation procedure. Nevertheless, the result is the same as for the simple adjustment: the ratio of estimated change of gravity to land uplift is of the order –0.16 ± 0.04 μGal/mm in agreement with a simple viscous flow model.

1.   INTRODUCTION

Since the middle of the 1960s, under the auspices of the Nordic Geodetic Commission, the Nordic national agencies in cooperation with universities have undertaken repeated precise gravity observations along four east–west land uplift gravity lines in Fennoscandia (Fig.1). The aim is to estimate the secular change of gravity and to study its relation to the present land uplift of the area. The observed land uplift (Fig.1) reaches a maximum of 9.2 mm/year in the upper part of the Bay of Bothnia. Adding an eustatic sea rise component of +1.0 mm/year and a geoid uplift of 0.7 mm/year (Sjöberg, 1982, 1983a) the total uplift peak is estimated to 10.9 mm/year.

Mäkinen et al. (1986) compiled and analysed the gravity measurements carried out for the 18 year period 1966–1984. Their study shows that merely the data from one

Fig. 1. Observed land uplift (uplift relative to mean sea level) in mm/year, and the land uplift gravity lines. From Mäkinen et al. (1986).

out of four gravity lines, the 63° latitude line, reveals a significant secular change of gravity. An analysis of the 63° line data was presented by Ekman et al. (1987). Further studies on this data set are given in Sjöberg et al. (1988) and Sjöberg (1989).

The present article will review the least squares adjustments carried out in the last three papers. Finally, a refined adjustment model including variance component estimation will be utilized.

2.   OBSERVED SECULAR CHANGE VERSUS A VISCOUS FLOW MODEL

If the Fennoscandian uplift process was not accompanied by any mass flow the change of gravity would be a pure free air change of about -0.31 μGal/(mm total uplift). A more realistic model assumes a linear flow of mantle material along with the uplift. Assuming that this flow occurs along an infinite (Bouguer) plate with mantle density 3.27 g/cm^3, one arrives at a gravity change of the order of -0.17 μGal/mm uplift. Based on a simple viscous flow model (Sjöberg, 1982, 1983a) a more refined model for the secular change of gravity was derived in Sjöberg et al. (1988):

$$\dot{g}_j = G\rho \iint_\sigma \{\dot{h}_i(\frac{1-s\,\cos\,\psi_{ji}}{1_{ji}^3} - \frac{2}{1_{ji}}) - \dot{H}_i \frac{3\delta_i \cdot \mu}{2L_{ji}}\} d\sigma - 0.3086\,\dot{H}_j, \qquad (1)$$

where G is the gravitational constant, ρ is the density of upper mantle, μ is the ratio ρ_{water}/ρ, σ is the unit sphere, δ_i is unity for the running point P_i at sea and zero for P_i on land, $\dot{H}_j$ is the observed land uplift rate and $\dot{h}$ is the total uplift rate (composed of $\dot{H}$, the uplift rate of the geoid and the eustatic rise). Furthermore

$$s = 1 - t/R$$
$$\ell_{ji} = \sqrt{1-2s\,\cos\psi_{ji}+s^2}$$
$$L_{ji} = \sqrt{2(1-\cos\psi_{ji})} \,,$$

339

where t is the depth of the crust and R is the mean Earth radius. The result of using this formula with the land uplift data of Fig. 1 is displayed in Fig. 2.

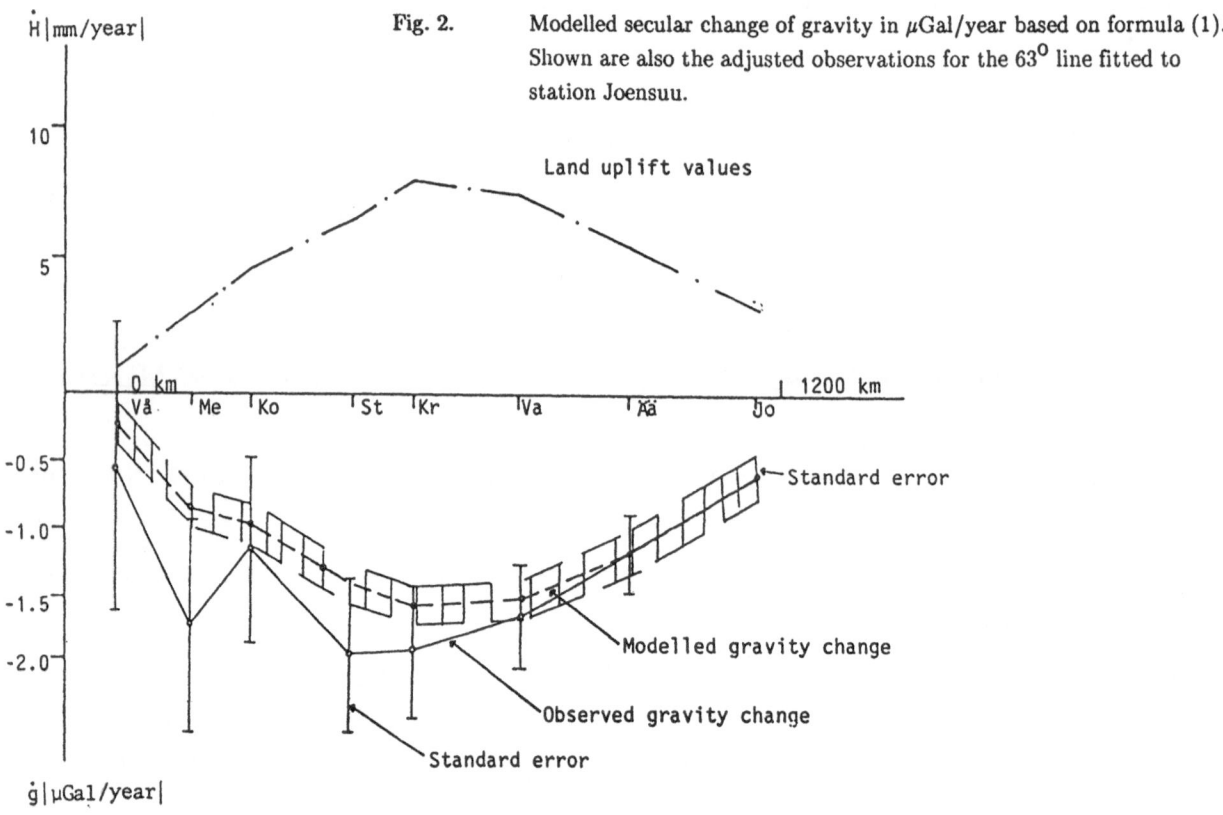

Fig. 2.   Modelled secular change of gravity in μGal/year based on formula (1). Shown are also the adjusted observations for the 63° line fitted to station Joensuu.

In Table 1 the observed and modelled rates of change of gravity along the 63° line are compared. As the observations were performed by relative gravimeters (La Coste & Romberg model G and D), the estimable rates are also relative and non—determined by a constant (bias). In the table and Figure 2 the observations are fit to the model at the station Joensuu. This fit yields a mean difference of —0.295 μGal/year for all stations. If the observations are fit to the mean of modelled data the differences become the residuals of the last column of Table 1.

Table 1. Comparison of observed and modelled change of gravity (ġ) in μGal/year for the 63° line. Residual = difference − mean difference

Station	Observed	Modelled	Difference	Residual
Joensuu	−0.60	−0.60	0	0.295
Äänekoski	−1.17	−1.17	0	0.295
Vaasa	−1.67	−1.50	−0.17	0.125
Kramfors	−1.90	−1.57	−0.33	−0.035
Stugun	−1.96	−1.40	−0.46	−0.165
Kopparå	−1.14	−1.16	−0.02	0.275
Meldal	−1.74	−0.67	−1.07	−0.775
Vågstranda	−0.63	−0.22	−0.31	−0.015

Mean difference:	−0.295	μGal/year	0
RMS difference:	0.44	μGal/year	0.33 μGal/year
Standard deviation			0.36 μGal/year

Except for the station Meldal all residuals are within the standard deviation. Hence we conclude that (except for Meldal) the observations fit the simple viscous flow model.

3.   LEAST SQUARES ADJUSTMENT OF OBSERVATIONS

In Table 2 we summarize the gravity observations along the 63° line for the western part (Vågstranda–Kramfors) and the eastern part (Vaasa–Joensuu).

Table 2. Measured gravity differences on the 63° line. From Ekman et al. (1987)

Western part		Eastern part	
Date	g–diff. in μGal	Date	g–diff. in μGal
1967.5	532.6 ± 8.3	1966.8	134.6 ± 1.9
1972.7	534.1 ± 6.1	1967.9	129.6 ± 6.4
1977.7	519.1 ± 3.9	1971.8	131.9 ± 7.4
1982.7	509.9 ± 7.0	1977.7	133.8 ± 3.3
		1979.8	143.3 ± 4.7
		1982.7	144.1 ± 2.3
		1984.7	151.0 ± 5.9

3.1 Adjustment by elements

Using a simple regression model between gravity difference (Δg) and time (t), i.e.

$$\Delta g_i = a + b(t_i - t_0) \tag{2}$$

where t_0 is a suitable reference epoch, Ekman et al. (1987) estimated the parameters a and b for the western and eastern part of the 63° line. By dividing these estimates of b by $\Delta \dot{H}_W = 7.0$ mm/year for the western part and by $\Delta \dot{H}_E = -4.3$ mm/year for the eastern part, they arrived at the ratio

$$\Delta \dot{g}/\Delta \dot{H} = \begin{cases} -0.23 \pm 0.07 \ \mu\text{Gal/mm (western part)} \\ \\ -0.22 \pm 0.06 \ \mu\text{Gal/mm (eastern part)} \end{cases}$$

where the two parts were separated by the Baltic.

This consistent result for both parts of the 63° line was obtained with equal weighting of all individual observations of gravity differences. Later Sjöberg et al. (1988) showed that the weighting is rather critical for the estimate. Starting from the model

$$\Delta \dot{g}_i = \{ {a_1 \atop a_2} \} + b \{ {\Delta \dot{H}_W \atop \Delta \dot{H}_E} \} (t_i - t_0) \ , \tag{3}$$

where the upper and lower formulas apply to the western and eastern parts, respectively, a joint solution for both parts is possible. The results of these computations with various weights are summarized in Table 3. It shows that the standard errors of the estimated ratios vary very little, while the estimates themselves vary considerably with respect to choice of weights! The ratio $b = \Delta \dot{g}/\Delta \dot{H} = -0.162 \pm 0.038$ μGal/mm, obtained for weights determined from individual standard errors (s_i) as s_i^{-2}, fits the data the best (i.e. has minimum standard error).

Table 3.   Estimates of the ratio $b = \Delta\dot{g}/\Delta\dot{H}$. Number of observations: Western part: 4, Eastern part: 7, joint solution: 11. s_i = standard error of Δg_i for epoch t_i. Units: μGal/mm.

Weights	Western part	Eastern part	Joint solution
Unweighted	−0.23 ± 0.07	−0.22 ± 0.06	−0.228 ± 0.039
s_i^{-2}	−0.26 ± 0.07	−0.15 ± 0.04	−0.162 ± 0.038
No. of gra−vimeters	−0.27 ± 0.08	−0.22 ± 0.07	−0.243 ± 0.048
$\sqrt{\text{No. of gra−vimeters}}$	−0.25 ± 0.07	−0.22 ± 0.07	−0.234 ± 0.044

3.2   Refined adjustment model

One reason for the sensitivity of $\hat{b}$ and unsensitivity of $s_{\hat{b}}$ to selected weights (Table 3) could be that the regression model is too simple to absorb data inconsistencies such as non−random errors. One such contribution might stem from erroneous a priori uplift rates for the two parts of the 63° line. In the previous section $\Delta\dot{H}_W$ and $\Delta\dot{H}_E$ were considered as given without errors. We will now modify the adjustment model to take possible uplift rate errors into account.

Introducing the errors $\epsilon_{\Delta g_i}$, ϵ_W and ϵ_E of Δg_i, $\Delta\dot{H}_W$ and $\Delta\dot{H}_E$, respectively, eqn.(3) is easily modified to

$$\Delta g_i - \epsilon_{\Delta g_i} = \binom{a_1}{a_2} + \{(b) + db\} \begin{bmatrix} \Delta\dot{H}_W - \epsilon_W \\ \Delta\dot{H}_E - \epsilon_E \end{bmatrix} \Delta t_i, \qquad (4)$$

where db is a correction to approximation (b) and $\Delta t_i = t_i - t_o$. From the 11 observations of Δg_i eqn.(4) yields the following system of equations of the type condition adjustment with unknowns X:

343

$$AX + B\epsilon = BL \, , \tag{5}$$

where

$$
\underset{(11.3)}{A} = \begin{bmatrix} 1 & 0 & \Delta\dot{H}_W\Delta t_1 \\ 1 & 0 & \Delta\dot{H}_W\Delta t_2 \\ \vdots & \vdots & \vdots \\ 1 & 0 & \vdots \\ 0 & 1 & \Delta\dot{H}_E\Delta t_{10} \\ 0 & 1 & \Delta\dot{H}_E\Delta t_{11} \end{bmatrix}
\qquad
X = \begin{bmatrix} a_1 \\ a_2 \\ db \end{bmatrix}
\qquad
\underset{(13.1)}{\epsilon} = \begin{bmatrix} \epsilon_{\bar{\Delta}\,g} \\ \epsilon_W \\ \epsilon_E \end{bmatrix}
$$

$$
\underset{(13.1)}{L} = \begin{bmatrix} \Delta\bar{g} \\ \Delta\dot{H}_W \\ \Delta\dot{H}_E \end{bmatrix}
\qquad \text{and} \qquad
\underset{(11.13)}{B} = \begin{bmatrix} & (b)\Delta t_1 & 0 \\ & \vdots & \vdots \\ \underset{(11.11)}{I} & (b)\Delta t_4 & 0 \\ & 0 & (b)\Delta t_5 \\ & \vdots & \vdots \\ & 0 & (b)\Delta t_{11} \end{bmatrix}
$$

Here $\Delta\bar{g}$ and $\epsilon_{\Delta\bar{g}}$ are the vector of gravity observations and its error. The least squares solution to eqn. (5) becomes

$$\hat{X} = (A^T C^{-1} A)^{-1} A^T C^{-1} BL \tag{6a}$$

with the covariance matrix

$$Q_{\hat{X}\hat{X}} = \sigma^2 (A^T C^{-1} A)^{-1}, \tag{6b}$$

where Q is the covariance matrix of the observations L, and σ^2 is the variance of unit weight and

$$C = BQB^T.$$

The following covariance model was used

$$\underset{(13 \times 13)}{Q} = \alpha_1 \begin{bmatrix} Q_1 & 0 \end{bmatrix} + \alpha_2 \begin{bmatrix} 0 & Q_2 \end{bmatrix}, \tag{7}$$

where α_1 and α_2 are unknown variance components to be estimated,

$$Q_1 = \text{diagonal } (\sigma_1^2, \sigma_2^2, \dots, \sigma_{11}^2) \, ,$$

including the a priori variances of observed gravity differences and

$$Q_2 = \text{diagonal } (\sigma_W^2, \sigma_E^2)$$

including the a priori variances of $\Delta \dot{H}_W$ and $\Delta \dot{H}_E$.

The adjustment was carried out iteratively for eqs.(6) and the variance components α_1 and α_2 of eqn. (7). For the estimation of α_1 and α_2 two methods were applied. One was using simple estimators of the form

$$\hat{\alpha}_1 = \frac{13}{10 \times 11} \sum_{i=1}^{11} (\hat{\epsilon}_i^2 / Q_{ii}) \tag{8a}$$

and

$$\hat{\alpha}_2 = \frac{13}{(10 \times 2)} \sum_{i=12}^{13} (\hat{\epsilon}_i^2 / Q_{ii}) \, . \tag{8b}$$

The other method was the MINQUE procedure. Its application to condition adjustment with unknowns was described in Sjöberg (1983b). In both cases the iteration yielded

$$\hat{\alpha}_2 = 0, \text{ (no errors of uplift rates)}$$

and

$$\hat{b} = -0.162 \pm 0.038 \ \mu\text{Gal/mm}$$

as in the previous section. Hence, for this data set the refined model does not improve the result.

4. CONCLUDING REMARKS

In conclusion the refined adjustment procedure does not improve the estimated ratio gravity change/uplift rate. The best estimate is -0.162 ± 0.038 μGal/mm. Considering the error bounds this estimate agrees with the simple viscous flow model applied in Fig.2. In addition Table 1 shows that the data and the model are consistent. The overall pattern of the uplift along the 63° gravity line is that of a glacial rebound model. An exception is station Meldal where other effects, such as ocean tide loading and plate tectonics are likely to interfere.

Today there is a rather common belief among geoscientists that the present land uplift phenomenon in Fenoscandia includes not only postglacial rebound but also a component related with tectonics and mantel convection. See e.g. Gregersen and Basham (1989, pp. 1–3). However in order to be able to study such inhomogeneities of uplift over Fennoscandia, significant uplift rates are needed for more than one gravity line. Hopefully a future gravity data bank will consist not only of relative gravity but also precise absolute space–time variation of gravity over the area.

REFERENCES

Ekman,M., J.Mäkinen, Å.Midtsundstad, O.Remmer (1987), Gravity change and land uplift in Fennoscandia. **Bull. Géod. 61**, 60–64.

Gregersen,S. and P.W. Basham (1989),(Eds.), Earthquakes at North–Atlantic Passive Margins: Neotectonics and Postglacial Rebound. NATO ASI Series C: Mathematical and Physical Sciences − Vol. 266, Kluwer Academic Publishers, Dordrecht.

Mäkinen,J., M.Ekman, Å.Midtsundstad and O.Remmer (1986), The Fennoscandian Land Uplift Gravity Lines 1966–1984. **Reports of the Finnish Geodetic Institute** 85:4, Helsinki.

Sjöberg,L.E. (1982), Studies on the land uplift and its implications on the geoid in Fennoscandia. **Department of Geodesy Rep. No. 14,** University of Uppsala.

Sjöberg,L.E. (1983a), Land uplift and its implications on the geoid in Fennoscandia. **Tectonophysics, 97,** 87–101.

Sjöberg,L.E. (1983b), Unbiased estimation of variance – covariance components in condition adjustment with unknowns – A MINQUE approach. **ZfV, 108,** No.9, pp. 382–387.

Sjöberg,L.E., H.Fan and E.Asenjo (1988), Studies on the secular change of gravity in Fennoscandia. **The Department of Geodesy Report,** No.14, The Royal Institute of Technology, Stockholm.

Sjöberg,L.E. (1989), The secular change of gravity and the geoid in Fennoscandia. In S.Gregersen and R.W.Basham (Eds.), see above, pp. 125–139.

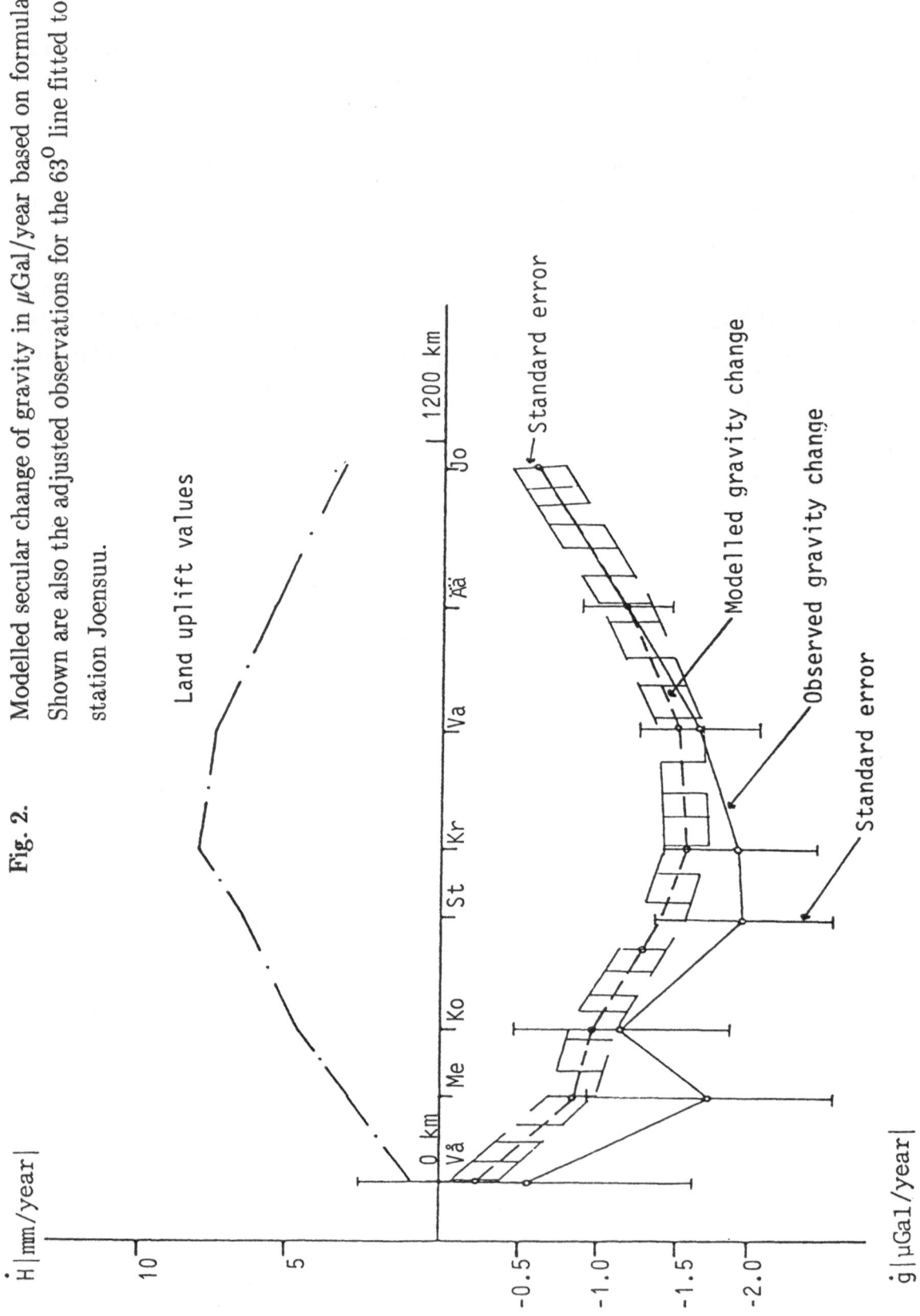

Fig. 2. Modelled secular change of gravity in μGal/year based on formula (1). Shown are also the adjusted observations for the 63^0 line fitted to station Joensuu.

Symposium 101 Author Index

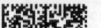